Scientific Testimony

Scientific Testimony

Its roles in science and society

MIKKEL GERKEN

OXFORD
UNIVERSITY PRESS

OXFORD
UNIVERSITY PRESS

Great Clarendon Street, Oxford, OX2 6DP,
United Kingdom

Oxford University Press is a department of the University of Oxford.
It furthers the University's objective of excellence in research, scholarship,
and education by publishing worldwide. Oxford is a registered trade mark of
Oxford University Press in the UK and in certain other countries

First Edition published in 2022

Published in the United States of America by Oxford University Press
198 Madison Avenue, New York, NY 10016, United States of America

British Library Cataloguing in Publication Data
Data available

Library of Congress Control Number: 2022933701

ISBN 978–0–19–885727–3

DOI: 10.1093/oso/9780198857273.001.0001

Printed and bound in the UK by
Clays Ltd, Elcograf S.p.A.

For Teo and Loa

Contents

Preface

This book is about the roles of scientific testimony within the scientific enterprise and in the wider society. Spoiler alert! I think that scientific testimony has vast significance in both realms. Thus, I will try to get a grasp on varieties of scientific testimony and begin to explore specific aspects of their roles in science and society.

It is widely appreciated that science is a forceful source of information about the world, in part because it is based on collaboration that is characterized by a fine-grained division of cognitive labor. Metaphorically speaking, scientists no longer stand on the shoulders of giants as much as they stand within an edifice built by their predecessors and contemporary peers. A central conclusion of this book is that scientific testimony is not merely an add-on to such collaborative science but a *vital part* of it. Given that scientific testimony contributes centrally to the epistemic force of science, it should be a central topic in philosophy of science. For the same reason, social epistemologists who theorize about the nature of testimony should regard scientific testimony as a central case. So, I will integrate philosophy of science and social epistemology in order to bring scientific testimony to the center stage as a research topic in its own right. While the book may be characterized as a treatise in the philosophy of science that draws on social epistemology, it is also true that philosophy of science informs foundational questions in the epistemology of testimony.

The significance of scientific testimony in the wider society is well recognized. In fact, an entire interdisciplinary field, *the science of science communication*, is devoted to it. As a philosopher, it has been inspiring to engage with this body of empirical research, and I hope that the book will introduce some of it to philosophers who are unfamiliar with it. Likewise, I hope that the book will exemplify how philosophical resources from philosophy of science and social epistemology may contribute to our understanding of scientific testimony and the obstacles that may hamper laypersons' uptake of it.

The book has been in the making for quite a while. One impetus to write it was my work on epistemic norms of assertion, which struck me as relevant to a principled characterization of scientific collaboration. Another entry point was my work on folk epistemological biases, which struck me as relevant to understanding science denialism. I wrote some articles on these issues but quickly realized that the topic called for a more sustained and coherent treatment. So, I applied for, and was awarded, the Carlsberg Foundation's *Semper Ardens* grant, which allowed me to dedicate the academic year 2018–19 to drafting the manuscript. Two more years and change were spent on revisions.

The final rounds of revision of a major book project tend to be painful. Each revision manifests a diminishing marginal improvement and an increasing urge to get the book out in the world. However, the revisions of this book took place during the coronavirus pandemic, which provided abundant examples of public scientific testimony—both good and bad. While the pandemic lockdowns with two wee rascals presented their own challenges, the infodemic accompanying the pandemic helped a worn-out writer retain the sense that the book deals with an important subject.

Acknowledgments

A book that extols the virtues of collaboration and the division of cognitive labor should be informed by the expertise and scrutiny of others. So, in order to ensure that I walk the collaborative talk, I have solicited and received a tremendous amount of help from a number of poor, unfortunate souls.

Many people commented on big chunks of the book and some commented on all of it. Notably, Kenneth Boyd and Uwe Peters commented extensively on every chapter and endured my mulling over details in our little research group which had the good fortune of being joined by Niklas Teppelmann for a semester. Likewise, I am very grateful to Haixin Dang and Jie Gao, who both commented on the entire first half, and to Andy Mueller who commented on the entire second half. During January 2020, I taught a course on testimony at Stanford in which I presented Chapter 6 and benefitted from the feedback of Sarah Brophy and Daniel Friedman. During May 2019, I was fortunate to visit VU Amsterdam for two weeks and have Chapters 1–3 subjected to careful criticism by a social epistemology research group consisting of Jeroen de Ridder, Tamarinde Havens, Thirza Laageward, Rik Peels, Christopher Ranalli, and René van Woudenberg. Chapter 4 was dealt with by Catharina Dutilh Novaes and her research group. The feedback from these sessions, and from the talk I gave during my stay, turned out to be extraordinarily helpful during the revision stage. Getting these perspectives on big chunks of the manuscript helped to ensure its overall coherence. However, I also received expert help on specific chapters. So, although I am dreading to have forgotten someone, I want to thank those who provided written feedback on chapter drafts:

Chapter 1: Ken Boyd, Haixin Dang, Jie Gao, Uwe Peters.
Chapter 2: Ken Boyd, Adam Carter, Haixin Dang, Jie Gao, Peter Graham, Uwe Peters.
Chapter 3: Ken Boyd, Haixin Dang, Jie Gao, Uwe Peters.
Chapter 4: Ken Boyd, Haixin Dang, Jie Gao, Hein Duijf, Uwe Peters.
Chapter 5: Ken Boyd, Andy Mueller, Uwe Peters.
Chapter 6: Ken Boyd, Andy Mueller, Uwe Peters, John Wilcox.
Chapter 7: Ken Boyd, Alexander Heape, Andy Mueller, Nikolaj Nottelmann, Uwe Peters.

For discussions and correspondence about articles that are precursors to parts of the book, I am especially grateful to Carrie Figdor, Bjørn Hallsson, and Karen

Kovaka, who helped me get my bearings early on. For discussion, I am grateful to all of those already mentioned as well as to Hanne Andersen, Line Edslev Andersen, Michel Croce, Jesus Vega Encabo, Pascal Engel, Axel Gelfert, Sanford Goldberg, Katherine Hawley, Christoph Kelp, Klemens Kappel, Arnon Keren, Søren Harnow Klausen, Martin Kusch, Krista Lawlor, Anna-Sara Malmgren, Boaz Miller, Anne Meylan, Esben Nedenskov Petersen, Asbjørn Steglich-Petersen, Ángel Pinillos, Melanie Sarzano, Samuel Schindler, Anders Schoubye, Mona Simion, Matthias Skipper, and Åsa Wikforss. I am especially grateful to Hannah Kim for stellar proofreading and insightful comments as well as to Lauren Thomas for valuable finishing touches regarding both grammar and style (both were funded by the Carlsberg Foundation).

Finally, I have presented the material as talks in a number of settings. Since 2018, when I started writing the book in earnest, relevant talks include: 2018: Danish Philosophical Society, Roskilde University; University of Southern Denmark; University of Copenhagen (twice); 1st SENE Conference, University of Oslo; University of St Andrews; University of Stockholm. 2019: University of Southern Denmark (twice); Stanford University; International Network for Danish Philosophers, University of Aarhus; Vrije Universiteit Amsterdam; Collage de France. 2020: Stanford University; Danish Philosophical Society, University of Southern Denmark; University of Zürich; University of Leeds (online).

Having written a book about testimony it is ironic that words cannot express my gratitude to my wife, Julie. The writing of my previous book, *On Folk Epistemology*, coincided with the arrival of our firstborn, Teo. So, I should have learned my lesson not to mix books and babies. But, as Hegel says, "We learn from history that we do not learn from history." So, I wrote this book when our second child, Loa, arrived with an agenda of her own. Once again, Julie marvelously anchored our family as the baby became a toddler, the coronavirus struck, and deadlines whizzed by. The fact that Teo and Loa were as cute as can be made the rough stretches of the writing process easier to get through. OK, maybe not exactly easier, but much more fun! I dedicate the book to Teo and Loa with love.

List of Figures

Introduction

Science and Scientific Testimony

The slogan of the Royal Society is *Nullius in verba*. While the exact translation and point of the slogan are debated, the core idea is, according to the Royal Society itself, that scientists should "take nobody's word for it" (Royal Society 2020). The key point is that science should be based on "facts determined by experiment" rather than on mere trust in authority.

This *Nullius in verba* sentiment is reflected in the philosophical foundations for an Enlightenment view of science. For example, Descartes's Rules for the Direction of the Mind explicitly forbids inquiring minds from relying on "what other people have thought" (Descartes 1628/1985: 13). Locke states, "In the Sciences, every one has so much as he really knows and comprehends: What he believes only and takes upon trust, are but shreads; which however well in the whole piece, make no considerable addition to his stock, who gathers them" (Locke 1690/1975: I.iv, 23). Kant, in turn, characterizes the Enlightenment itself in terms of the ability to understand things oneself without relying on others: "Enlightenment is man's emergence from his self-incurred immaturity. Immaturity is the inability to use one's own understanding without the guidance of another" (Kant 1784/1991: 54).

This book is not a historical treatise. If it were, it would reveal a more complex picture of scientific testimony in the philosophical foundations of the Enlightenment than these quotes might suggest. Likewise, critical historical studies suggest that the individualistic, anti-testimonial ethos of early scientists did not accurately reflect their scientific practice (Shapin 1994). Nevertheless, it is important to address the simplified picture that opposes testimony in science. Often, simplified pictures are more forceful than complex ones in influencing our *folk theory of science*.

For example, it is natural to think of science as an enterprise that produces "first-hand knowledge" from careful analysis of meticulous observation rather than mere "second-hand knowledge" from the testimony of someone else. According to this line of thought, the rest of us may defer to scientists' testimony precisely because the scientists themselves are autonomous in the sense that they base their views on observation rather than deferring to someone else's say-so (Dellsén 2020). Thus, a natural folk theory of science may well encompass an inarticulate yet influential *science-before-testimony* picture. According to this

Scientific Testimony: Its roles in science and society. Mikkel Gerken, Oxford University Press. © Mikkel Gerken 2022.
DOI: 10.1093/oso/9780198857273.003.0001

picture, scientific testimony's place in the scientific practice is after its conclusions have been established.

I venture to guess that many philosophers of science would now reject such a *science-before-testimony* picture and agree with Lipton's dictum: "Science is no refuge from the ubiquity of testimony" (Lipton 1998: 1). Philosophers of science have highlighted the importance of scientific collaboration and division of cognitive labor in philosophy of science. Yet, despite notable exceptions, philosophy of science features comparatively little work on the roles of scientific testimony. This is startling if scientific testimony is an important part of science rather than an add-on. In contrast, social epistemologists spend their days and nights theorizing about testimony, but they often do so without thematizing *scientific* testimony. Consequently, a central ambition of this book is to situate scientific testimony as a primary topic of investigation by drawing on both philosophy of science and social epistemology. I will try to not merely reject the *science-before-testimony* picture by articulating negative arguments against it. I will also begin to articulate a principled *testimony-within-science* alternative according to which scientific testimony is not merely a *product* of science but a *vital part* of it. This picture is painted by mixing philosophy of science and social epistemology. Developing a positive alternative picture that highlights the significance of scientific testimony is important because it helps us understand the nature of science. But it is also important because it puts us in a better position to ameliorate the role of science in society.

One main theme of the book will be the significance of what I call *intra-scientific testimony*, which is scientific testimony from a scientist that has collaborating scientists as its primary audience and which aims to further future scientific research. I will argue that intra-scientific testimony and the norms governing it are as vital to collaborative science as scientific norms governing observation, data analysis, theorizing, etc. While observations may be the building blocks of the scientific edifice, scientific testimony is required to unify them. In slogan: *Scientific testimony is the mortar of the scientific edifice.*

While the slogan provides a metaphorical contrast to the *Nullius in verba* slogan, I will make the metaphor more tangible by developing concrete norms of intra-scientific testimony. For example, I will propose an epistemic norm for providing intra-scientific testimony as well as a norm for its uptake in the context of scientific collaboration.

One reason to develop a positive alternative to the *science-before-testimony* picture is that it may continue to hold some sway in folk conceptions of science insofar as many laypersons may share the misconception that scientific progress owes to an autonomous individual genius. Just think about the image conveyed by the TED talk—the immensely popular platform for science communication—which is built around a solitary presenter musing in the spotlight. Likewise, history of science and science education often focuses on individual geniuses such as Galileo, Newton, Darwin, and Einstein (Allchin 2003). But science is not well represented by focusing on individual efforts of great white males who, after a *Eureka!* moment, produce an entirely novel theory and prove it by a crucial experiment. A focus on such a narrative may give rise to what I call a *great white man fetish*, which is both misguided and likely to sustain structural injustices. In particular, I will argue that it may sustain *testimonial* injustice, which is an important species of epistemic injustice (Fricker 2007).

Folk misconceptions of science that are influenced by the *science-before-testimony* picture are relevant for another main theme of the book: *public scientific testimony*, which is scientific testimony that is primarily directed at the general lay public or select members of it such as policy makers. To address public scientific testimony, philosophical resources will be integrated with empirical work on laypersons' uptake of public scientific testimony in the novel interdisciplinary field called *the science of science communication* (Jamieson et al. 2017). In doing so, I focus on general norms of public scientific testimony that apply to both scientific expert testifiers and to science reporters. But I also articulate more specific guidelines that may inform their public scientific testimony. I develop such norms and guidelines on the basis of philosophical reflection on the nature of scientific testimony and its proper role in societies that pursue ideals of deliberative democracy. But although I pursue a principled account, I do not presuppose that the nature and role of scientific testimony are eternal truths that may be uncovered by a priori reflection alone. Rather, I also draw heavily on empirical research on the social context of public scientific testimony and of laypersons' psychological obstacles to a reasonable uptake of it. This engagement with the relevant empirical work is *critical* insofar as I criticize some proposals and articulate some important conceptual distinctions. But it is also *constructive* in that I draw on the empirical research to formulate working hypotheses on clearly empirical questions regarding folk misconceptions of science, cognitive biases, and strategies for overcoming these obstacles.

Given that I seek to address the significance of scientific testimony both within scientific practice and in the wider society by drawing on a broad range of philosophical and empirical resources, a couple of brief methodological preliminaries are in order.

Methodological Considerations

There are three methodological aspects of the book that readers should prepare themselves for: Reliance on approximate characterizations of paradigm cases, reliance on substantive background assumptions, and efforts to integrate disparate discussions. Let me say a bit about each.

Definitions vs Characterization of Paradigm Cases

The book concerns complex phenomena such as expertise, collaboration, groups, collective belief, public deliberation, and so forth. Since each of these phenomena is a self-standing research topic which involves debates over definitions, I must frequently work with approximate characterizations. This is no less true of the two core components of scientific testimony—namely, *science* and *testimony*.

The attempt to characterize science well enough to distinguish it from pseudoscience is a long-standing ambition of philosophy of science (Lakatos 1978; Hansson 2017). Likewise, contemporary epistemology and philosophy of language feature intense debates about the characterization of testimony. I have sought to reflect these debates without getting stuck in the pursuit of reductive analyses. In fact, I suspect that many of the phenomena are too basic to admit of reductive analyses. Rather, I will try to uncover, in a non-reductive manner, some principled, and sometimes constitutive, relations between the relevant phenomena (see Gerken 2017a for elaboration of such an *equilibristic* methodology). So, rather than hunting for necessary or jointly sufficient conditions, I will often provide a characterization in terms of some hallmark properties of paradigm cases and then restrict the discussion to such cases. Of course, some cases are hard to capture in this manner. For example, testimony given at a research conference that is open to the public lies in the gray zone between intra-scientific testimony and public scientific testimony. But although some cases are hard to categorize, many cases are clear enough. For example, the case of an epidemiologist who gives an interview on the radio may still be discussed as a paradigm case of public scientific testimony. So, I follow Kripke's *hard cases make bad law* methodology of beginning with the paradigm cases (Kripke 1979/2011: 160). When things go well, the account of paradigm cases may eventually be extended to harder peripheral or derivative cases. But this is not always an ambition of the present exploration. In sum, while the book contains a good deal of conceptual clarification, I have sought to strike a balance between working with clear characterizations and adopting approximations that will do for the purpose at hand.

Background Assumptions

Space and focus also dictate that I assume some substantive theoretical views without much argument. For example, I adopt a broadly realist background stance according to which approximating truth is an actual and reasonable aim of scientific theories (Psillos 1999; Godfrey-Smith 2003; Chakravartty 2011). While this remains a controversial assumption in the philosophy of science, it will be a working hypothesis that I will adopt with very little defense. Consequently, the parts of the investigation that rest on this stance may not speak to some scientific anti-realists. On the other hand, one way to motivate a philosophical framework is to adopt it and consider whether a particular issue may be fruitfully investigated within it. This is my approach with regard to scientific realism in the present investigation. In other cases, I allow myself to rely on arguments that I have given elsewhere. For example, I will rely on previous criticisms of the knowledge-first program (see Gerken 2011, 2012a, 2014a, 2015b, 2017a, 2018a). There are other debates which I regard as very important, but which I have had to sidestep due to space and focus. One example is the debates concerning the value-free ideal of science—roughly, the idea that at least some central parts of the scientific enterprise should aim to be as neutral as possible with regard to non-cognitive values (Douglas 2009, 2015; Brown 2020). The value-free ideal has been the subject of intense debate, which bears on scientific testimony. In this case, I do not speak to the grand debate about the value-free ideal in scientific research (although I do have views on it). Rather, I rely on the much less controversial assumption that the practical ramifications of public scientific testimony bear on the conditions under which it is appropriate to assert it.

Integrative Efforts

Perhaps the most striking methodological aspect of the book is its close integration of related fields that sometimes fail to draw on each other. One such integration is between philosophy of science and (social) epistemology. The book is written from the conviction that an understanding of the significance of scientific testimony must be based on foundational theorizing in the epistemology of testimony. On the other hand, I have repeatedly found that reflecting on issues and cases in the philosophy of science informs fundamental issues about the nature of testimony. Scientific testimony is an area in which philosophy of science and (social) epistemology may be mutually illuminating. I hope that integrating these two adjacent subdisciplines of philosophy, which are often conducted in relative isolation from each other, will shed light on scientific testimony. More generally, I hope that the discussion will exemplify how philosophy of science and social epistemology may benefit from further integration.

Another important integration is between philosophy and the empirical *science of science communication*. For a philosopher, this interdisciplinary field is obviously valuable in providing empirical warrant for empirical assumptions. But I have also found it to be a treasure trove of novel ideas and perspectives on public scientific testimony. That said, the book is by no means an attempt to naturalize philosophy. On the contrary, I aim to provide both critical and constructive philosophical contributions to the debates. Often, they consist in foundational concepts, arguments, or distinctions between, for example, types of scientific testimony. Furthermore, empirically informed philosophical reflection may provide substantive theses about the nature of scientific testimony, the norms that apply to it, and its role in scientific collaboration as well as in the wider society. Thus, I hope that the investigation will indicate that philosophy has a lot to contribute to the understanding of the significance of scientific testimony in science and society.

An Overview

The book consists of seven chapters and a brief coda. It is organized in four parts. The first three parts each consist of two chapters, and the final part consists of a concluding chapter and the coda.

Part I: Philosophical Foundations of Scientific Testimony. The first part of the book approaches its subject matter by some principled characterizations and by taxonomizing varieties of scientific testimony. Moreover, I articulate and motivate substantive theses about scientific testimony, epistemic expertise, scientific collaboration, etc. So, Part I contributes to the conceptual foundations for the investigation of scientific testimony.

In **Chapter 1**, I start the investigation with some conceptual clarifications and a provisional taxonomy of types of scientific testimony. Notably, this includes the distinction between *intra-scientific testimony*, which takes places between collaborating scientists, and *public scientific testimony*, which is directed at laypersons and comes in two varieties. *Scientific expert testimony* is characterized by the testifier being a scientific expert. *Science reporting*, in contrast, is public scientific testimony by testifiers, such as journalists, who often lack scientific expertise. Given this initial clarification of scientific testimony, I consider its relationship to prominent themes in philosophy of science. These include scientific expertise, scientific collaboration, and the division of cognitive labor. In discussing these themes, I articulate conceptual and empirical arguments that scientific collaboration contributes immensely to the epistemic force of science and that intra-scientific testimony is a vital part of such collaboration.

Chapter 2 opens with a discussion of the nature of testimony as a speech act and an epistemic source. This discussion draws on foundational epistemological work involving, for example, the internalist/externalist debate and the reductionist/anti-reductionist debate. Relatedly, I consider the senses in which testimony may and may not be said to transfer epistemic warrant from testifier to recipient. Specifically, I argue for a negative principle, *Non-Inheritance of Scientific Justification*, according to which the kind or degree of scientific justification that the testifier possesses is typically *not* transmitted to the recipient—even when the testimonial exchange is epistemically successful. I will often view scientific testimony through the lens of norms. Consequently, Chapter 2 also includes a brief discussion of *norms*, which I consider as objective benchmarks of assessment, and *guidelines*, which are more concrete directives that scientific testifiers may follow.

Part II: Scientific Testimony within Science. The two chapters that make up Part II address the nature of scientific testimony and the roles it plays *within* the scientific practice. On the basis of a characterization of scientific testimony, I focus on intra-scientific testimony's role in truth-conducive scientific collaboration.

Chapter 3 provides a characterization of scientific testimony that differentiates it from other types of testimony. Specifically, I articulate and defend a characterization of scientific testimony as testimony that is properly based on scientific justification. Further specification of this characterization is provided by way of a discussion of some of the central properties of scientific justification. These include its being gradable, its being discursive, and the senses in which it is and is not epistemically superior to non-scientific justification. Likewise, I discuss what being properly based on scientific justification amounts to. Apart from helping to clarify what scientific testimony is, these arguments help to specify why intra-scientific testimony contributes to the epistemic force of collaborative science. Likewise, they help to specify why public scientific testimony may serve as a central epistemic authority in society.

In **Chapter 4**, I continue the overarching argument that intra-scientific testimony is a vital part of the scientific practice by articulating some norms for it. The first one is a *Norm of Intra-Scientific Testimony* (NIST), according to which a scientist who provides intra-scientific testimony within a scientific collaboration must base it on a contextually determined degree of scientific justification. I then turn from the producer side to the consumer side and develop a *Norm of Intra-Scientific Uptake* (NISU). According to NISU, a collaborating scientist receiving intra-scientific testimony should, as a default, believe or accept it insofar as he has strong and undefeated warrant for believing that the testimony is properly based on scientific justification. In developing this duo of norms of the production and consumption of intra-scientific testimony, I argue that they partly but centrally contribute to explaining the truth-conduciveness of scientific collaboration. This reflects the book's general attempt to replace a *science-before-testimony* picture with a *testimony-within-science* alternative according to which intra-scientific testimony is not an add-on to scientific practice but a vital part of it.

Part III: *Scientific Testimony in Society*. In Part III, I turn to *public scientific testimony* and its roles in society. In particular, I will propose a number of norms and guidelines for *scientific expert testimony* and *science reporting*, respectively. My approach is informed by empirical research on the psychology of laypersons' uptake of public scientific testimony.

Chapter 5 concerns scientific expert testimony. It begins by surveying empirical research on psychological challenges for the public's uptake of public scientific testimony. On the basis of this work, I articulate a novel norm for scientific expert testifiers: *Justification Expert Testimony* (JET). According to JET, scientific expert testifiers should, whenever feasible, include appropriate aspects of the nature and strength of scientific justification, or lack thereof, in their testimony for the scientific hypothesis in question. I furthermore argue that JET motivates a more specific guideline concerning *scientific expert trespassing testimony* which occurs when a scientific expert testifies on matters in a domain of epistemic expertise other than her own. According to this *Expert Trespassing Guideline*, a scientific expert who provides expert trespassing testimony should, in some contexts, qualify her testimony to indicate that it does not amount to expert testimony. So, Chapter 5 exemplifies the gradual movement from foundational research on general norms to applied research on more specific ameliorative guidelines.

Chapter 6 is devoted to *science reporting* and begins with a critical assessment of some prominent principles of science communication that appeal to scientific consensus, recipient values, etc. This serves as the background for my own proposal, *Justification Reporting*, which has it that science reporters should seek to include appropriate aspects of the nature and strength of scientific justification in science reporting. I consider the prospects and limitations of this norm in light of empirical research on laypersons' uptake of public scientific testimony. The chapter concludes with a more ameliorative perspective. Specifically, I consider the journalistic principle of *Balanced Reporting* according to which science reporters should seek to report opposing hypotheses in a manner that does not favor any one of them. By an application of *Justification Reporting*, I set forth an alternative, *Epistemically Balanced Reporting*, according to which science reporters should seek to report opposing hypotheses by indicating the nature and strength of their respective scientific justifications.

Part IV: The Significance of Scientific Testimony. Part IV consists of Chapter 7 and a short Coda. In Chapter 7, I draw the previous sub-conclusions together in arguments for general conclusions about the significance of intra-scientific testimony and public scientific testimony, respectively. The Coda briefly relates the central themes of the book to cognitive diversity and epistemic injustice.

Chapter 7 begins with arguments for two theses concerning intra-scientific testimony. The first thesis, *Methodology*, is the claim that the distinctive norms governing intra-scientific testimony are vital to the scientific methods of collaborative science. The second thesis, *Parthood*, is the claim that intra-scientific testimony is a vital part of collaborative science. Jointly, these two theses help to replace the *science-before-testimony* picture with a *testimony-within-science* alternative. I then turn to arguments for two theses about public scientific testimony. The first thesis of this duo, *Enterprise*, has it that public scientific testimony is critical for the scientific enterprise in societies pursuing ideals of deliberative democracy. The second thesis, *Democracy*, is the claim that public scientific testimony is critical for societies pursuing ideals of deliberative democracy. In light of these two theses, I discuss the role of public scientific testimony in the *societal* division of cognitive labor. In particular, I argue that it is an important societal task to secure a social environment in which laypeople may acquire epistemically entitled testimonial belief through *appreciative deference* to public scientific testimony. This results in a novel norm for laypersons' uptake of public scientific testimony.

Coda. The brief Coda indicates how scientific testimony relates to (cognitive) diversity and epistemic injustice. After characterizing these notions, I consider how cognitive diversity bears on intra-scientific testimony. I argue that it has good epistemic consequences in virtue of adding critical perspectives but also bad consequences in virtue of complicating intra-scientific communication. Relatedly, I note that cognitively diverse minorities' intra-scientific testimony is particularly liable to be received in epistemically unjust ways. Turning to public scientific testimony's relationship to cognitive diversity and epistemic injustice, I suggest that a social environment characterized by an appreciative deference to scientific testimony may help minimize some types of epistemic injustice for cognitively diverse or epistemically disadvantaged groups. On this basis, I suggest that social and institutional initiatives combating epistemic injustice for cognitively diverse groups should be central to the pursuit of the broader goal of aligning scientific expertise and democratic values.

Stylistic Notes

I label cases by italicized full capitalization. For example: As the case *WIND SPEED* exemplifies...

I label principles by upper and lower case italics. For example: According to the principle *Distinctive Norms*, science relies...

I label acronymized principles by full capitalization. For example: The principle NIST is one which...

I use single quotes to mention words and sentences. For example: The word 'testimony' which occurs in the sentence 'scientific testimony is important' is a controversial one.

I use double quotes for real or imagined quotations and occasionally to indicate metaphors or to introduce novel terminology.

I use italics for emphasis and occasionally to indicate quasi-technical phrases.

PART I

PHILOSOPHICAL FOUNDATIONS OF SCIENTIFIC TESTIMONY

Part I of the book consists of two chapters concerning fundamental debates which are about, or relevant for, understanding scientific testimony. Thus, Part I contributes to laying the conceptual foundations for more specific arguments and theories about scientific testimony. It does so by surveying some of the relevant debates in philosophy of science and social epistemology. However, along the way, I will contribute to these debates by providing conceptual clarifications, making distinctions, and articulating substantive theses and principles.

In Chapter 1, I distinguish among some central kinds of scientific testimony and consider it in relation to themes in philosophy of science, such as scientific expertise and scientific collaboration. On this basis, I begin to develop an account of the roles of scientific testimony in scientific collaboration that is characterized by a high degree of division of cognitive labor.

In Chapter 2, I characterize the fundamental features of testimony in general, and as an epistemic source in particular. For example, I address central epistemic features of testimony by relating them to some foundational epistemological debates, such as the internalist/externalist debate and the reductionist/anti-reductionist debate. Finally, I consider how scientific testimony may be characterized via the epistemic norms governing it.

1
Testimony and the Scientific Enterprise

1.0 The Roles of Scientific Testimony

A study of scientific testimony involves considering the relationship between two phenomena: science and testimony. Consequently, I will begin with provisional characterizations of the relevant kinds of testimony and move on with some select points about the relevant aspects of the scientific process.

In Section 1.1, I will provide some core distinctions in a taxonomy of scientific testimony that I will examine. In Section 1.2, I distinguish among some varieties of scientific expertise at the individual level. In Section 1.3, I move to the social level by highlighting the collaborative aspects of the scientific process and method. Section 1.4 continues this theme by focusing on the division of cognitive labor that characterizes scientific work. In Section 1.5, I draw on these discussions to argue that the division of cognitive labor characteristic of science depends on distinctive *norms* of intra-scientific testimony. Thus, the chapter concludes by initiating arguments for a broad *testimony-within-science* picture.

1.1 Kinds of Scientific Testimony: Intra-Scientific and Public

Testimony is a varied phenomenon, and in order to provide some classification of the various types of *scientific* testimony, a bit of an overview is called for. So, I will briefly consider testimony in general before focusing on scientific testimony.

1.1.a Testimony in general: For the purposes of this book, I will think of testimony in a fairly broad manner as an assertive expression which is offered as a ground for belief or acceptance on its basis. Utterances or writings are central examples of testimony although they are not exhaustive. For example, representational depictions, maps, or icons may count as types of testimony—including types of scientific testimony. Likewise, nods, hand waves, and grimaces may qualify as testimony. However, I will focus on familiar written and spoken forms of propositional scientific testimony that purport to convey a worldly fact. I also construe testimony broadly as to include assertions that *p* that include a justification or explanation for *p*. Consider, for example, the assertion: "The meeting will be postponed. It makes no sense without the investor, and she is

Scientific Testimony: Its roles in science and society. Mikkel Gerken, Oxford University Press. © Mikkel Gerken 2022.
DOI: 10.1093/oso/9780198857273.003.0002

delayed." I take this to qualify as testimony that the meeting will be postponed although a rationale is given for this. Given the broad conception of testimony, it is all the more important to zoom in on scientific testimony and its species. There will be plenty of zooming in and out throughout the book. In this opening section, I will simply draw some basic distinctions and settle on some terminology.

Although the term 'testimony' has solemn and austere connotations, it may just consist in an everyday assertion. When Teo tells me that he had rye bread for lunch, he is testifying in this relaxed sense of the term. When I believe him, I form a *testimonial belief*. Likewise, when I read that FC Barcelona won El Clásico 5–0, I read a testimony and my resulting belief is a paradigmatic testimonial belief (I elaborate in Chapter 2.1.a–b). Scientific testimony may have the same relaxed character. My testimonial belief that nothing travels faster than light may be formed much like my testimonial belief that Barça won El Clásico 5–0. So, testimony need not occur in courtrooms or in formal pronouncements.

Terminologically speaking, we may follow Coady in distinguishing between *formal* testimony, such as in a courtroom, and *natural* testimony, such as Teo's one about rye bread.[1] While the distinction is helpful, many of the cases that will be discussed are situated in a gray zone between these categories. For example, an assertion in response to a question at a scientific conference has aspects of both natural and formal testimony. Likewise, a scientist's quotes in a semi-structured interview for a newspaper have aspects of both natural and formal testimony.

Turning to the recipient's side, the idea of a *minimal background case* is a useful one that I will rely on: In a minimal background case, the recipient has minimal information about the testifier and the testifier's epistemically relevant properties, such as his competence, reliability, and sincerity. Minimal background cases also involve minimal warrant for beliefs about the broader informational environment. Of course, the recipient will always have some background information (Audi 2006: 27–8). So, minimal background cases are limiting cases that contrast with cases with richer background information. A good example is that of an epistemologically naïve recipient, such as a young child who believes an unfamiliar testifier.

Let us fix some terminology: I use "warrant" as a genus of epistemic rationality which harbors two species.[2] The first species of warrant is called "justification" and may be generally characterized as a warrant that constitutively depends, for its warranting force, on the competent exercise of a subject's faculty of reason. The warrant for a conclusion-belief on the basis of reasoning is a central example. The second species of warrant is called "entitlement" and does not depend on reason in this manner. The basic warrant for perceptual belief is a central example of entitlement. Entitlement is an epistemically externalist type of warrant that

[1] Coady 1992: 38. See also Shieber 2015: 10ff.; Gelfert 2014: 14ff.
[2] Burge 2003; Graham 2012a; Gerken 2013a, 2013b, 2020a.

partly depends on environmental conditions that the individual needs no cognitive access to. In contrast, justification may be said to be epistemically internalist. One subspecies of justification—discursive justification—is important for scientific testimony since it requires that the subject be capable of *articulating* aspects of the warrant as epistemic reasons (Gerken 2012a). I take "epistemic reasons" to consist of propositional contents that may provide truth-conducive support for believing other propositions, whereas I regard "epistemic grounds" as environmental circumstances that may provide truth-conducive support. I will return to these issues in Chapter 2 and beyond. Now I move on to the main topic of scientific testimony.

1.1.b Scientific testimony and its varieties: It is not a trivial matter to distinguish scientific testimony from other types of testimony. So, to get things moving, I simply present my view, which I will elaborate on and argue for (in Chapter 3.1): What makes a given testimony a *scientific* testimony is the fact that it is properly based on scientific justification.

Scientific testimony is often more formal than everyday testimony, but this is not a defining feature of it. Consider a scientist informing a colleague that the abnormality in their data was due to a defective instrument, or a postdoc emailing the principal investigator that there was a significant effect in the pilot study. Such testimonies exemplify scientific testimony among collaborating scientists that I call "intra-scientific testimony." Yet they are no more formal than the testimony from a realtor who writes her client that the buyer has now signed off on the contract. Likewise, a newspaper may report a study finding that inadequate sleep dramatically increases the risk of traffic accidents in a format that does not differ from a report on policy or sports. Nevertheless, such a report would exemplify another type of scientific testimony—namely, the type I call "public scientific testimony." Yet more specifically, it would exemplify the subspecies that I call "science reporting." Another subspecies of public scientific testimony that I call "scientific expert testimony" occurs when scientific experts testify in some context of scientific communication to laypersons. For example, a particular scientific expert on sleep and sex drive might testify during a public presentation that the two are correlated. The final type of scientific testimony that I will mention is labelled "inter-scientific testimony." It communicates the results of scientific investigation to the general scientific community. This tends to be quite formal since it typically takes the form of publications, such as a journal article.

One thing that all these species and subspecies of scientific testimony have in common is that they are all properly based, in importantly different ways, on scientific justification. In Chapter 3.1, I will argue that this is no coincidence since being properly based on scientific justification is what makes a testimony a scientific testimony. A nicety of this way of looking at things is that *pseudo*-scientific testimony may be derivatively characterized: Pseudo-scientific testimony is testimony that purports to be scientific although it is not because it is not properly based on scientific justification. Merely non-scientific testimony is also not properly based on scientific justification, but, in contrast with pseudo-scientific testimony, it does not purport to be scientific testimony.

Perhaps a map will be helpful. Figure 1.1 shows the central types of scientific testimony just mentioned:

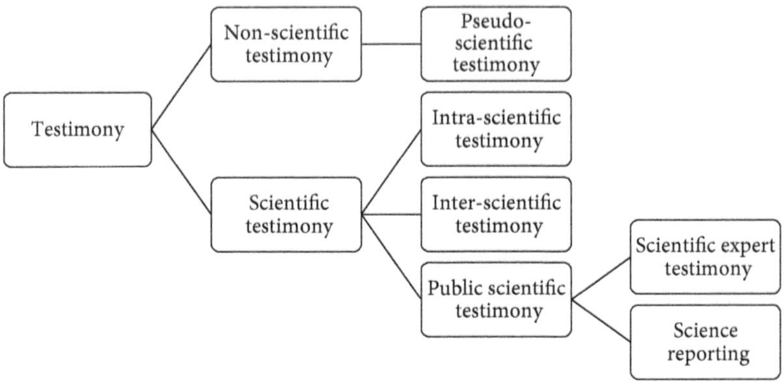

Figure 1.1 Types of testimony

I hasten to note that the overview is not comprehensive. There are further subcategories, as well as hybrids and overlaps, among the mapped categories. Consider, for example, an influential scientist who provides expert scientific testimony to a prominent news platform that a classic study has failed to replicate. In some cases, she might be simultaneously testifying to the lay public and her colleagues via a public news channel. Other examples are "breaking scientific news" conferences or press releases in which major findings are simultaneously communicated to the general public and, in a preliminary form, to the scientific community. Likewise, some scientific experts have a side hustle with scientific outreach in popular science media, and their testimonies may therefore be situated in the intersection of *scientific expert testimony* and *science reporting*.

Such hybrid scientific testimonies and borderline cases are important to bear in mind. But they hardly compromise the distinctions insofar as there are reasonably clear and paradigmatic cases of each category. The best way to illustrate intra-scientific, inter-scientific, and public scientific testimony is to consider these categories in turn.

1.1.c Intra-scientific testimony: Intra-scientific testimony may be approximately characterized as "scientific testimony from a scientist that has collaborating scientists as its primary audience and which aims to further future scientific research" (Gerken 2015a: 570). According to this characterization of intra-scientific testimony, it is partly distinguished in terms of, first, its primary audience of (collaborating) scientists and, second, its central aim of furthering future research. These two components are related insofar as it makes sense to communicate to collaborating scientists if one aims to further future research. All in all, intra-scientific testimony concerns science in the making in the daily hustle and bustle of lab meetings, emails, watercooler talk, internal memos, and progress reports, etc.[3]

The characterization is not a reductive analysis which captures all cases. The two components are neither individually necessary nor jointly sufficient for intra-scientific testimony. For example, some cases of intra-scientific testimony obstruct future research or promote past research. However, the characterization captures paradigm cases well enough. For example, it dissociates intra-scientific testimony from standard cases of scientific expert testimony to laymen due to its component concerning the primary audience. Similarly, it dissociates intra-scientific testimony from scientific testimony that is aimed at application in, for example, public policy.

The characterization in terms of primary audience and aim also allows for an initiation of the extended argument that intra-scientific epistemology is not merely a *product of* science but rather a *vital part of* the scientific process. Yet, the characterization remains a rather broad one, and once intra-scientific testimony takes the center stage in Chapter 4, some subspecies of it will be distinguished between. Here, my main aim has primarily been to identify the phenomenon and distinguish it from public scientific testimony, to which I now turn.

1.1.d Public scientific testimony: Public scientific testimony is scientific testimony that is primarily directed at the general lay public or select members of it, such as policymakers. Given this broad characterization, there is an enormous variety of public scientific testimony. Public scientific testimony will take center stage in Part III. Here, I will just draw a couple of rudimentary distinctions that I will need to get going.

Some public scientific testimonies are directed at the lay population at large for the purpose of general information. A common example is a scientist's testimony that is quoted in an interview for a public media platform. Such public scientific testimony reflects an important enlightenment ideal of a scientifically informed public (Jasanoff 1990; Kitcher 2011: 85). However, public scientific testimony may also be directed at highly select stakeholders in the layperson population, and these may include political decision makers. A scientific report commissioned by a ministry or scientific expert testimony in legal proceeding are examples.

[3] For a classic and a recent report, see Latour and Woolgar 1979/2013 and Cho 2011.

I draw the general distinction between *scientific expert testimony* and *science reporting* in terms of source and, derivatively, in epistemic terms. What characterizes scientific expert testimony is that its immediate source is a scientific expert in the relevant domain. In contrast, science reporting is typically mediated by someone, such as a journalist, who is a non-expert in the relevant domain. Note that the phrase 'science reporting' may be used in a broad way that denotes discussion of scientific practice, for example, "scientists relocate resources to develop a coronavirus vaccine." But I will be more concerned with a use that qualifies as scientific *testimony* in which a hypothesis or finding is presented as true, for example, "COVID-19 has a longer median incubation period than influenza." Like ordinary testimony, science reporting may be qualified as to indicate the epistemic status of the hypothesis, and I will argue that science reporters should often include such epistemic qualifications (Chapter 6).

To recap, the central difference between scientific expert testimony and public scientific testimony is whether the testifier has relevant scientific expertise. I will argue that scientific expertise standardly involves epistemic expertise. Hence, science reporting has epistemic force since its ultimate source is scientific expert testimony. For example, a science journalist may base their report on a press release, on interviews with scientists, or even by consulting some of the relevant scientific publications. However, given the indirectness of the ultimate source, there are distinctive pitfalls for science reporting that may render it less reliable than scientific expert testimony. For example, even dedicated science journalists tend to be laypersons when it comes to the highly specialized science they report on (Goldman 2001; Figdor 2010, 2018). The additional link in the communication chain is a distinct source of fallibility. Moreover, journalists work in an attention economy in which accessibility, novelty, and other news criteria may trump accuracy and reliability.[4]

1.1.e Inter-scientific testimony: scientific publications and scientific reports: An important type of scientific testimony that I will not thematize, although it will figure occasionally, is inter-scientific testimony. Roughly, this is scientific testimony which aims to communicate the results of scientific investigation to the general scientific community (I owe the label to Dang and Bright forthcoming). A central type of inter-scientific testimony is scientific publications which are, as the name indicates, ways of making scientific findings and theories public. Examples include articles in scholarly journals, academic books, conference proceedings, and so forth. These are public venues, but their primary audience is typically other scientists. Scientific publications are central to the scientific practice and, therefore, governed by both explicit conventions and implicit

[4] Valenti 2000; Miller 2009; Nisbet and Fahy 2015; Figdor 2017; Gerken 2020d.

disciplinary norms. As an example of explicit conventions, consider the *Publication Manual of the American Psychological Association* (American Psychological Association 2009). As an example of fucking implicit disciplinary norms, consider the use of redundant profanity in this sentence. The reason why it is jarring is that redundant profanity violates implicit disciplinary norms of academic writing.

Inter-scientific testimony may include scientific reports which are distinct from science reporting in virtue of being directed to, and often commissioned by, policymakers or other stakeholders in need of scientific assessment. So, scientific reports are typically instances of formal testimony, and for that reason they are subject to more explicit, and often highly idiosyncratic, aims and norms. For example, a scientific report may have to be written in a manner that is apt for basing legislation on it. However, some scientific reports have other scientists as their primary audience. For example, reports from the WHO are resources for health scientists and policymakers alike. Likewise, IPCC reports are also resources for both climate scientists and policymakers. So, while some scientific reports are best classified as public scientific testimony, others are best classified as inter-scientific testimony, and many are in the gray zone between these categories. Likewise, there are gray zones between intra- and inter-scientific testimony. A central difference is that inter-scientific testimony does not have *collaborating* scientists as the primary audience. But social norms and conventions that determine whether another scientist is collaborating in the relevant sense may leave some cases open. Nevertheless, reflection on such norms may provide some principled help in distinguishing intra- and inter-scientific testimony. For example, a scientist may be required to tell a collaborator about the outcome of the pilot study, but she may be required to withhold this information in communicating with non-collaborators from a competing research group.

Both scientific publications and science reports are important types of scientific testimony. In the case of publications, this is because of their dual role of making scientific work public and contributing to future scientific research. In the case of scientific reports, this is because they help apply scientific work to concrete problems. In doing so, they legitimize, and thereby sustain, the scientific enterprise. So, although these types of scientific testimony are not the primary phenomena of investigation here, their importance ensures that they will both make their occasional return.

1.1.f Concluding remarks on varieties of scientific testimony: The distinctions drawn and the associated terminology do not come close to a comprehensive taxonomy of scientific testimony. However, they do mark out some important categories, and the brief discussion of the various types of scientific testimony begins to reveal its wide-ranging significance.

1.2 Aspects of Scientific Expertise

With some preliminary distinctions and rudimentary terminology concerning testimony in hand, it is time to turn to the other part of our explanandum of scientific testimony—namely, the *science* part. I will begin at the individualistic level by characterizing the kinds of expertise in virtue of which individuals are to be regarded as scientists. Both *expertise* and *scientific expertise* are substantive research topics in their own right.[5] So, I will approach the issues by providing some broad characterizations and by drawing some distinctions that are particularly relevant for discussing scientific testimony.

1.2.a Scientific expertise as epistemic expertise: Since science aims to be, and often succeeds in being, an epistemically forceful mode of cognition, scientific expertise is distinctively epistemic. However, as will transpire, this is a truth with important qualifications concerning the collaborative nature of science. So, it would be too fast to infer from the epistemic force of science to the idea that the individual scientists are invariably epistemic experts. On the other hand, it would be rash to ignore the epistemic aspects of scientific expertise. This is particularly so when it comes to assessing the epistemic credentials of scientists *qua* testifiers in public discourse (Baghramian and Croce 2021). Generally speaking, the epistemic authorities in a given domain are the best scientific experts in that domain—at least if the domain falls within a reasonably mature science (see Chapter 3.3 for elaboration and qualifications). Consequently, I will begin the discussion of scientific expertise with a characterization of epistemic expertise.

Provisionally, the property of *possessing expertise* may be broadly, albeit not reductively, characterized as an acquired specialized competence that enables the expert to excel in a specific domain. This is, of course, very broad insofar as it encompasses everything from being an expert foxtrot dancer to being an expert logician. The narrower category of cognitive expertise, then, involves an acquired specialized competence that enables the expert to excel in a specific domain of cognition. But this remains a rather broad characterization given how broadly the term 'cognitive' applies. So, for the purpose of investigating scientific testimony, the following provisional characterization of the property of possessing *epistemic* expertise may be of use:

> *Epistemic Expertise*
>
> S possesses epistemic expertise in a domain, D, that consists of a set of propositions iff S has acquired a specialized competence in virtue of which she is likely to possess or be able to form, in suitable conditions, extraordinarily reliable judgments about members of D.

[5] See, for example, papers in Selinger and Crease 2006; Quast and Seidel 2018.

Epistemic Expertise is a characterization of possessing epistemic expertise as opposed to being an epistemic expert. Possessing epistemic expertise is to possess expert competence and I take this to be necessary but not sufficient for being an epistemic expert. A further condition on being an epistemic expert might be that of properly applying one's epistemic expertise in minimally suitable conditions. Consider, for example, an epidemiologist, S, who has the generic competence to gather and analyze register data about a range of domains, D^1–D^n. But assume that S lacks any access to evidence about a certain domain D^7. Given this assumption, there is at least a sense in which she is not an expert regarding D^7, although she possesses the expertise to become one. Since ordinary language is unlikely to be a reliable guide on such a subtle matter, I am making a terminological choice in drawing the distinction between possessing expertise and being an expert. Although the two categories will be co-extensional in most of the cases that I will discuss, it is important to recognize the distinction.

Notice that *Epistemic Expertise* is consistent with the idea that epistemic expertise comes in degrees since reliability comes in degrees and since the (potential) judgments may cover varying numbers of members of D. I take "extraordinarily reliability" to involve a comparative component. Goldman considers adding an objective threshold on the proportion of propositions in D with regard to which S needs to be extraordinarily reliable (Goldman 2018: 5). This would have the consequence that seismologists are not epistemic experts with regard to judgments about future earthquakes since they predict earthquakes with extremely low reliability (Jordan et al. 2011; thanks to Jie Gao for suggesting the case). However, if seismologists consistently outperform laypersons in virtue of their acquired specialized competence, it seems reasonable to regard them as possessing epistemic expertise in the domain. But since epistemic expertise comes in degrees, the seismologists may be ascribed a low degree of epistemic expertise in virtue of their comparative advantages. For the present purposes, I will leave these issues open but focus on cases in which the scientists have a comparatively extraordinary and objectively respectable degree of reliability with regard to a respectable proportion of members of D.

In any case, an objectively high degree of reliability does not suffice for epistemic expertise (Goldman 2018 agrees with this much). Almost everybody is extremely reliable in their judgments about whether seawater is salty, but this does not make almost everybody an epistemic expert on the matter. So, a comparative component is required. Of course, a comparative component admits of considerable gray zones in part because the comparison class may be unclear. So, often the cut-off for expertise is vague.

However, there are clear cases in which someone's judgments are extraordinarily reliable in virtue of an acquired cognitive competence. A meteorologist with access to the relevant data (i.e., in suitable conditions) will be extraordinarily reliable in predicting when the storm will make landfall both in the objective sense and the comparative sense. The binary formulation 'extraordinarily reliable' is useful as it allows us to say that she possesses epistemic expertise outright or, if the further conditions are met, that she is an epistemic expert simpliciter. However, in borderline cases, *Epistemic Expertise* is consistent with saying that someone has more expertise than most laypersons or that he is approximating expert status.

Epistemic Expertise builds on (Goldman 2001, 2018). But it diverges from the suggestion that "S has the capacity to help others (especially laypersons) solve a variety of problems in D or execute an assortment of tasks in D which the latter would not be able to solve or execute on their own" (Goldman 2018: 4). Likewise, Coady characterizes an epistemic expert as "someone laypeople can go to in order to receive accurate answers to their questions" (Coady 2012: 30). In contrast, I do not think that the *pedagogical* ability to address laypersons should not be a defining feature of *epistemic* expertise (Croce 2019b). However, *Epistemic Expertise* explains why epistemic experts may be helpful to epistemic laypersons in terms of their capacity to provide extraordinarily reliable testimony.

The 'in suitable conditions' qualification in *Epistemic Expertise* is required by the fact that the apparatus, observation conditions, or resources may eliminate the opportunity for putting the competence to work. A chemist needs a lab to conduct her experiments, a developmental psychologist needs some babies to observe, and a philosopher needs copious amounts of coffee. The 'is likely' qualification is required to account for cases in which an expert subscribes to a warranted but false theory and, therefore, forms unreliable judgments about D—even compared to laypersons.

While provisional, *Epistemic Expertise* allows us to derivatively characterize S as possessing epistemic expertise with respect to a particular proposition, *p*, as follows: S possesses epistemic expertise with respect to a proposition, *p*, iff *p* is a member of D and S possesses epistemic expertise in D. As above, possessing epistemic expertise in D does not suffice for being an epistemic expert with respect to *p* even if *p* is a member of D. In order to be an expert with respect to *p*, S must have made a judgment regarding *p* that is properly based on her expertise in D. Finally, *Epistemic Expertise* allows us to derivatively characterize expert judgments as judgments that are properly based on epistemic expertise.

These characterizations have the—I think reasonable—consequence that epistemic expert judgment is domain restricted. Scientific expertise, in particular, is distinctively domain restricted due to the hyper-specialization of science. Of course, the borders of epistemic expertise may be vague or porous. Expertise in one domain may carry over to other domains. As philosophers are aware, epistemic expertise in logic and argumentation may be applied in a wide variety of contexts. The same is true for epistemic expertise in statistics. Likewise, many scientists have acquired generic competences that are widely applicable. These include the ability to reflect on study designs, data gathering, the strength of evidence, etc. Such general epistemic expertise is also widely applicable and important for providing and receiving scientific testimony.

Nevertheless, there are clear and paradigmatic cases in which epistemic expertise in D1 does not entail any epistemic expertise in D2. Indeed, many such cases are found in science. One can be an expert in paleontology while remaining a layperson about microeconomics and vice versa. Every scientist who has attended an interdisciplinary conference is familiar with the humbling experience of finding oneself in the position of a layperson or novice. Even within a discipline an epistemic expert in one subfield may be closer to a layperson than to an epistemic expert with respect to another subfield. Consider, for example, social psychology and vision science, which are subfields of psychology. So, it is reasonable to characterize epistemic expertise in a domain-specific manner and then account for spillover cases within this framework. One case may be accounted for in terms of the domains D1 and D2 intersecting. In other cases, a high degree of epistemic expertise with regard to D1 may result in a low or partial epistemic expertise in D2.

A general account of epistemic expertise would require a full treatise. For example, *Epistemic Expertise* is restricted to domains of propositions and, hence, it is an acquired cognitive competence that may generate accurate or reliable *sub-propositional representations*. Moreover, the idea of "extraordinarily reliability" must be specified (cf. Goldman 2018). So, my approach will be to mainly consider cases in which the expert is both objectively highly reliable and comparatively more reliable than laypersons since this is highly relevant to *scientific* expert testimony.

In sum, *Epistemic Expertise* is central to scientific expertise but by no means a reductive definition of it. Scientific expertise is a multifaceted affair with facets that go beyond epistemic expertise. For example, it includes varieties of knowhow. Someone who has merely memorized a large number of the propositions that make up a domain may qualify as an epistemic expert in that domain. But if the individual is not capable of contributing to a scientific investigation, there is a strong sense in which he lacks scientific expertise. Consider, for example, an amateur historian who has memorized most of the publicly available information about an era but lacks the ability to synthesize the information or provide historical analysis of it. Or consider a butterfly connoisseur who has memorized a fantastic amount of butterfly trivia but completely lacks understanding of the basics of biology. Such individuals appear to lack an important aspect of scientific expertise (see also Croce 2019a, 2019b). Thus, reflection on such cases has given rise to the influential idea of contributory expertise.

1.2.b Contributory expertise: Whereas I take scientific expertise to centrally involve epistemic expertise, there is more to scientific expertise. In their endeavors to provide a "Periodic Table of Expertise," Collins and Evans have articulated a distinction between contributory expertise and interactional expertise (Collins and Evans 2007, 2015). My ambition here is not to provide a taxonomy—and much less a Periodic Table. But the popular contributory/interactional distinction may provide a helpful perspective on scientific expertise. Very roughly, contributory expertise consists in the ability to contribute to some area, whereas interactional expertise merely requires the ability to discuss it.

However, there is an active debate on how the contributory/interactional distinction is best drawn (Goddiksen 2014; Reyes-Galindo and Duarte 2015). Collins and Evans themselves have provided a number of non-equivalent characterizations of contributory expertise. In a recent paper, they initially characterize contributory expertise as "the ability to contribute to an area of practical accomplishment" (Collins et al. 2016: 104). Transposing this general characterization to the case of scientific expertise, contributory scientific expertise consists in the ability to contribute to an area of scientific accomplishment.

Although contributory expertise often overlaps with *Epistemic Expertise*, they are distinct because *contribution* tends to be articulated in sociological or functional terms rather than in epistemological terms (cf. Collins 2004: 128). Often, but not invariably, someone with contributory expertise with regard to a particular domain, D, is also an epistemic expert in D or a domain D* that is a subset of D. But since the types of qualifications that permits an individual to contribute to a scientific process are extremely varied, contributory expertise does not *entail* epistemic expertise.

The distinction between contributory and interactional expertise is unclear in ways that matter for how it bears on scientific expertise and, consequently, for scientific testimony. Collins, Evans, and Weinel note one type of trouble cases: "peer-reviewers and committee members who are understood to be primarily interactional experts but clearly contribute to the technical domain" (Collins et al. 2016: 104). Likewise, project managers, and even principal investigators, may solely in virtue of their organizational position become co-authors on publications outside their domain of *epistemic* expertise. Such individuals should be characterized as lacking contributory expertise in the relevant domain. But according to the broad characterization, they possess it. While Collins et al. recognize this problem, they talk down its significance: "perhaps it is one of those borderline problems that are philosophically irritating but which do not pose any serious real world problems" (Collins et al. 2016: 104).

In contrast, I suspect that the philosophical irritation indicates real world problems. For example, interdisciplinary collaboration involves many contributors who lack any epistemic or otherwise substantive expertise within the domain of inquiry (Hvidtfeldt 2018). But although a taxonomy that aspires to be analogous to a Periodic Table should aim for more exactness, I will work with the broad characterizations. However, I will seek to invoke them only in clear cases. For example, a developmental psychologist who is competently running a competently designed study of her own in a baby lab clearly has contributory expertise. In contrast, an unspecialized science journalist who interviews the developmental psychologists about early signs for autism will at most possess a variety of interactional expertise. In such cases, the distinction may be illuminating.

This does not mean that the problems surrounding the distinction are negligible. For example, it is unclear whether contributory expertise is useful for a characterization of scientific expertise. Even if the problematic cases are peripheral, they indicate the principled point that *Epistemic Expertise* is, in an important sense, a primary aspect of scientific expertise. Here is one reason why: Contributory expertise very frequently *derives* from epistemic expertise. For example, a biologist may contribute to an interdisciplinary project on the impact of pesticides on insect diversity in virtue of being extraordinarily reliable in discriminating among kinds of insects. Thus, she meets the criterion for contributory expertise in virtue of exercising her epistemic expertise. This contrasts starkly with the project manager, who contributes merely in virtue of exercising his managerial expertise. Clearly, the biologists should be regarded as possessing scientific expertise in the investigation-relevant domain, and the manager should not. The key difference is whether their contributions obtain in virtue of exercise of an epistemic expertise. This gives a reason to think that contributory expertise, broadly characterized, is only a proxy for scientific expertise insofar as it tends to be derived from *Epistemic Expertise.*

However, contributory expertise does illuminate an aspect of scientific expertise which is not captured by *Epistemic Expertise.* Recall the amateur historian with encyclopedic knowledge of a particular era but no ability to expand, criticize, or analyze scientifically justified assumptions about it. She qualifies as an epistemic expert, but her lack of contributory expertise is precisely what allows us to characterize her as falling short of an expert in the science of history. Despite her epistemic expertise, she lacks scientific expertise in virtue of lacking contributory expertise. A general lesson is that scientific expertise paradigmatically consists of a combination of contributory and epistemic expertise (see also Croce 2019b). Although contributory expertise is, in a sense, secondary because it typically arises from exercising an epistemic expertise, it does not *reduce* to epistemic expertise and may therefore play a limited but important role in characterizing scientific expertise.

1.2.c Interactional and T-shaped expertise: It is also controversial how to characterize interactional expertise. Collins and Evans have gone through several non-equivalent characterizations such as the following one: “expertise in the language of a specialism in the absence of expertise in its practice” (Collins and Evans 2007: 28). Applied to scientific expertise, the core idea is that someone possessing interactional expertise may communicate with scientists in a discipline without being able to contribute to their research. For example, sociologists of science may, over time, become capable of discussing scientific work on a particular domain without being able to contribute to such scientific research (Collins 2004; Collins and Evans 2015). Given the focus on communicative abilities, interactional expertise is relevant for scientific testimony. For example, science journalists often exemplify people who possess interactional expertise but lacks contributory expertise.

Whereas contributory expertise is often explained by epistemic expertise, an individual who has interactional expertise (but no contributory expertise) with regard to a scientific domain, D, may frequently lack epistemic expertise with regard to D or subsets thereof. For example, a sociologist of science may learn the lingo of a discipline, its journal hierarchy, main figures, etc. But she may be so focused on these social structures that she lacks epistemic expertise concerning what the lingo is about, what is published in the journals, and what the main figures think.

However, interactional expertise does not *entail* the lack of epistemic expertise or contributory expertise. Recognizing this, Collins et al. coin the phrase "special interactional expertise," which is interactional expertise without contributory expertise (Collins et al. 2016). Although this is an important category, scientists frequently possess interactional expertise because they are epistemic experts. Moreover, since contributory expertise is often associated with proper scientific expertise, it is inadequately appreciated that interactional expertise is also an important aspect of scientific expertise in collaborative science (but see Collins and Evans 2015; Collins et al. 2016). In fact, interactional expertise may partly explain why a scientist has contributory expertise. After all, her ability to contribute often depends on her ability to collaborate, which, in turn, partly depends on interactional expertise.

Often, scientists who are valuable in collaboration possess what is sometimes called T-shaped expertise, which is, roughly, the combination of broad superficial (interactional) expertise and domain-restricted deep (epistemic) expertise (Oskam 2009; Enders and de Weert 2009; Conley et al. 2017). It is called T-shaped expertise since the broad superficial expertise may be represented by the horizontal bar of a 'T', whereas the narrow deep expertise may be represented by the vertical part of the 'T'.

T-shaped expertise may be thought of as the type of expertise that enables interdisciplinary collaboration. Interdisciplinary collaboration is characterized by disciplinary *integration* of terminology and methods.[6] This contrasts with mere multidisciplinary collaboration, in which different disciplines are brought to bear on a research problem without any such integration (Oskam 2009; Holbrook 2013). When a biologist takes a water sample and hands it over to a chemist who analyzes it

[6] Oskam 2009; Rossini and Porter 1979; Klein 2005; O'Rourke et al. 2016.

and conveys the results to a medical scientist, the collaboration is merely multidisciplinary because each scientist contributes within their isolated domains.

The idea of T-shaped expertise is used more frequently in HR and management theory than in the philosophy of science. But although the T-shape metaphor has its limitations, it is apt to illustrate a type of expertise that many scientists possess. Scientists tend to possess hyper-specialized domain-restricted epistemic expertise that allows them to contribute to their field as well as interactional expertise that allow them to collaborate with other hyper-specialized scientists. It is in part due to this combination of expertise that they are capable of contributing to interdisciplinary collaborations. So, despite their limitations, the ideas of interactional expertise and T-shaped expertise help to make it vivid that there is more to scientific expertise than what an individualistic conception such as *Epistemic Expertise* might suggest. Moreover, I will argue that they are central for public scientific testimony.

1.2.d Expertise and the reputation of expertise: Some sociological examinations of expertise appear to conceive of expertise as a social construction—i.e., as a concept without a substantive basis over and beyond social conventions. According to one constructivist approach, someone has expertise of a certain kind insofar as she is treated as an expert of this kind in accordance with social conventions. Indeed, a standard criticism of Collins's and Evans's various characterizations of expertise is that they are often more sociological than substantive (see, e.g., Collins 2004: 128). But these proposed criteria of contributory expertise are better thought of as contingent social indicators of it (Goddiksen 2014).

In contrast, I clearly distinguish between indicators of a given type of expertise and the presence of that type of expertise. Indeed, this distinction is required to ask and answer important questions concerning scientific testimony. For example, the problem of pseudo-scientific testimony requires that we be able to clearly distinguish between pseudo-scientific experts and genuine scientific experts. It is a real problem that pseudo-scientific experts manage to succeed in being conventionally treated as scientific experts and fulfill the roles that genuine scientific experts are supposed to fulfill (Dunlap et al. 2011; Oreskes and Conway 2010). However, if scientific expertise is reduced to such conventional social role fulfillment, the pseudo-scientific experts will too often be categorized as scientific experts. Such an approach to the problem of pseudo-scientific testimony obscures the important problem of expert identification (Goldman 2001; Martini 2014; Grundmann forthcoming).

Consequently, I will clearly distinguish between genuine scientific expertise and social conventions or prevalent representation of it. Of course, this stance is a controversial one within some quarters but here I rest content with clarifying the distinction and sketching a (putatively question-begging) motivation for heeding it.

1.2.e Concluding remarks on expertise: As with the other phenomena introduced in this chapter, expertise is an incredibly varied phenomenon calling for a study in its own right. And the complexity is not reduced much by focusing on scientific expertise. So, I have continued my practice of introducing the core distinctions that I will need in order to proceed.

1.3 Science as Collaboration among Scientific Experts

The structure of scientific collaboration is where our two topics *science* and *testimony* intersect. Collaboration among diverse scientists would not be possible unless the scientists could communicate effectively, and the testimonial norms and practices shape the nature of scientific collaboration.

1.3.a The rise of collaboration in science: While the history of science often focuses on individual geniuses, science has developed to be a collaborative affair.[7] Collaboration makes it possible for scientists to investigate areas that would otherwise be impracticable, or even impossible, to investigate. Moreover, both collaboration within research teams and within the larger scientific enterprise increases the accuracy and reliability of scientific investigations. Arguably, most scientific research currently produced could not have been produced by a lone genius. In slogan: *Scientists no longer stand on the shoulders of giants as much as they stand within an edifice built by a myriad of their predecessors and contemporary peers.*

From a sociological point of view, collaboration has become the norm of most scientific research. Measures of collaboration in terms of co-authorship clearly indicate that scientific collaboration in the natural and social sciences has been on the rise throughout the twentieth century (Thagard 1999; Wray 2002, 2015). For example, co-authored publications in the natural sciences from 1920 to 1929 amounted to 49 percent and 6 percent for the social sciences. But in 1950–9 these numbers had risen to 83 percent and 32 percent, respectively.[8] Recent work indicates that the trend toward collaboration has continued (Wuchty et al. 2007; Sonnenwald 2007).

The humanities "show lower growth rates in the fraction of publications done in teams, yet a tendency toward increased teamwork is still observed" (Wuchty et al. 2007: 1037). However, bibliometric analyses of scientific collaboration only focus on co-authorship. This is an imperfect proxy for scientific collaboration. Even in the humanities, where co-authorship remains limited, there is ample

[7] Hardwig 1985; Thagard 1997, 2006; Fallis 2006; Tuomela 2013; Wagenknecht 2016; Miller and Freiman 2020.

[8] Wray 2015 summarizing Zuckerman and Merton 1973.

formal and informal scientific collaboration. Typically, a publication in the humanities has been vetted at conferences, colloquia, research groups, and by individual readers who are experts in the area. (The present book is no exception.)

Following Andersen and Wagenknecht, we may distinguish between *unilateral* epistemic dependence, which concerns some individual scientists' dependence on other scientists, and *multilateral* epistemic dependence, which concerns a group of scientists' epistemic dependence on each other (Andersen and Wagenknecht 2013; Rolin 2020). While the humanities do not exhibit nearly the same degree of multilateral epistemic dependence as the natural sciences do, unilateral epistemic dependence pervades virtually all contemporary scientific work.

Why has collaboration come to dominate science? There are many reasons for this, but a basic one is that collaboration has simple practical advantages with great epistemic payoffs. These include the ability to get research done more effectively and quickly. For example, gathering evidence is often more efficient when done by a group than by an individual, and this may help increase statistical power. Whenever sophisticated technology is required, the numbers of collaborators tend to grow dramatically. The outcome is a phenomenon that is sometimes labelled "radically collaborative research" (Winsberg et al. 2014). The current record is the 5154-author article which presents the evidence for the Higgs boson (Aad et al. 2015). This case, and many less dramatic ones, indicate that the necessity of collaboration typically goes beyond merely having enough hands. It takes a lot of hands to run a network of particle accelerators. But it also takes a lot of minds.

In fact, many scientific fields have become so sophisticated and specialized that it is psychologically impossible for a single individual to acquire all the expertise required to advance science—even if the resources to conduct experiments were available to her.[9] In consequence, collaboration is required insofar as it is a way to assemble the conceptual resources, skills, and methods that are required for contributing to scientific progress. There are also more subtle reasons to divide cognitive labor. For example, Hackett has argued that if a research leader meticulously controlled her collaborators, she might impede their independent creativity (Hackett 2005).

In general, it is important to realize that scientists collaborate for many reasons. As Thagard has argued, scientific collaboration is an important institutionalized form of on-the-job training for scientists as part of their education (Thagard 1999). Other prudential reasons for collaboration may be that it is a way to expand networks, amplify influence, secure funding, share physical resources, and signal competence or status (Wray 2002; Jasanoff 2004b). However, a *central*

[9] Thagard 1997; Winsberg et al. 2014; Huebner et al. 2018; Klausen 2017.

reason for scientific collaboration is that it is conducive to the epistemic goals of science, and this will be my focus.

1.3.b Scientific collaboration as a truth-conducive social structure: As noted, some philosophers of science highlight that scientists collaborate for many reasons, some of which are non-epistemic (Thagard 1999). But although this is an important insight, I will highlight the epistemic reasons for scientific collaboration by arguing that the nature of scientific testimony and the norms that govern it are keys to explain the truth-conducive structure of scientific collaboration (Chapters 4 and 7).

Practical advantages of collaboration, such as enabling hyper-specialization, utilizing complex scientific instruments, and including methodologically diverse approaches, all have epistemic motivations. Each of these advantages is motivated by benefits to at least one of two epistemic dimensions: scope or quality. Let us consider these in turn.

First of all, scientific collaboration may extend the scope of scientific investigations (Sonnenwald 2007). Due to the complexity of the experiments and apparatus required to investigate the domain, collaboration may be required for observation. For example, observing a galaxy far, far away requires extensive scientific collaboration due to the complexity of the instruments involved. Moreover, in many domains, a hypothesis can only be justified by combining methods from distinct disciplines. For example, investigating a hypothesis concerning how CO_2 omissions impact coral reefs will presumably require interdisciplinary collaboration among, for example, chemistry, atmospheric composition research, marine biology, oceanography, etc. Moreover, the partial investigations in these areas call for intra-disciplinary collaboration in larger research units. An individual marine biologist cannot measure the decay of the Great Barrier Reef. So, without interdisciplinary as well as intra-disciplinary collaboration, the scientific investigation of the hypothesis would be epistemically compromised. Hence, collaboration may increase the epistemic force of science in the sense that it extends the *scope* of human cognition.

Secondly, scientific collaboration may increase the epistemic force of science in the sense that it may increase the degree of scientific justification that a theory or hypothesis enjoys. The increase in scientific justification may come about in many ways. A relatively straightforward way occurs when a larger pool of evidence is gathered and analyzed. This may improve the inductive justification for a hypothesis. For example, an instance of radically collaborative research in genomics resulted in a paper with 1014 authors (Leung et al. 2015). Over nine hundred of the co-authors were undergraduate students who manually corrected and annotated a stretch of the fruit fly genome. According to the lead author, this required extensive data analysis and, hence, a "significant intellectual contribution" on part of each student (Woolston 2015). Even if granting the students co-authorship is controversial, the collaborative process increased the strength of justification for the hypothesis. However, collaboration may also boost the strength of scientific justification in other ways. For example, it may help to minimize performance mistakes or idiosyncratic biases of individual authors (Longino 1990, 2002). Moreover, collaboration may consist in methodological triangulation—i.e., the use of independent methods that may jointly provide a stronger justification than a single method may produce (Kuorikoski and Marchionni 2016; Heesen et al. 2019). Mixed-methods approaches may combine qualitative and quantitative methodologies. But other combinations are possible. For example, research in anthropology may combine evidence from two qualitative methods, such as structured interviews and participant observation (Creswell and Clark 2017). Since the state of the art of each method requires considerable specialization, collaboration is required. So, given that collaboration is required to reap the epistemic advantages of triangulating mixed methods, collaboration may be epistemically beneficial (Miller and Freiman 2020; Dang 2019).

The idea that scientific collaboration is epistemically advantageous has also been explored empirically by investigating correlations between co-authorship and impact. For example, empirical research indicates that collaborative research—as measured by co-authorship—is more epistemically impactful—as measured in terms of citations—than non-collaborative research.[10] Citations are a reasonable proxy for impact, and some studies control for self-citations.

Citation rates are not clear-cut proxies for quality, but some empirical work supports a correlation. For example, studies have found that patent citations correlate with market value (Hall et al. 2005). More pertinent to scientific quality, findings indicate that citation rates correspond reasonably well with authors' self-assessment of the scientific contribution—modulo perceived overcitation of survey articles (Aksnes 2006). So, the robust finding that scientists are more likely to cite co-authored papers than single authored ones may be taken as a fallible measure of trust, which may, in turn, fallibly reflect epistemic quality (Korevaar and Moed 1996 corroborates this assumption in mathematics).

Some empirical research addresses the concern that increased citations are explained by an effect of network and increased visibility rather than by a quality effect. For example, Franceschet and Costantini examined the referees' quality ratings of about 18,500 research products and found that in "8 out of 14 areas ... the average peer rating increases monotonically with the number of authors" (Franceschet and Costantini 2010: 546). Interesting exceptions include mathematics and computer science, where the best-rated papers had two authors (Franceschet and Costantini 2010: 547).

[10] Hull 1988; Wray 2002; Beaver 2004; Wuchty et al. 2007; Schmoch and Schubert 2008; and Franceschet and Costantini 2010, which surveys parts of the literature.

However, it is challenging to provide empirical evidence of epistemic quality precisely because it is controversial how to measure it (Schmoch and Schubert 2008). Nevertheless, the existing empirical research and the noted methodological reflections provide reason to assume that collaboration can improve both the scope and strength of justification of scientific research. More specifically, the considerations motivate the thought that scientific collaboration contributes immensely to science's epistemic force—i.e., to its capacity for generating warrant for hypotheses about a wide range of issues. By saying that the epistemic contribution of scientific collaboration is "immense," I mean that it is among the main contributors to the epistemic force of science, but not that it is the sole or the primary contributor. Given these qualifications, the preceding discussion may be summed up in the following doctrine:

Collaboration's Contribution

Scientific collaboration contributes immensely to the epistemic force of science.

This thesis will serve as a premise in arguments for assuming that intra-scientific testimony is a vital part of science and that the norms governing it are vital to the scientific methods of collaborative science (Chapter 7). So, I will return to it and justify it from various angles as the investigation unfolds.

1.3.c Intra-scientific testimony as a vital part of scientific collaboration: Given that the epistemic force of science partly derives from collaboration among hyper-specialized scientists, communication between them is paramount. Arguably, there would be little point in collaborating if the collaborators did not communicate with one another, at least centrally in the form of inter-scientific testimony. Lipton puts this point as follows: "At least most of the theories that a scientist accepts, she accepts because of what others say. The same goes for almost all the data, since she didn't perform those experiments herself. Even in those experiments she did perform, she relied on testimony hand over fist" (Lipton 1998: 1; Rolin 2020).

Given that inter-scientific testimony plays such a role in scientific collaboration, it is aptly characterized as a vital part of scientific collaboration in the sense that it is a metaphysically contingent, but nevertheless principled, part of scientists' practice of collaboration. Thus, the idea of *vitality* contrasts with, on the one hand, *essence* and, on the other hand, *enabling conditions*. For example, my hip joint is a vital part of my body that is central to a principled account of my pedestrian practices. However, it is a contingent part of my body insofar as it could—perhaps sooner than I would like to think about—be replaced with a hip prosthesis. Likewise, intra-scientific testimony is a vital part of scientific collaboration in virtue of contributing in a principled manner to its pursuit of its epistemic functions. I will elaborate on the notion of vitality in Chapter 7.1.b.

But at this point of the investigation, I will rely on the notion to draw a sub-conclusion from the preceding discussion of scientific collaboration.

Testimony's Contribution

Intra-scientific testimony is an epistemically vital part of scientific collaboration.

The preceding reflections on the epistemic functions of scientific collaboration and intra-scientific testimony's role in it motivate *Testimony's Contribution*. Scientific collaboration partly consists in communicating findings, analyses, criticisms, ideas, and perspectives among collaborators and much of such communication takes the form of intra-scientific testimony. Thus, intra-scientific testimony is part of what scientists do when they collaborate. Given the assumption, *Collaboration's Contribution*, that scientific collaboration contributes immensely to the epistemic force of science, it is reasonable to suppose that intra-scientific testimony is, moreover, an epistemically *vital* part of scientific collaboration. As science is in fact practiced, the epistemic contribution of scientific collaboration would be seriously compromised if intra-scientific testimony were not a part of it.

This thesis, *Testimony's Contribution*, will also serve as a premise in an argument for the more controversial further conclusion that intra-scientific testimony is a vital part of collaborative science (Chapter 7.2.b). However, *Testimony's Contribution* is a substantive sub-conclusion in its own right. So, I will both invoke it and motivate it further throughout the book.

1.3.d Scientific collaboration and scientific testimony: I opened this section by claiming that scientific collaboration is where *science* and *testimony* intersect. Hopefully, some dimensions of this intersection are now a bit clearer. In particular, I have highlighted that intra-scientific testimony is a vital part of scientific collaboration in virtue of contributing significantly to its truth-conducive nature.

However, there is a further dimension to this line of thought. Whereas collaboration has obvious epistemic and practical advantages, it is not clear what is distinctive of *scientific* collaboration. But although it may not be a defining trait, it is clear that scientific collaborations are characterized by a fine-grained division of cognitive labor. So, in order to consider the roles of scientific testimony, the division of cognitive labor requires some discussion.

1.4 The Division of Cognitive Labor

Scientific collaboration involves hyper-specialization, and when it is epistemically superior to other kinds of collaboration, this is a central reason why. The division of cognitive labor is often taken to be central to the scientific method and the epistemic success of science. I will argue that scientific testimony is critical in structuring the division of cognitive labor in science and between science and society. So, in this section, I will briefly survey select research on the division of cognitive labor and clarify how it relates to the questions concerning varieties of scientific testimony.

1.4.a The structure of science and invisible hands: Theorizing about the structure of scientific collaboration has been on the rise over the last half century or so. An important impetus was Kuhn's claims that dogmatism on part of individual scientists in periods of normal science is a requirement for effective collaboration (Kuhn 1962/2012). However, Kuhn also emphasizes that dogmatism leads to the accumulation of anomalies, which may eventually lead to a scientific revolution. So, according to Kuhn, the dogmatism of individual scientists helps to explain that science as a communal enterprise is dynamic over time. This relationship between individual dogmatism and communal dynamism suggests a tension between individual rationality and the rationality of the scientific community (Kuhn 1977). Consequently, Kuhn's account of the structure of science has been compared to Smith's *invisible hand* account of economic markets, given the tension between dogmatic and narrow-minded individual scientists who form a dynamic open-minded scientific community (cf. Godfrey-Smith 2003: 99ff.; Rowbottom 2011; Strevens 2011).

Merton's sociology of science added the assumption that the distribution of *recognition* for being the first to publish a novel finding is the central reward system in science (Merton 1973). The focus on science as a norm-governed reward system alongside invisible hand dynamics has shaped much work on the division of cognitive labor in science. For example, both ideas figure prominently in Hull's influential discussion of scientific processes as driven by the individual scientist's desire for recognition in the form of use of their ideas and citations (Hull 1988). According to Hull, such incentives structure scientific collaboration, and this bears on testimony's role in science. Specifically, Hull argues that scientific collaboration requires trust and that breaches of such trust are met with extremely harsh sanctions (Hull 1988).

Kitcher develops this interplay between the rational choices of individuals and a reasonable overall distribution of individuals across different scientific programs as a problem of coordination of the scientific community's resources (Kitcher 1990, 1993). He highlights that pursuing potentially groundbreaking research programs is a high-risk/high-benefit strategy, whereas engaging in what Kuhn would regard as *problem-solving* is a low-risk/low-benefit strategy. According to Kitcher, researchers pursue the research projects where they can rationally expect their contribution to be most significant. The individual maximization of expected payoff helps to ensure that scientists specialize in narrow domains and that scientists are distributed reasonably among competing research programs. So, according to Kitcher, the incentive structure of science yields a distinctively fine-grained *division of cognitive labor* (Kitcher 1990, 1993; see also Muldoon 2013).

However, according to a development by Strevens, the benefits of partaking in a successful research program are not evenly distributed but rather distributed in accordance with the individual scientist's contribution (Strevens 2003). Specifically, Strevens argues that this reward system explains the *Priority Rule*, according to which scientific findings are primarily valuable the first time they are produced, and that novelty is, therefore, prized extraordinarily highly in science (Strevens 2003). This, in turn, helps explain the division of cognitive labor in science (Strevens 2003, 2011, 2017). Kitcher's and Strevens's approaches have been dubbed the *Marginal Contribution/Reward model* (Muldoon 2013). This approach has been criticized on various grounds. For example, Solomon has warned against an overly idealized rationalization of the scientists' choices by highlighting a theme that I will return to: Scientists' choices are driven by cognitive heuristics that must be understood in relation to the relevant social environment (Solomon 1992, 2006a). Moreover, alternative formal models of cognitive diversity have emerged, such as Muldoon and Weisberg's *Epistemic Landscape* approach.[11]

Despite importantly different theoretical approaches to the division of cognitive labor, the body of research on it generally substantiates the idea that scientific collaboration helps increase the truth-conduciveness of science. Thus, reflection on the division of cognitive labor augments the motivation for the thesis, *Collaboration's Contribution*, according to which scientific collaboration contributes immensely to the epistemic force of science. Given the further assumption that an epistemically efficient, fine-grained division of cognitive labor must involve intra-scientific testimony, the considerations also augment the motivation for *Testimony's Contribution*. That is, the broad image of scientific collaboration as epistemically forceful in virtue of a fine-grained division of cognitive labor supports the thesis that intra-scientific testimony is an epistemically vital part of scientific collaboration.

The idea that scientific collaboration is epistemically forceful due to its fine-grained division of cognitive labor may also be cast in terms of Mandevillian intelligence—i.e., the idea that "individual cognitive vices (i.e., shortcomings, limitations, constraints and, biases) are seen to play a positive functional role in yielding collective forms of cognitive success" (Smart 2018; Peters forthcoming a). In pursuing this idea, it is important to avoid overly idealized assumptions about the rationality and self-interestedness of scientists and the resulting collective rationality. For example, the replication crisis in social psychology may be partly explained by incentives to be the first to publish striking findings. So, a general

[11] Weisberg and Muldoon 2009; Muldoon 2013. For criticism, see Alexander et al. 2015.

lesson may be that neither cognitive diversity nor invisible hand dynamics automatically guarantee truth-conduciveness (Solomon 2006a; Intemann 2011; Eagly 2016). Rather, the truth-conduciveness of scientific collaboration and epistemic benefits of diversity hinges on the general social environment and, in particular, the norms and incentives that structure scientific collaboration. Indeed, norms of providing and receiving intra-scientific testimony are required for effective scientific collaboration given that it is characterized by a fine-grained division of labor (Strevens 2011). Likewise, the norms governing the reward schemes that help structure scientific collaborations have consequences for scientific testimony. So, considering scientific norms brings the considerations about the division of cognitive labor to bear on the theme of scientific testimony.

1.4.b The inter-scientific, intra-scientific, and societal division of cognitive labor: To avoid equivocations, it is important to distinguish three types of division of cognitive labor: *inter*-scientific, *intra*-scientific, and *societal* division of cognitive labor. This distinction is broadly mirroring the trifold distinction among inter-scientific, intra-scientific, and public scientific testimony.

A good deal of the research surveyed above concerns the *inter*-scientific division of cognitive labor, which may be provisionally characterized as the division of cognitive labor of the scientific community as a whole. A prominent example is the allocation of scientific resources (including scientists) to research projects. While this is a very important topic, I will pay more attention to the *intra*-scientific and *societal* divisions of cognitive labor. The intra-scientific division of cognitive labor may be provisionally characterized as the divisions of cognitive labor that take place within the scientific practice. This is most vividly present in multi- and interdisciplinary collaborations (Wagenknecht 2016; Miller and Freiman 2020). But specialization can be equally important in complex collaborations within a single discipline. This is important to recognize because conclusions concerning inter-scientific division of cognitive labor do not automatically hold for its *intra*-scientific counterpart. For example, Kitcher's and Strevens's focus on competition among research projects does not clearly apply to intra-scientific division of cognitive labor in a collaborative research project that is up and running. For example, the *Priority Rule* may govern the question of which collaboration or role in collaboration that scientists select. But it is less clear that it governs their collaborative practice once it is under way. Thus, the intra-scientific division of cognitive labor should be investigated in its own right, and I will begin to do so by considering the question: How does intra-scientific testimony relate to the intra-scientific division of cognitive labor? This is a question that intersects philosophy of science and social epistemology.

The *societal* division of cognitive labor may be broadly characterized as the divisions of cognitive labor that take place within society at large. However, since science plays a distinctive and prominent epistemic role in society, the societal division of cognitive labor also raises a number of questions: How may the epistemic authority of science be reconciled with democratic values? Should science be representative of the cognitive diversity in society? And, most pertinently for the present treatise: How should scientists disseminate their research to the public? These questions intersect philosophy of science and political philosophy.

The distinctions among the inter-scientific, intra-scientific, and societal division of cognitive labor are not always clearly drawn, although they are occasionally noted in differing terminologies.[12] Such variation in terminology and fine-grainedness of distinctions is unavoidable. The key point is that the different types of division of cognitive labor raise importantly different questions about scientific testimony (Whyte and Crease 2010).

For example, reflection on the intra-scientific division of cognitive labor raises questions about the roles and norms of intra-scientific testimony (Chapter 4). Another important aspect of intra-scientific division of cognitive labor has been brought to the stage by studies that suggest that diversity in a group of investigators is epistemically beneficial. For example, Fishkin and Luskin found that groups of managers, jurors, and students *outperformed their best member* in tasks ranging from mathematics to political matters (Fishkin and Luskin 2005; but see Eagly 2016). Likewise, Hong and Page argued for a *diversity trumps ability* theorem according to which: "a random collection of agents drawn from a large set of limited-ability agents typically outperforms a collection of the very best agents from that same set" (2004: 16,386). While the theorem has received criticism, subsequent work has extended this broad idea to various models of scientific collaboration.[13] For example, it has been argued that it is epistemically valuable to pursue an epistemically diverse community of scientists (Grim et al. 2019; Singer 2019; but see Eagly 2016 for important reservations). Relatedly, Thoma distinguishes between *explorers* who follow "approaches that are very different from those of others" and *extractors* who pursue work "that is very similar to but not the same as that done by others" (Thoma 2015). Thoma then argues, via a version of Weisberg and Muldoon's *epistemic landscape* model, that groups including both "explorer" and "extractor" scientists outperform groups composed of either one type of scientist.[14]

Interestingly, some of the formal work on cognitive diversity aligns broadly with arguments for diversity from feminist philosophy of science. Feminist

[12] For example, Kitcher's notion of division of epistemic labor is roughly equivalent to what I call the societal division of cognitive labor (Kitcher 2011: 20ff.). In contrast, his notion of cognitive division of labor appears to straddle what I call inter-scientific and intra-scientific division of cognitive labor, although his discussions mainly concern the former (Kitcher 1990; 2011: 249 fn. 3). See also Wilholt 2013; Origgi 2015; Keren 2018.

[13] Weisberg and Muldoon 2009; Zollman 2010; Muldoon 2013; Stegenga 2016; Bright 2017.

[14] Thoma 2015. For further developments, see Zollman 2010; Stegenga 2016; Bright 2017.

philosophers of science have long emphasized the epistemic benefits of including a plurality of perspectives in science. For example, Longino argues that there are epistemic advantages to increasing the pool of ideas and perspectives in science (Longino 1990, 2002). Moreover, standpoint theorists have argued that marginalized groups may be in a privileged epistemic position with regards to specific domains of inquiry (Harding 1991, 2004). However, it has been highlighted that the epistemic benefits of cognitive diversity depend on the general social environments and its characteristic norms (Solomon 2006a, 2006b; Intemann 2011). But despite substantive differences in feminist philosophies of science, a common denominator is a plea for cognitive diversity and social conditions in which it may flourish. So, generally, an important aspect of both inter-scientific and intra-scientific division of cognitive labor pertains to epistemic diversity among scientists. It is reasonable to suppose that the epistemic benefits of intra-scientific collaboration between cognitively diverse scientists require norm-governed intra-scientific testimony (Longino 2002: 129ff.). If so, this body of work may also support the thesis *Testimony's Contribution*, according to which intra-scientific testimony is an epistemically vital part of scientific collaboration. However, *Testimony's Contribution* is compatible with the idea that cognitive diversity is not *invariably* beneficial. In fact, it is important to acknowledge that it may, in some cases, even be epistemically detrimental, e.g., via weakening social cohesion (Ely and Thomas 2001; Galinsky et al. 2015; Eagly 2016).

How does the societal division of cognitive labor bear on the roles of scientific testimony? Questions arising from reflection on the societal division of cognitive labor do not solely concern the supplier side—i.e., scientific experts and science reporters. They also concern the consumer side—i.e., elected decision-makers and the general lay public. Hence, a sensible societal division of cognitive labor does not merely consist in a reasonable distribution of scientists with contributory expertise within science and within society. It also involves a reasonable distribution of individuals with *interactional* expertise and an *appreciative deference* on part of laypersons—i.e., an appreciation of the fact that they are standardly in an inferior epistemic position compared to the state-of-the-art science (I elaborate on appreciative deference in Chapter 7.4.c). Thus, the societal division of cognitive labor raises questions about the proper roles of scientific testimony in the public sphere. So, an account of public scientific testimony must consider both the testifiers and the audience (Keren 2018). Doing so requires integration with the empirical science of science communication. Hence, the pursuit of such integration is among this book's central methodological aims.

The debates concerning cognitive diversity also impact the societal division of cognitive labor. For example, Hong and Page's *diversity trumps ability* theorem has been invoked in defenses of democracy (Landemore 2012, 2013). Philosophers of science may consider whether the elitist and meritocratic scientific institutions adequately represent the diverse lay public. Such considerations are important given that the epistemic status of many of our beliefs is dependent on our social environment. Much of our everyday knowledge is epistemically dependent in a diffused manner which, roughly, consists in our epistemic position being dependent on social facts that go beyond direct testimony (Goldberg 2011; Gerken 2013b). For example, distributed credibility monitoring underwrites the phenomenon of *coverage*, which occurs when we form abductively warranted beliefs on the basis of absence of testimony (Goldberg 2011; Pedersen and Kallestrup 2013). Public scientific testimony plays important roles in credibility monitoring and coverage. These roles include myth-busting and rebutting pseudo-scientific testimony. One consequence is that if science is not representative of the cognitive diversity in the lay population, some groups will be in a significantly worse epistemic position than others. This may be regarded as a kind of epistemic injustice (Fricker 2007, 2013; Gerken 2019, 2020e).

1.4.c Concluding remarks on the division of cognitive labor: Research on the intra-scientific division of cognitive labor helps explain why scientific collaborations are central to the epistemic success of science and underscores the thesis, *Testimony's Contribution*, that intra-scientific testimony is an epistemically vital part of scientific collaboration. Thus, I have begun to argue that the hyper-specialized division of cognitive labor that characterizes scientific collaboration requires intra-scientific testimony. Finally, I have emphasized how reflection on the *societal* division of cognitive labor raises hard questions about the balance between respect for epistemic diversity in society and the epistemic authority of public scientific testimony. So, all in all, the prominent discussions about scientific collaboration and the division of cognitive labor raise complex questions about the nature and roles of various types of scientific testimony. One way to get clearer on the roles that varieties of scientific testimony play consists in reflecting on the norms governing it. In the final section of the chapter, I will initiate such reflection.

1.5 Norms of Scientific Testimony

In this section, I initiate an approach that will be prominent throughout the book. This is the approach of viewing scientific testimony through the lens of the norms that govern it. In doing so, I set forth a substantive thesis, *Distinctive Norms*, that the epistemic contribution of scientific collaboration depends on distinctive norms of intra-scientific testimony.

1.5.a How norms, guidelines, and incentives structure scientific collaboration: The term 'norm' is both polysemous and controversial. So, let me gloss over how I use the term in a preliminary way (which I will elaborate in Chapter 2.4.c). I take *norms* to be objective benchmarks of assessment that the agent need not have any cognitive access to. So, even though scientific norms are social and contingent, they are typically tacit, and they need not be conceptualized by scientists. In contrast, *guidelines* are met only if they are, in some sense, sufficiently appreciated to be *followed* by the agent (for elaboration, see Gerken 2017a: ch. 6.1). A guideline may be a simplified approximation of the norm which is feasible to conceptualize and follow. Indeed, guidelines may figure in textbooks. For example, there are textbooks of how to conduct structured interviews or perform statistical analysis. In contrast, an aspiring scientist often develops a sensitivity to norms by taking part in a scientific practice.

Norms are frequently postulated on the basis of reflection on scientific *practice*, or, more ambitiously, reflection on the nature and aims of science. But the norms are typically tacit and, hence, distinct from *guidelines* that the scientists explicitly consider. Nevertheless, such methodological norms may indirectly structure scientific practice, and this is visible in cases of norm violation which are associated with implicit or explicit sanctions (Bicchieri 2006; Graham 2015b; Fricker 2017). To some extent, the risk of such sanctions regulates scientific practice. Moreover, the distribution of incentives such as prestige, funding, and such also reflect scientific norms. That is, norms of science reflect both the sticks of sanctions and the carrots of rewards. Hence, it is a central task for philosophy of science to articulate the norms distinctive of scientific practice.[15] A more ameliorative project consists in articulating norms and guidelines that scientists ought to follow. With regard to intra-scientific testimony, I will primarily articulate the operative norms of scientific practice, and with regard to public scientific testimony, I will have more ameliorative ambitions.

1.5.b A thesis concerning norms of intra-scientific testimony: Since understanding the scientific practice and methodology partly consists in identifying the norms governing it, scientific testimony may also be investigated by considering the norms governing it. Hence, I will focus on norms for both intra-scientific testimony and public scientific testimony. As noted, this approach extends general philosophy of science, and it has also influenced some discussions of scientific testimony.[16] Note that this ambition does not entail that distinctive norms of science amount to a demarcation criterion or that they are individually necessary for science. Rather, the thesis is a fairly modest one:

[15] Hull 1988; Longino 2002; Strevens 2003, 2017; Gerken 2015a.

[16] See, for example, Fricker 2002; Douven and Cuypers 2009; Olsson and Vallinder 2013; Mayo-Wilson 2014; Gerken 2015a, 2018; Zollman 2015.

> ***Distinctive Norms***
> The epistemic contribution of scientific collaboration depends on distinctive norms of intra-scientific testimony.

To say that the norms of intra-scientific testimony are distinctive amounts to the idea that they go beyond generic normative requirements that apply to all types of testimony. *Distinctive Norms* is best motivated by way of concrete norms of intra-scientific testimony. Consequently, I will postpone my main defense of it until I have articulated some such norms. Specifically, I will articulate concrete norms of production as well as the uptake of such norms in Chapter 4 and thereby motivate *Distinctive Norms* by example.

However, *Distinctive Norms* may gain a more general motivation from the general idea that a practice depends on norms governing aspects of it.[17] Rescorla articulates the idea as follows: "Every practice is associated with "internal" standards of normative assessment codified by norms dictating how to execute the practice correctly" (Rescorla 2009: 101). The fact that norms govern correct engagement in a practice is central to the distinction between practices and activities: "The contrast between practices and mere activities suggests the following formulation: a norm is constitutive of a practice iff one must obey the norm to engage correctly in the practice" (Rescorla 2009: 101). The epistemic norms of assertion are often taken to be (partly) constitutive of this speech act insofar as they help characterize it and distinguish it from other speech acts (Williamson 2000). Williamson's knowledge norm has been roundly criticized.[18] But the broader assumption that the epistemic norm of assertion is constitutive of assertion is widely shared. In comparison, *Distinctive Norms* is more modest given that it is compatible with, but does not require, that the norms in question are constitutive.

This general motivation for *Distinctive Norms* may be augmented by analogy with norms of scientific collaboration that do not directly involve testimony. For example, the norms that govern the choice of statistical test and the norms that restrict *p*-hacking generally contribute to a truth-conducive data analysis (UCLA Statistical Consulting Group 2021). Similarly, the practice of blinding procedures in many types of experiments is governed by social norms distinctive of the relevant scientific discipline. For example, randomized controlled trials are standardly double blinded in the biomedical sciences. That is, the participants and the researchers monitoring the effects of a treatment are unaware whether participants are undergoing treatment or are in a control condition receiving an inert placebo. In many disciplines, the norm of double blinding is so established that it

[17] Rescorla 2009; Burge 2011; Graham 2015b; Bicchieri 2017; Fricker 2017.
[18] For my line of criticism and positive alternatives, see Gerken 2011, 2012a, 2014, 2017a, 2017b, 2018d, 2020b, 2021.

figures as an explicit guideline in textbooks (e.g., Aschengrau and Seage 2020: 196). Like many other social norms, the epistemic norms of intra-scientific testimony tend to be tacit (Bicchieri 2006, 2017; Graham 2015b). But they may nevertheless contribute to the epistemic force of scientific collaboration as *Distinctive Norms* has it. The assumption that they do so is motivated by the idea that norms of intra-scientific testimony are part of what allows scientific collaborators to communicate effectively. Without such norms, the collaboration will be impeded in a manner that compromises its epistemic contribution. So, reflection on scientific collaboration and the intra-scientific division of cognitive labor motivates the assumption that norms govern intra-scientific testimony. Likewise, reflection on norms of speech acts motivates the idea that these norms are distinctive in the sense that they go beyond what testimony generally requires and contribute to the epistemic aims of science. In Chapter 4, I will give concrete examples of such distinctive scientific norms.

1.5.c Concluding remarks of norms of scientific testimony: Since understanding scientific practice and methodology partly consists in understanding the norms governing it, scientific testimony may also be investigated by considering those norms. Hence, I will consider norms for both intra-scientific testimony and public scientific testimony. As noted, this approach extends general philosophy of science, and it has also influenced some discussions of scientific testimony.[19]

It is a methodological ambition of this book to integrate research on norms in philosophy of science with research on norms on testimony in epistemology and philosophy of language. An effort to consider the norms of scientific testimony may benefit from a closer integration with epistemological work on testimony and, more generally, assertive speech acts, which is also revolving around the pursuit of epistemic norms.[20]

Philosophy of science and social epistemology may also have ameliorative ambitions. At any rate, I have. Consequently, I will also seek to articulate norms that *ought* to govern the practice of scientific testimony as well as more concrete guidelines for testifying in accordance with such norms. This ambition primarily concerns public scientific testimony where the operative social norms of, for example, science journalism often differ considerably from objective epistemic norms. Consequently, the latter cannot be identified solely by reflection on the practices of science reporting. Rather, a philosophical investigation with ameliorative ambitions must be informed by empirical research on how various types science communication fare with regard to laypersons' uptake. Therefore, the

[19] See, for example, Fricker 2002; Douven and Cuypers 2009; Olsson and Vallinder 2013; Mayo-Wilson 2014; Gerken 2015a, 2018a; Zollman 2015.

[20] Brown and Cappelen 2011; Littlejohn and Turri 2014; McHugh et al. 2018. Gerken and Petersen 2020 gives a survey whereas Johnson 2015 criticizes the idea that norms of assertion should inform the epistemology of testimony.

philosophical investigation must be integrated with the empirical science of science communication.[21]

1.6 Concluding Remarks

The key ambition of this chapter has been to put on the table central pieces of the puzzle about the role and nature of scientific testimony. These pieces include some varieties of scientific testimony, varieties of scientific expertise, varieties of scientific collaboration, and various divisions of cognitive labor. Moreover, I have emphasized how the structure of scientific collaboration is governed by norms that help ensure a truth-conducive division of cognitive labor. Both inter-scientific and public scientific testimony are governed by such norms. Thus, one ambition of what is to come is to articulate the norms that govern, or should govern, varieties of scientific testimony.

My aim in this opening chapter, however, has primarily been to introduce the issues to be discussed, to draw some conceptual distinctions, and to introduce some associated terminology: the distinction between intra-scientific and public scientific testimony (and, within the latter, the sub-distinction between scientific expert testimony and science reporting); the distinctions among epistemic, contributory, interactional, and T-shaped expertise; the distinction between unilateral and multilateral epistemic dependence; the idea of radically collaborative research; the distinctions between norms and guidelines; the distinctions between intra-scientific and societal division of cognitive labor, etc.

Apart from putting these pieces of the puzzle on the table, I have sought to indicate the agenda by articulating substantive proposals. For example, I have set forth the thesis *Collaboration's Contribution*, that scientific collaboration contributes immensely to the epistemic force of science. Likewise, I have articulated and begun to motivate the thesis *Testimony's Contribution*, that intra-scientific testimony is an epistemically vital part of scientific collaboration, and the thesis *Distinctive Norms*, that the epistemic contribution of scientific collaboration depends on distinctive norms of intra-scientific testimony. These theses will be invoked in arguments for a broad *testimony-within-science* picture, which will be developed from various angles throughout the book and culminate in Chapter 7.

[21] Burns et al. 2003; Fischhoff 2013; Dunwoody 2014; Kahan 2015a; Jamieson et al. 2017; John 2018.

2
The Nature of Testimony

2.0 Testimony as a Source of Epistemic Warrant

The main purpose of this chapter is to consider some of the general foundational properties of testimony as a source of epistemic warrant and knowledge. This is a major research area in social epistemology (Gelfert 2014 and Shieber 2015 provide fine overviews). In consequence, my selective discussion will focus on the aspects of testimony that are important for understanding scientific testimony. For example, some debates that dominate epistemology will be treated rather superficially, and I will not be shy about making assumptions without defending them at length. In short, I seek to say only what needs to be said about testimony in general in order to home in on scientific testimony.

In Section 2.1, I consider the nature of testimony, testimonial belief, and testimonial acceptance. In Section 2.2, I discuss how some prominent debates in epistemology bear on scientific testimony. In Section 2.3, I consider the social basis for testimonial warrant and the social environment that is relevant for scientific testimony. In Section 2.4, I consider the interplay between the recipients' personal vigilance and the larger social environment. This includes a return to the idea that testimony and its uptake are governed by norms.

2.1 Testimony, Testimonial Belief, and Acceptance

To understand the varieties of scientific testimony, it will be useful to situate them in a more general account of the nature of testimony. So, in this section, I will briefly address this question. On this basis, I will consider what a testimonial belief is and note an important distinction between belief and acceptance. I conclude the section by going from an individualistic perspective to a social one by considering how we should think about group testimony, belief, and acceptance.

2.1.a Testimony as a speech act: Given that testimony is an important source of belief and knowledge, it is no surprise that epistemology and the philosophy of language feature debates about how testimony is best characterized. It is no less surprising that it has proved hard to provide a reductive definition.

Scientific Testimony: Its roles in science and society. Mikkel Gerken, Oxford University Press. © Mikkel Gerken 2022.
DOI: 10.1093/oso/9780198857273.003.0003

A natural account has it that (verbal) testimony is the linguistic expression of a speaker's belief. Such an account is sometimes called the "broad view."[1] However, the broad view is criticized on the grounds that it classifies all lying as non-testimony (Cullison 2010; Shieber 2015). Coady sets forth an alternative "narrow view" which requires, among other conditions, that a testifier "has the relevant competence, authority, or credentials to state truly that P" (Coady 1992: 42). But this is misguided because it runs together the conditions concerning epistemically permissible testimony with the conditions for what testimony is (Fricker 1995; Graham 1997).

Lackey distinguishes between speaker and hearer testimony and provides a disjunctive account encompassing both: "S testifies that *p* by making an act of communication A if and only if (in part) in virtue of A's communicable content, (1) S reasonably intends to convey the information that *p* or (2) A is reasonably taken as conveying the information that *p*" (Lackey 2008: 35–6; Lackey 2006b). But Cullison argues that speaker testimony, supposedly captured by the first disjunct, (1), cannot account for cases in which a speaker testifies against her intention under pressure (Cullison 2010; see Saul 2002 for general considerations against intentional approaches). Cullison's account retains (2) but Faulkner argues that it is too inclusive since S's communicable content may reasonably be taken to convey propositions that the speaker does not testify (Faulkner 2011). Shieber illustrates this by noting that a racially insensitive assertion may reasonably be taken as conveying that the speaker is racially insensitive, although he clearly does not assert that (Shieber 2015: 15–16). Consequently, Shieber prefers an account that requires that the recipient takes S to convey information that *p* by performing A. But this does not appear to be able to account for unsuccessful testimony. If I assert "we had a ball" and you think we had a spherical recreational device when I was talking about a party, my assertion was still a testimony.

This brisk tour of the debates illustrates the difficulties of providing a reductive definition of testimony. But, as is often the case, there are lessons to be learned from reflection on unsuccessful definitions. One such lesson is that a characterization of testimony should be clearly distinguished from the epistemic constraints on testimony. Another putative lesson is that both intention-based and recipient-based views face challenges.

However, a characterization of (linguistic) testimony may be approximated from a speech act theoretic perspective (Bach and Harnish 1979: 1.3; Graham 1997, 2015a). Often testimony is characterized as a type of assertion that is offered as a ground for belief. For example, Gelfert provides the following gloss: "the assertion, by a speaker (or author), of a spoken (or written) sentence, which is offered as a ground for belief to the hearer (or reader)" (Gelfert 2014: 232; see also

[1] See Lackey 2008, who associates it with Fricker 1995: 396–7; Audi 1997: 406; Sosa 1991: 219.

Hinchman 2005). While I think this is on the right track, I will broaden the characterization in two ways. First, the notion of *belief* should be construed broadly to include less committal attitudes such as credences. Second, testimony may also offer a ground for *acceptance*, which is a less committal attitude to be discussed shortly (Cohen 1992). This characterization raises questions about what it takes for an assertion to be offered as a ground for belief or acceptance. One answer is a broadly Gricean intention-based account (Graham 2015a). A standing concern with such an account is that it hyper-intellectualizes testimony and testimonial belief. However, a Gricean account may be developed in a manner that only requires fairly primitive cognitive competences (Bach and Harnish 1979; Moore 2017, 2018). An alternative view is that the conversational context determines whether an assertion is offered as a ground for belief or acceptance. I will not attempt to settle this subtle matter, as testimony is closely related to assertion in either case. I also leave open whether assertion is a broader notion that may serve communicative purposes other than offering a ground for belief or acceptance.

Thus, the following speech act theoretic gloss is a reasonable approximation of the type of testimony that will be most relevant to an investigation of scientific testimony:

Testimony

Testimony that *p* is an assertion that *p*, which is offered as a ground for belief or acceptance that *p* on its basis.

Importantly, this gloss does not provide a reductive analysis of testimony but a working characterization of the type of testimony that I will focus on. For example, the gloss leaves out non-propositional forms of testimony, such as gestures or pictures (Lackey 2008). Moreover, a characterization of the terms 'testimony' and 'testify' should reflect that both terms are polysemous in ordinary language and that further technical uses are found in law, religion, and elsewhere (Graham 2015a). I will focus the discussion on cases in which a testimony is offered as a ground for believing its *content*, and this is reflected in the present characterization. The philosophy of language features further subtleties concerning, for example, conversational implicatures (Grice 1989), impli*cit*ures (Bach 1994), and explicatures (Sperber and Wilson 1986/1995; Carston 2002). There are debates over the epistemic norms for asserting that *p* with the dominant communicative function of conveying that *q* (Fricker 2012; Gerken 2017a: chs. 7–8). Likewise, complexity arises when a public scientific testimony that *p* predictably leads to other consequences than the audience coming to believe or accept that *p* or increase their credence in *p*. Some of these thorny issues will become relevant in the discussions of public scientific testimony. But, unless I make explicit qualifications, I will focus on the core phenomenon of testimony that *p*, which has conveying that *p* as its primary communicative function, and this is what is glossed by *Testimony*.

2.1.b Testimonial belief and acceptance: A characterization of testimonial belief may be approached by considering some basic cases. In Chapter 1, I assumed that the case in which Teo said "I had rye bread for lunch" and I took his word for it to exemplify a testimonial belief that Teo had rye bread for lunch. While such cases are reasonably clear, it is not easy to provide a general characterization. For example, the idea of *taking someone's word* stands in need of explication.

The first thing to note is that testimonial belief is not just belief that is caused by a testimony. For example, S's testimony "I drank all the rum" may cause my belief that S speaks English or that S is going through a breakup. But my belief is not a testimonial belief. Likewise, I may falsely testify "I am a tenor" and you may, due to my distinctively baritone voice, come to believe that I am a baritone. In this case, you are not forming a testimonial belief. Even if I truly testify "I am a baritone" but you believe me solely because of the tone of my voice, your belief is not testimonial (the baritone examples are variations of Audi 1997: 420; Graham 2015a enlists further examples).

In these cases, beliefs are not based on the testimony in the manner required for instances of testimonial belief. Thus, the cases indicate that testimonial belief requires a *basing relation* between the speaker's testimony and the recipient's belief that differs from mere causation. Rather, the basing relation requires, at least typically, that the recipient has a minimally competent uptake of the content of the testimony and forms the belief on the basis of this uptake (Pritchard 2004; Shieber 2015: 8ff.; Peet 2019). That said, I will mainly discuss cases without deviant causal links between the speaker's testimony that *p* and the audience's belief that *p*, although these complexities will resurface occasionally. This exemplifies the modus operandi of starting with simple paradigmatic cases and using the account as the basis for discussing more complex cases. When a scientist testifies "we found no trace of mercury in the samples" to a collaborator who forms a corresponding belief on the basis of this testimony, the collaborator forms a fairly clear-cut case of testimonial belief. Likewise, when a scientist testifies "we have found increasing levels of mercury in yellow fin tuna" in a radio interview and a listener believes him, the listener forms a fairly clear-cut case of testimonial belief.

However, there is another type of attitude that may be based on testimony that requires discussion in a study of scientific testimony. This is the attitude called "acceptance."[2] The distinction between belief and acceptance was brought to prominence by van Fraasen, who argued that belief involves commitment

[2] Cohen 1992; Engel 2000, Tuomela 2000; Gilbert 2002. See also Fleisher 2021.

to the truth of the theory whereas "acceptance of a theory involves a belief only that it is empirically adequate" (van Fraassen 1980: 12—italics removed). For van Fraassen, the notion of permissible acceptance plays an important role within a constructive empiricist framework according to which science aims to provide empirically adequate, rather than true, theories (van Fraassen 1980). But in the debates about collective knowledge, it is often invoked without a commitment to constructive empiricism. For example, Cohen provides the following generic characterization: "to accept that *p* is to have or adopt a policy of deeming, positing, or postulating that *p*" (Cohen 1989: 367). Thus, the central difference between belief and acceptance does not pertain to the basing relation but to the nature of the doxastic attitude. Belief involves a commitment to the truth of *p* whereas acceptance only involves a commitment to treating *p* as true or likely in specific contexts (Toumela 2000; Wray 2001; Gilbert 2002).

The notion of acceptance is important for an analysis of scientific testimony since a scientist may accept theories and hypotheses without believing them. In fact, the attitude of acceptance may be central to scientific practice (Cohen 1992; Engel 2000). If so, an account of scientific testimony that ignored testimonial acceptance would be highly incomplete. Moreover, the notion of acceptance plays a major role in characterizing scientifically important types of groups as well as collective scientific belief, knowledge, and testimony.

2.1.c Collective acceptance, belief, knowledge, and testimony: Scientific collaboration and hyper-specialization raise hard questions about *who* forms a judgment that a scientific hypothesis is true and *who* comes to know it when things go well.[3] Naturally, these questions bear on the question "who is the scientific testifier?"

A natural approach is a *summative* account according to which a group believes that *p* just in case all or most members of the group believe that *p* (Fagan 2012: 825). However, many theorists now favor non-summative approaches, such as Gilberts's account, according to which individuals expressing willingness to act in a particular manner constitutes a *joint commitment* which, in turn, constitutes the group as a *plural subject* (Gilbert 1989, 2000, 2002). Plural subjects and joint commitments are holistic phenomena that are irreducible to individual commitments. A plural subject is, roughly, a "corporate body" constituted by joint commitments that may be ascribed collective belief, acceptance, and knowledge. According to Gilbert, this account helps explain phenomena such as scientific change (Gilbert 2000).

Further work has focused on the specific conditions for collective knowledge, belief, and acceptance. For example, Wray adopts Cohen's appeal to *acceptance* in

[3] Tuomela 2000, 2004; Tollefsen 2007, 2015; Lackey 2021; Brady and Fricker 2016.

accounting for group acceptance and group knowledge on the basis of organic solidarity in a research group (Wray 2001; see also Tuomela 2000). Many theorists have thought that *collective scientific knowledge* obtains in some cases where the scientific justification for a hypothesis is generated by a large network of collaborating scientists.[4]

Rolin extends this approach by arguing that collective scientific knowledge may be ascribed to the larger scientific community, and not merely to a research group.[5] Given that scientific testimony is often collective in nature, these discussions have been extended to address the following question:

The Question of Collective Scientific Testimony

How may a collective scientific testifier be characterized?

The idea of joint commitment has been used to answer this question. For example, Fricker conceives of a "group testifier as constituted, at least in part, by way of a *joint commitment to trustworthiness as to whether p* (or whatever range of *p*-like questions might delineate the body's expertise, formal remit, or informal range of responsibility)" (Fricker 2012: 272; Faulkner 2018 criticizes). In contrast, Lackey provides a deflationary account, according to which talk about group testimony does not require postulating group knowers or believers (Lackey 2015, 2021). Criticizing this account, Faulkner suggests, drawing on Burge, that since the justification may be distributed, we may consider collective knowledge and belief without postulating collective knowers and believers (Burge 2013; Faulkner 2018). Another option, however, is to acknowledge collective knowers, believers, and testifiers without reifying them to be a collective subject that is constituted by joint commitments. One example of such an approach is due to Bird, who argues that the fact that "certain people have mutually interacting jobs or roles may be sufficient to bind them together" (Bird 2014: 59). So, Bird concludes that collective scientific knowers and testifiers are constituted by "mutual dependence arising from, above all, the division of labour" (Bird 2014: 54).

There is no consensus on these vexed issues. On occasion, I will sidestep them by focusing on the testimony of individual scientists who speak on their own behalf. Furthermore, I will, when considering collective knowledge and collective acceptance, tend to focus on simple cases in which most or all group members accept or believe the hypothesis in question. When discussing harder cases, I will treat them in a piecemeal manner. However, I assume that cases of collective scientific knowledge, acceptance, and testimony must at least meet a mutual

[4] de Ridder 2014a, Wray 2015; Miller 2015; Boyer-Kassem et al. 2017. For criticism, see Giere 2002, 2007.

[5] Rolin 2008, 2010. See also Longino 1990; Bird 2010; List and Pettit 2011; Cheon 2014, 2016; Wilholt 2016; Kallestrup 2019; Lackey 2021.

dependence constraint (Bird 2014). Likewise, I assume that collective scientific knowledge is at least partly characterized by distributed scientific justification (Burge 2013; de Ridder 2014a).

An important qualification is that even when the justificatory process is extended in scientific collaboration, we may still epistemically assess the cognitive states and processes of the various individuals who make up the collective. Failing to do so would come with a loss of explanatory power since occurrences of testimony between collaborating scientists represent epistemologically important types of possible failure (Gerken 2014b). So, even in cases of distributed justification, it is often relevant to focus on individual scientists' intra-scientific and public scientific testimony.

2.1.d Concluding remarks on testimony, testimonial belief, and acceptance: The working characterization of testimony and the discussion of testimonial belief and acceptance do not amount to reductive analyses. Likewise, I have not given substantive answers to the questions about collective knowledge, belief, acceptance, and testimony. However, the discussions have clarified the difference between characterizing phenomena such as testimony or testimonial belief in their own right and characterizing testimony as a source of warrant. Likewise, the belief-acceptance distinction may be recognized even if neither phenomenon is reductively defined. So, hopefully, the provisional working characterizations of testimony, testimonial belief, and acceptance developed in this section put us in a better position to turn to the epistemological side of the matter.

2.2 Testimony as an Epistemic Source

Testimony is not merely a source of belief and acceptance; it is also an *epistemic* source of warranted belief and acceptance as well as knowledge. The sense in which testimony is an epistemic source may be approached by considering principles reflecting the idea that testimony transmits the testifier's warrant and knowledge to the recipient. Such principles are staples of social epistemology (Gelfert 2014; Levy and Alfano 2019). But in this section, I will merely draw some fairly basic, but nevertheless important, conclusions concerning the type of warrant that *scientific* testimony provides.

2.2.a Necessary conditions for testimonial transmission: How is testimonial warrant and knowledge transmitted from testifier to recipient? A common idea is that someone who forms a testimonial belief may, in some epistemologically important sense, come to acquire the testifier's warrant or knowledge. After all, we provide and accept testimony because it conveys the speaker's knowledge and warranted beliefs, and thereby improves the recipient's epistemic position.

Moreover, the idea is motivated by the thought that testimony can at best convey existing warrant and knowledge, but not improve on it or generate new knowledge. Lackey articulates a principle that captures an aspect of these ideas (with a slight notational change from Lackey 2006a: 79—the 'N' is for 'necessary condition'):

> ***Transmission of Epistemic Properties-N***
>
> For every speaker, S, and hearer, H, H's belief that *p* is warranted (justified, known) on the basis of S's testimony that *p* only if S's belief that *p* is warranted (justified, known).

Lackey attributes versions of the principle to several philosophers.[6] To avoid exegesis, and because there are important differences between transmission principles for knowledge and warrant, I will begin the discussion with a simple principle:

> ***Testimonial Knowledge Requires Testifier Knowledge***
>
> H acquires testimonial knowledge that *p* through S's testimony that *p* only if S knows that *p*.

A prominent case type directed against *Testimonial Knowledge Requires Testifier Knowledge* is a non-believer case in which S testifies that *p* on the basis of strong epistemic warrant, although she does not believe that *p*. Hence, S does not know that *p*. In some such cases, many philosophers take it that the recipient may acquire testimonial knowledge. An influential version concerns a teacher who is a creationist but knows evolutionary science and accurately testifies a truth about evolution to a group of students despite not believing it.[7] However, Burge objects that testimonial knowledge does not require that it is the "recipient's intermediate interlocutor" who knows (Burge 2013: 256; see also Faulkner 2011). Moreover, Wright notes that the case would not compromise an analog principle concerning warrant since the teacher possesses strong warrant for the contents of her testimony (Wright 2016). Moreover, he argues that a transmission principle for warrant may explain why the students acquire knowledge once they form the testimonial belief that their teacher lacks (Wright 2016, 2018).

However, Graham has devised a different non-believer case, *FOSSIL*, in which our creationist teacher, while on a field trip, "discovers a fossil that proves that ancient humans once lived in this area (itself a surprising discovery no one knew

[6] Hardwig 1985, 1991; Burge 1993, 1997; Williamson 2000; Audi 1997, 2006; Owens 2000; Adler 2002; and Schmitt 2006. Due to subtle variations in formulation, some attributions may be questioned. Cf. Burge 2013: 256. See also Keren 2007; Wright 2016, 2018.

[7] Lackey 1999, 2006, 2008; Graham 2000, 2006; Carter and Nickel 2014.

before)" and testifies this to her students.[8] So, in *FOSSIL*, *no one* in the epistemic network that is responsible for the justifying evolutionary and paleontological theory believes the specific proposition that *p*—i.e., that ancient humans lived in the area. Thus, it cannot be objected that someone in a simple chain of testimonies knows that *p*. However, it remains true that the testifier is so well warranted that she would know if it were not for her lack of belief.

Reflection on *scientific* knowledge and warrant suggests that non-believer cases may be fairly prevalent in science. In radically collaborative research, individual researchers may not believe certain hypotheses but merely *accept* them. In some such cases, a scientific testifier's lack of belief that *p* or active disbelief (i.e., belief that not *p*) will be a defeater of the testifier's knowledge that *p*. So, if the audience is nevertheless capable of acquiring knowledge by the scientist's testimony, scientific testimony provides an important class of *actual* counterexamples to *Testimonial Knowledge Requires Testifier Knowledge*. These include examples of scientific expert testimony that is directed at public deliberation or policymaking.

Moreover, for collaborative research in which scientific justification is distributed across an epistemic network, the individual testifier may lack scientific justification for the hypothesis, *p*. Yet, *p* may only be said to be scientific knowledge if it is backed by all the available scientific justification or by significantly more scientific justification than the testifier possesses. If the audience may acquire testimonial knowledge in some such cases and the individual scientist is the testifier, rather than a mere spokesperson for the network, there is yet another class of counterexamples to *Testimonial Knowledge Requires Testifier Knowledge*.

2.2.b Does testimony transmit kind and degree of warrant?: A different type of transmission principle reflects the idea that recipients may, in an epistemologically important sense, come to acquire the testifier's warrant or knowledge. For example, Faulkner articulates this idea as follows: "If the audience is warranted in forming a testimonial belief, then whatever warrant in fact supports a speaker's testimony continues to support the proposition the audience believes" (Faulkner 2000: 591).

If we apply this idea to scientific warrant, the principle has it that the testifier's scientific justification continues to support the recipient's testimonial belief. There is room for interpretation of the phrase 'continues to support.' More generally, the idea of transmission of warrant may be explicated in many ways (Gelfert 2014; Wright 2016, 2018). Here, I will be concerned with an explication according to which the recipient acquires the same kind and strength of warrant that the testifier possesses when the testimonial transaction is successful. The idea is motivated by the thought that testimony cannot produce novel warrant but only

[8] Graham 2016: 176 from Graham 2000. See Carter and Nickel 2014 for further varieties.

transmit already existing warrant. To reflect that not all who defend a transmission view are committed to this explication, I will label it "the inheritance view." Here, I will set aside other ways of explicating *transmission* in order to argue that the inheritance view fails since this is an assumption that I will rely on down the line. So, to avoid exegesis and to make my commitments explicit, I will target the following general principle:

> *Inheritance of Warrant*
>
> Whenever H acquires testimonial warrant through S's testimony that *p*, S's testimony that *p* transmits the kind or degree of warrant possessed by S to H.

An initial consideration against *Inheritance of Warrant* is that the warrant for testimonial belief partly depends on the recipient's exercise of truth-conducive cognitive competences in the uptake of the testimony (Graham 2020). These include the cognitive competences involved in comprehending the content of the testimony. In virtue of this different epistemic basis for the recipient's testimonial warrant, it is at least partly different in kind from the testifier's warrant.

The problems with *Inheritance of Warrant* may also be illustrated via a twin case: Assume that S has deduced that there were more than ten visible meteors in Hornstrandir on October 18th on the basis of astronomical and meteorological data. In contrast, S* was in Hornstrandir on October 18th and visually observed more than ten meteors. So, S and S* have very different epistemic warrants for their beliefs. Assume, finally, that S testifies to H and S* testifies to H* that there were more than ten visible meteors in Hornstrandir on October 18th in a *minimal background case* in which the recipient has minimal information about the testifier's epistemically relevant properties (Chapter 1.1.a). Hence, H and H* have no ideas of the epistemic basis for the respective testimonies. Given that H and H* inhabit the same general social environment, there is good reason to think that their testimonial beliefs are on a par, epistemically speaking.[9] However, the assumption required to compromise *Inheritance of Warrant* is only that H does not acquire inferential justification and that H* does not acquire perceptual entitlement that there were more than ten visible meteors in Hornstrandir on October 18th. So, H and H* do not inherit the kind of warrant for the testifiers' beliefs. Moreover, assume that S** is another astronomer with access to far better data than S has and has superior models for predicting meteors than S does. Hence, S**'s scientific justification is superior to S's. Assume now that both S and S** testify to H and H**, respectively, that there were more than ten visible meteors in Hornstrandir on October 18th. Assume, again, that the case is a *minimal background case* and that the general environmental conditions are

[9] Gerken 2013b. Goldberg 2010a argues otherwise and I respond in Gerken 2012b, 2014b.

held fixed. In such a case, their respective recipients, H and H**, will be equiwarranted (Gerken 2013b). Thus, both kind and degree of testimonial warrant are standardly lost in transmission (Peet and Pitcovski 2017 provide further arguments). So, *Inheritance of Warrant* must go.

This is a modest but important lesson about the nature of testimonial warrant that it is not typically inherited in the sense that it is similar to the kind and strength of the testifier's warrant. It is mistaken to suppose that the specific epistemic properties of the *particular* speaker are invariably central the recipient's testimonial warrant. Rather, what matters are the general features of the relevant social environment. I will substantiate this assumption in the discussion of epistemic externalism below (Section 2.3.a), but for now I will note this commitment with the following epistemological principle:

> ***Non-Inheritance of Warrant***
>
> Unless the nature of the warrant that S possesses for believing that *p* is articulated, or otherwise clear, to H, S's testimony that *p* does not transmit the kind or degree of warrant possessed by S to H.

Non-Inheritance of Warrant articulates a negative point. I state it as an epistemic principle in its own right because it consists of a necessary condition of obtaining the kind or degree of testimonial warrant that the testifier possesses. However, articulating the nature of the warrant is far from sufficient to ensure that the recipient obtains the kind or degree of testimonial warrant that the testifier possesses, and this point will become important once we turn to public scientific testimony in Chapters 5 and 6.

Moreover, *Non-Inheritance of Warrant* may help explain why a recipient may acquire testimonial knowledge from an immediate interlocutor who does not know. It is partly because testimony provides a type of warrant that is distinct from the testifier's warrant that there may be conditions which defeat the testifier's knowledge without defeating the recipient's knowledge. I avoid talking about testimony "generating" knowledge and warrant because we have yet to see cases without warrant in the antecedent testimonial chain and network. Both in non-believer cases and cases of distributed warrant, the immediate interlocutor draws on antecedent warrant. This characteristic feature of *scientific* knowledge and warrant is highlighted by Burge addressing the case, *FOSSIL*: "In an abstract sense, the knowledge resides in the antecedent chain, including the teacher (with his/her knowledge of the fossil). We do talk this way about complex, collective scientific or mathematical work. Knowledge of a theoretical explanation or a proof resides in a group, even though, because of the complexity of the content of knowledge, no individual has full control of the explanation of the proof" (Burge 2013: 257).

Thus, both non-believer cases and distributed scientific justification cases connect the considerations about collective knowledge and distributed scientific justification to epistemological principles of testimony. Given that my focus is scientific testimony, I will set forth a limited non-inheritance principle which concerns scientific justification specifically:

> ***Non-Inheritance of Scientific Justification***
>
> Unless the scientific justification that S possesses for believing that *p* is articulated or independently clear to H, S's testimony that *p* does not transmit the kind or degree of scientific justification possessed by S to H.

I take cases of distributed scientific justification to motivate *Non-Inheritance of Scientific Justification*. In such cases, the scientific justification that supports the speaker's testimony does not typically transfer in the sense that the recipient obtains the same type and degree of warrant that the testifier possesses. Likewise, *Non-Inheritance of Scientific Justification* may also be motivated by reflection on the elaborate and specialized scientific justification for believing or accepting that *p* that scientists tend to possess. An expert in astrophysics may calculate from a complex set of evidence that it will be about 6,800 years before the comet NEOWISE returns to the inner solar system. If she testifies this conclusion to a layperson, the layperson does not acquire the same type of warrant that the testifier possesses. Moreover, the expert testifier's warranted belief about the return of NEOWISE is robust with regard to a number of defeaters. For example, the expert may rebut pseudo-scientific arguments that NEOWISE will return in sixty-eight years. But the recipient will not be able to do so on the basis of the testimony. If so, the recipient lacks testimonial warrant of a strength that matches the strength of the testifier's warrant. So, scientific experts cannot transmit all the rich scientific justification or its degree to a layperson simply by testifying that *p*. If they could, our educational system's aim of making students acquire and be able to articulate the justification for scientific hypotheses, rather than merely believe them, would be redundant and, hence, entirely misguided. But that aspect of our educational systems is not entirely misguided. So, recipients of scientific testimony cases do not invariably, or even typically, acquire *scientific* justification. Similar points have been made regarding transmission of understanding (Gordon 2016; but see also Boyd 2017).

Although basic, this negative point raises an important question: What is the nature of the testimonial warrant that the recipient obtains when things go well? In some minimal background cases, recipients may only acquire a basic testimonial entitlement. However, there is enough variation in minimal background cases to suggest that recipients may acquire different types or strengths of testimonial warrant depending on the case and formation of testimonial belief. Another key point is that cases of scientific testimony are rarely minimal background cases. The conversational context or the content of the testimony often enables the audience to infer that the testifier is a scientific expert. In such cases, reflective recipients may—in addition to the default basic entitlement—acquire a higher-order non-scientific justification that *p* is scientifically justified. And when environmental conditions are in place, even unreflective recipients may acquire an entitlement for presupposing that *p* is scientifically justified. So, although *Non-Inheritance of Scientific Justification* is a negative principle, it suggests some positive lessons about public scientific testimony.

2.2.c Concluding remarks on testimony and transmission: I conclude that the warrant that recipients acquire through testimony is typically different in kind from the warrant possessed by the testifier. In particular, recipients do not by default acquire scientific justification from scientific testimony. This is important for the study of scientific testimony because the recipients' testimonial warrants may be vulnerable to defeaters that the testifiers' scientific justification may rebut. The point bears on the kind of testimonial warrant: As a general rule, whenever a scientist possesses an epistemically internalist *justification* for the content of her testimony, she possesses a type of epistemic warrant that is robust against certain defeaters, such as lack of consensus among laypersons. But the layperson recipient often only acquires an epistemically externalist *entitlement* for his testimonial belief that may be defeated for example by lack of consensus among his epistemic peers. This point also bears on the strength of testimonial warrant given that a warrant which is robust to misleading defeaters is *ceteris paribus* stronger than a warrant that is not. So, *Non-Inheritance of Scientific Justification* has some notable ramifications. However, it should not lead us to discount the value of testimonial entitlement from public scientific testimony. To elaborate on this point, I will briefly address the internalism-externalism debate in the epistemology of testimony.

2.3 Foundational Debates in the Epistemology of Testimony

In this section, I will briefly address some foundational debates in the epistemology of testimony: the debates between epistemic internalists and externalists and the related, but ultimately different, debates between reductionists and anti-reductionists.

2.3.a Internalism vs. externalism in the epistemology of testimony: The internalist-externalist debate is a foundational debate in epistemology. As a rough initial approximation, epistemic internalists emphasize the importance of *personal* epistemic properties, whereas epistemic externalists emphasize the importance of *environmental* epistemic properties (Gerken 2018c, 2020a).

In Chapter 1.1.a, I drew—without much argument—a distinction between an epistemically externalist species of warrant, entitlement, and an internalist species, justification. Given that a recipient of scientific testimony based on scientific justification often only acquires testimonial entitlement, the distinction is important for understanding the epistemic properties of scientific testimony. Hence, I will clarify the distinction between entitlement and justification, although I will not engage in the internalism-externalist dispute in a manner that will satisfy opponents.[10] Rather, my aim is to connect the internalism-externalism distinction to questions regarding *scientific* testimony. I characterize entitlement and justification by way of a *Reason Criterion*:

Reason Criterion (Justification)

S's warrant, W, for her belief that *p* is a justification if and only if W constitutively depends, for its warranting force, on the competent exercise of S's faculty of reason.

Reason Criterion (Entitlement)

S's warrant, W, for her belief that *p* is an entitlement if and only if W does not constitutively depend, for its warranting force, on the competent exercise of S's faculty of reason.

In this manner, the entitlement-justification distinction is drawn by reference to a sophisticated cognitive faculty—the faculty of reason. Lower-level cognitive competencies, such as perception, involve non-propositional representations and yield entitlement for the resulting beliefs. In contrast, the cognitive faculty of reason is responsible for high-level, although not necessarily conscious, propositional thought that is attributable to the individual and yields justification (Gerken 2020a). Here, I assume that epistemic reasons are propositional. Given this assumption, there is a constitutive relation between the nature of epistemic justification and epistemic reasons. Even if the faculty of reason may yield warrant that is not based on epistemic reasons, it involves the capacity for basing epistemic warrant on such reasons. A paradigmatic case is that of *reasoning*, in which the premise-beliefs serve as epistemic reasons for the conclusion-belief (Gerken 2013a, 2020a; Burge 2020; Graham 2020).

It might be objected that the *Reason Criterion* rules out the possibility of testimonial entitlement on the basis of linguistic testimony, which is propositional and, hence, dependent on the recipient's exercise of the faculty of reason. In response, I call attention to the qualification 'for its warranting force' in the

[10] But see Gerken 2012b, 2013b, 2014b, 2018d, 2020a, 2020b.

Reason Criterion. In some minimal background cases, such as when a child believes a testifier, its faculty of reason only plays an *enabling* role in the comprehension of the testimony and not a *warranting* role. In such cases, the child only acquires an entitled testimonial belief. The enabling-warranting distinction is developed in (Burge 1993). Here he mistakenly invokes the distinction in defense of a priori testimonial warrant (Christensen and Kornblith 1997; Malmgren 2006). Burge has retracted this claim (Burge 2013). However, the distinction between warranting and enabling roles remains important. For example, it may apply to cases in which the faculty of reason is operative in comprehending the testimony and monitoring defeaters or background conditions. Even if this yields justification for the belief that there are no defeaters, for example, this may remain an enabling condition for the entitlement of the first-order belief that *p* if it does not contribute to it. However, given that the distinction between cognition that is based on the faculty of reason and more primitively based cognition is not firm, neither is the entitlement-justification distinction (Graham 2018b, 2020; Gerken 2020a).

The internalist-externalist distinction is more commonly drawn in terms of accessibility (Bonjour 1992: 132ff.). However, I have argued that accessibilist criteria face challenges in distinguishing between mere lack of access and inaccessibility (Gerken 2020a). Moreover, inaccessible reasoning that involves unconscious belief as premises (and, thus, epistemic reasons) is best categorized as yielding justification (Gerken 2020a). So, the *Reason Criterion* is also motivated by being able to classify cases in a reasonable manner.

However, accessibility may help to differentiate different subspecies of justification. Having access to reasons for one's belief is epistemically beneficial insofar as one's belief is more robust against certain defeaters. So, I will use the phrase "accessibilist justification" for this species of justification. For the following discussion, it is important to recognize an even more demanding species of justification, which I call "discursive justification." This involves the ability to *articulate* one's warrant for belief as epistemic reasons. Here is a statement of it:

Discursive Justification

S's warrant for believing that *p* is a discursive justification if and only if S is able to articulate some epistemic reasons for believing that *p* (Gerken 2012a: 385).

I will argue, in Chapter 3, that such articulability is an important aspect of scientific justification. Here, I only note that it is a socially important type of warrant insofar as it allows the discursively justified individual to defend her views and engage in social cognition (Gerken 2020a). Some might think that the ability to articulate only *some* reasons is too weak a requirement. Recall, however, that justification is gradable. So, if S cannot articulate other epistemic reasons than "she said so," S is *minimally* discursively justified, although he might be quite well entitled. If S can articulate "she said so and she is a world-class scientist in the area," he has much better discursive justification.

A presupposition of the present distinctions is the pluralist standpoint that there are several types of genuinely epistemic warrant. This becomes vivid when we reflect on recipients of scientific testimony. A recipient who is aware that the testifier is a scientific expert in the relevant domain may acquire both entitlement and justification. He may acquire the basic default entitlement for his testimonial belief that *p* and may even know that *p* in virtue of it. However, reflection on the testifier and the testimonial context may provide the recipient with a higher-order justification that *p* is scientifically justified. If the recipient exercises the faculty of reason well enough to acquire such justification, he has moreover acquired some access-justification—i.e., access to the nature of his epistemic warrant for belief. Finally, if he is also capable of articulating the warrant as *epistemic reasons* for his belief, he has acquired discursive justification that enables him to convey not merely the content of his testimonial belief but also some epistemic reasons for holding it. So, several distinct warrants may coexist (Gerken 2013b, 2020a).

Such a pluralism about warrant remains something that needs arguing. In fact, the internalist-externalist dispute in epistemology has been characterized by *monist* presuppositions—i.e., the presupposition that there is only one genuinely epistemic type of warrant. For example, epistemic internalists have often argued that epistemically externalist conceptions of warrant, such as reliabilism, are misguided (Bonjour 1985; Fumerton 1995). Epistemic externalists, in turn, have suggested that externalist conceptions of epistemic warrant should *replace* what they regard as hyper-intellectualized Cartesian internalist conceptions (Kornblith 2002).

I have argued against monism about warrant in a number of ways (Gerken 2013b, 2020a). One of these is a twin argument: A naïve recipient, four-year-old Muddy, is in an epistemically hospitable social environment, Chicago. Here, the speakers generally testify sincerely, and only when they are well-warranted. This contrasts with another naïve recipient, four-year-old twin-Muddy, in an epistemically inhospitable environment, Crete. Here, the speakers lie whenever they think they can get away with it, and speak vaguely when they can't. Consequently, mature recipients on Crete rarely accept testimony at face value. However, twin-Muddy is, like Muddy, naïve in that he accepts testimony in a fairly uncritical manner. The epistemic twins have the same track record of forming true testimonial beliefs because Muddy has been unlucky and twin-Muddy has been lucky in the past. Generally, the thought experiment involves the assumption that all of Muddy's and twin-Muddy's experiences with testimony and reflection on it are individualistically indiscernible (see Gerken 2013b for elaboration).

I take it that Muddy and twin-Muddy are epistemically on a par in terms of justification. But it is also plausible that the difference in their general environment makes them differ epistemologically in terms of entitlement. Thus, two kinds of distinct warrant may accrue to testimonial belief. This conclusion is supported via two separate arguments. Here is the argument for entitlement against internalist monists:

Twin Argument for Entitlement
E1: If Muddy and twin-Muddy differ in overall degree of warrant, and everything except the social and environmental facts is held fixed, a distinctive social externalist kind of warrant accrues to Muddy's testimonial belief.
E2: Muddy and twin-Muddy differ in degree of overall warrant for believing that *p*.
E3: Everything except the social and environmental facts is held fixed.
E4: A distinctive social externalist kind of warrant accrues to Muddy's testimonial belief.

And here is the argument for justification against externalists monists:

Twin Argument for Justification
I1: If there is a genuinely epistemic standard according to which Muddy and twin-Muddy are on a par, a distinctively epistemic internalist kind of warrant may accrue to testimonial belief.
I2: There is a genuinely epistemic standard according to which Muddy and twin-Muddy are on a par.
I3: A distinctively epistemic internalist kind of warrant may accrue to testimonial belief.

I motivate and defend the premises for each argument in (Gerken 2013b). Here, I assume the arguments to be sound and use the case to reflect on the nature of internalist and externalist testimonial warrant and their connections to scientific testimony.

One central lesson is, crudely put, that the social environment partly determines entitlement. More precisely, systematic patterns of relations between a recipient of testimony and the general social environment that he is embedded in matter for the epistemic status of his testimonial belief.[11] There are two externalist aspects to this account. The first is that the social environment that goes beyond the individual testifier and recipient matters for epistemic status. The second is the fact that it does so even though the recipient lacks cognitive access to the nature of the general environment (Gerken 2017b, 2018d, 2020a, 2020c). The fact that the general social environment is central to testimonial entitlement is important for several aspects of

[11] Graham 2012a, 2020b; Burge 2013; Gerken 2013a, 2013b, 2014b; 2018d.

scientific testimony that I will explore. For example, the warranting force of public scientific expert testimony to laypersons is dependent on a social environment in which scientists play certain testimonial roles and testify in accordance with the relevant epistemic norms (Chapters 5 and 7). Consequently, I will argue that laypersons may, in the right social environment, acquire entitled testimonial belief through *appreciative deference*. Roughly, this is the uptake of public scientific testimony in virtue of sensitivity to the idea that it is trustworthy in virtue of being scientific testimony (I elaborate on this notion in Chapter 7.4.c).

In general, to understand important aspects of scientific testimony as an epistemic source, it is important to recognize externalist warrant (entitlement) for testimonial belief. However, the step from recognizing externalist conceptions of epistemic rationality to rejecting the relevance of more traditional internalist conceptions is misguided. Justification—including, in particular, discursive justification—is central to an account of scientific warrant. Consequently, I will argue that scientific testimony has a distinctively internalist flavor insofar as it is paradigmatically based on a species of discursive justification (Chapter 3.5).

In sum, the internalism-externalism debate has important ramifications for an account of scientific testimony as an epistemic source. In particular, the role of the social environment that has been highlighted by epistemic externalists matters in characterizing the conversational contexts of scientific testimony. On the other hand, epistemic internalists have highlighted the ability to articulate epistemic reasons and the importance of the recipients' ability to assess the testifier. The debate about the epistemic importance of the latter has led to a related debate concerning whether testimonial warrant may *reduce* to the recipients' non-testimonial warrant for believing the testifier.

2.3.b Reductionism vs. anti-reductionism: Reductionists about testimonial warrant argue that it may be reduced to non-testimonial warrants from perception, induction, and reflection, etc. We may characterize the view as follows:

> ***Reductionism***
>
> Testimonial warrant requires, and reduces to, other types of warrant that are ultimately non-testimonial.

Reductionism comes in a global variety which requires a general justification of testimonial beliefs on the basis of non-testimonial warrants, and a local one according to which a testimonial belief reduces locally on the basis of case-specific warrants.[12] Anti-reductionist views have it that testimony is a basic source of epistemic warrant. It is basic because it is a *sui generis* epistemic source and

[12] Fricker 1995. For discussions, see Kusch 2002; Gelfert 2009; Graham 2018a.

because our epistemic reliance on it is unmediated. Hence, testimonial warrant does not reduce to other types of warrant. This view may be articulated as follows:

> ***Anti-Reductionism***
>
> Testimony is a basic, and hence irreducible, source of warranted belief and knowledge.

Often, anti-reductionism aligns with epistemic externalist theories, whereas reductionism aligns with epistemically internalist theories. However, the internalism/externalism distinction and the reductionism/anti-reductionism distinction are orthogonal insofar as anti-reductionism may combine with traditional epistemic internalist requirements (cf. Faulkner 2000). On the other hand, someone might argue that testimonial warrant reduces to perceptual entitlement by claiming that it is based on the recipient's mainly subconscious perception of cues of sincerity and reliability of the testifier. While there is little to say in favor of this view, it exemplifies a combination of epistemic externalism and reductionism.

Anti-reductionist views are often traced back to Reid, but much of the contemporary debate has revolved around varieties of a principle articulated by Burge:

> ***Acceptance Principle***
>
> A person is entitled to accept as true something that is presented as true and that is intelligible to him, unless there are stronger reasons not to do so. (Burge 1993: 467)

Several varieties of such a principle have been articulated.[13] However, its central idea is that a recipient of intelligible testimony may become epistemically entitled by default even though he lacks warranted beliefs about the specific testifier or the general social environment. Thus, the *Acceptance Principle* motivates *Anti-Reductivism.* To explain the principle, Burge articulates a transcendental argument according to which the intelligibility of a testimony is a sign of a rational, and hence truth-conducive, source (Burge 1993, 2013; Graham 2018b). In contrast, I take the general features of the social environment that the interlocutors are embedded in to be more central (Gerken 2013b: fn. 14). A congenial approach focuses on the norms operative in the social environment (Faulkner 2011; Graham 2015b; Simion forthcoming).

A central reductionist criticism is that the sort of testimonial belief that the *Acceptance Principle* sanctions amounts to a kind of gullibility which is

[13] Cf. the similar *Presumptive Right* principle in Faulkner 2000: 46. See also Gelfert 2014: ch. 5; Shieber 2015: chs. 3–4.

inconsistent with epistemic rationality (Fricker 1994: 145). Fricker also criticizes a central motivation for the *Acceptance Principle*—roughly, the assumption that not all recipients who acquire testimonial warrant can critically assess the testifier. However, Fricker clarifies that when recipients form judgments about a testifier's trustworthiness, "the specific cues in a speaker's behaviour which constitute the informational basis for this judgment will often be registered and processed at an irretrievably sub-personal level" (Fricker 1994: 150).

In response, anti-reductionists argue that non-reflective testimonial belief may be epistemically rational.[14] Often, such a defense of anti-reductionism aligns with a version of epistemic externalism. But, as mentioned, it need not do so. Moreover, research in developmental psychology suggests that while young children can track accuracy in some types of testimony, they lack the conceptual resources that reductionism would require. Hence, an extensive debate revolves around this empirical issue.[15]

My inclination is to side with the anti-reductionists, although I take the explanatory scope of the *Acceptance Principle* to be more limited than some of its defenders. This is particularly so with regard to most varieties of scientific testimony. So, I accept that testimony is a basic source of warrant and that something like the *Acceptance Principle* holds in some minimal background cases. But I also take it to be less relevant to analyzing many cases of scientific testimony which tend to feature a recipient who is involved in *some* degree of assessment of the testifier. This is true of intra-scientific testimony in which the recipient tends to rely on background assumptions about the testifier's role in the epistemic network (Chapter 4.3). But it is also true in public scientific testimony where laypersons often rely on background assumptions to the effect that the testimony is science-based (Chapters 5 and 6).

But even though scientific testimony is paradigmatically based on scientific justification and often scrutinized more carefully than more mundane types of testimony, it is important to recognize the relevance of the social environment in which it takes place. For example, I uphold the epistemically externalist view that differences in the general social environment may yield differences in the epistemic position of recipients of scientific testimony.

2.3.c Concluding remarks on internalism, externalism, reductionism, and anti-reductionism: Despite the brevity of my discussion, several lessons for an account of scientific testimony may be drawn. One lesson is that there are distinctively externalist types of testimonial warrant—entitlements—and that the social

[14] See, for example, Goldberg 2008; Gelfert 2009; Graham 2010, 2018; Burge 2013; Gerken 2013a; Kallestrup 2020; Simion and Kelp 2020a; Simion forthcoming.

[15] Goldberg and Henderson 2006; Goldberg 2008; König and Harris 2004, 2007; Clement 2010; Cole et al. 2012; König and Stephens 2014; Michaelian, 2010, 2013; Graham 2018a. For responses, see Audi 2006; Lackey 2005; Fricker 2006a, 2006b, 2006c; Shogenji 2006. See also Moore 2017, 2018.

environment is a significant determiner. Another lesson is that scientific testimony is often a highly intellectualized form of testimony. Scientific testifiers paradigmatically possess an internalist kind of warrant—scientific justification—for the content of their testimonies. Moreover, recipients of scientific testimony often assess scientific testimony in comparatively reflective manners that involve background information. Consequently, internalist warrant—justification—is relevant to an analysis of scientific testimony.

So, an overarching lesson from these foundational epistemological debates is that an account of scientific testimony should address both the role of the broader social environment as well as the types of individual vigilance that recipients of scientific testimony may engage in. In the following section, I will examine each of these aspects of scientific testimony.

2.4 Individual Vigilance and Social Norms

One set of questions concerns the recipient of testimony: What is the relevance of the recipient's background assumptions and assessments of the testifier's trustworthiness? What resources do the recipients have available for assessing testifiers? Etc. Another set of questions arises from the patterns of relations among testifiers, recipients, and the general social environment in which they are embedded: What are the relevant features of the social environment? What is the interplay between social norms governing testimony and the relevant social environment? Etc.

I will first consider how recipients of testimony are capable of critically assessing a testifier and then consider whether these capabilities carry over to scientific testimony. I then turn to the social environment and specify how it bears on recipients' epistemic position. Finally, I briefly discuss the role of social norms in structuring the social environment and their ramifications for trusting various kinds of scientific testimony.

2.4.a Varieties of vigilance: To assess recipients' cognitive resources for assessing the trustworthiness of a scientific testifier, it is worth considering the general cognitive basis for testimonial uptake. An influential treatment has it that a "disposition to be vigilant is likely to have evolved biologically alongside the ability to communicate in the way that humans do" (Sperber et al. 2010: 360). Moreover, Sperber et al. argue that the epistemic vigilance that recipients exercise is central to an explanation of why testifiers tend to be non-deceptive, and they note that this outlook aligns well with a reductionist approach (Sperber et al. 2010: 461).

However, on the basis of empirical work on deception, Michaelian argues that the type of epistemic vigilance that is not too cognitively costly to deploy continuously is not particularly effective at detecting cues of deception (Michaelian 2013). Moreover, Michaelian argues that other factors help ensure that testifiers are rarely deceptive. So, he concludes that individual epistemic vigilance only plays a lesser role (Michaelian 2013; Sperber 2013 responds).

How does this debate bear on *scientific* testimony? One thing to note is that the resources for epistemic vigilance, which are adapted for mundane testimony, cannot be assumed to be effective for assessing scientific testimony. In fact, I will argue that folk epistemological heuristics adapted for ordinary testimony may systematically mislead laypersons' assessment of scientific testifiers. What's worse, I will argue that reliance on such folk epistemological resources may be a source of mistaken skepticism about scientific testimony among laypersons (Chapter 5 and 6).

The debates over vigilance point to the important lesson that we rely heavily on cognitive cost-minimizing heuristics as recipients of testimony. This is no surprise as laypersons in general rely heavily on folk epistemological heuristics (Gerken 2017a). So, although science is a sophisticated subject matter, laypersons are likely to rely on some of the same cognitive heuristics in assessing sources of scientific testimony. Of course, the complexity of a scientific testimony may tend to trigger more reflective responses. But evidence suggests that even such judgments will be shaped by fairly basic biases. This evidence includes general reflections on human cognition and evidence that is specific to scientific testimony.[16] Taken together, the empirical evidence reinforces the assumption that laypersons are often biased in their assessment of public scientific testimony. For example, ubiquitous biases such as confirmation bias and motivated cognition are likely to affect laypersons' uptake of public scientific testimony (Chapter 5.2–3). Furthermore, the impact of cognitive heuristics occurs much prior to the assessment of the testimony in, for example, the more or less voluntary selection of testimonial sources. In consequence, I will argue for the view that both testifier reliability and recipient vigilance should be addressed on the social level rather than merely on the individual level.[17] Specifically, I will promote a *Testimonial Obligations Pyramid* structure according to which the bulk of the obligations that pertain to public scientific testimony are societal ones whereas the individual lay recipients are subject to the least demanding obligations (Chapter 7.4.d).

According to this view, any attempt to provide guidelines for public scientific testimony should pay careful attention to the psychological consequences of evolved dispositions for epistemic vigilance among laypersons. So, my approach to public scientific testimony will thematize the heuristics and associated biases in laypersons' uptake—or lack of uptake—of public scientific testimony.

[16] For empirical work regarding cognition generally, see Gigerenzer 2008; Evans 2010; Kahneman 2011; Stanovich 2011, Gerken 2017a. For research on scientific testimony, see Fischhoff 2013; Dunwoody 2014; Kahan 2015a; Jamieson et al. 2017.

[17] Gerken 2013b, 2020a, 2020e; Dutilh Novaes 2015; Guerrero 2016; de Melo-Martín and Intemann 2018; Contessa forthcoming.

Scientists may be expected to have a more sophisticated approach to the evaluation of scientific testimony due to their general scientific literacy. However, scientists are only human, and, therefore, reliant on cognitive heuristics in many assessments of the testimony of fellow scientists. This fact is important for understanding the role of intra-scientific testimony in scientific collaboration. Moreover, scientists often have vested interests in the relevant scientific testimony, and their judgments may be biased by such practical interests. A particular long-standing worry is that scientists are prone to varieties of *confirmation bias* in all areas of scientific practice ranging from study design over observation gathering to data analysis.[18] Grand proposals about the nature of science, such as Popper's critical rationalism, are partly motivated by the idea that scientists are disposed to look for evidence confirming their hypotheses (Popper 1934, 1963). One may appreciate this rationale without accepting Popper's framework. Indeed, it is widely accepted that scientists are incapable of eliminating even fairly basic cognitive biases by themselves. Consequently, many of the methodological rules for proper science may be seen as curbing the problems associated with confirmation bias. Some concrete examples involve the scientific practices of coding interviews, providing placebos, and preregistration of studies. These scientific practices are reasonably understood as addressing confirmation bias (Bird 2019 provides examples from medicine).

Given that cognitive biases in scientific practice are ubiquitous and tenacious, it would be surprising if such biases were not found when scientists provide scientific testimony in public or within scientific collaborations. Likewise, it would be surprising if scientists' uptake of intra-scientific testimony were entirely unbiased. Thus, empirically informed reflection on the cognitive heuristics involved in the assessment of scientific testifiers is paramount for the analysis of public scientific testimony and, perhaps more surprisingly, intra-scientific testimony.

While these are important lessons, a limitation of them is that they primarily concern the individual layperson's and scientist's psychology. However, epistemic vigilance may also be seen as taking place at a communal level. This is particularly so in science where it is the scientific community that enforces sanctions against individual scientists who violate the relevant norms. Thus, individuals may, to a significant extent, outsource epistemic vigilance to the wider social community. Consequently, the debates concerning epistemic vigilance connect to another important lesson from the foundational epistemological debates—namely, the significance of the background social environment in which the testimony takes place. So, I will turn the focus from individual social competences and direct it at the social environment.

[18] Bacon 1620; Mynatt et al. 1977; Nickerson 1998; Hergovich et al. 2010; Bird 2019; Peters forthcoming a.

2.4.b The social environment of scientific testimony: In the discussion of testimonial entitlement, I concluded that it is partly determined by systematic patterns of relations between testifiers and recipients and the general social environment that they are embedded in. An important dimension of the social basis of testimonial warrants consists in the general reliability and sincerity of the testifiers in the relevant social environment. However, varieties of scientific testimony introduce further epistemic factors and complicate the basic ones concerning reliability and sincerity. So, we should not assume without argument that scientific testimony can be subsumed without qualifications under extremely general epistemological doctrines.

For example, the issue of sincerity is complicated by that fact that scientific hypotheses are often accepted without being believed by members of scientific communities (Wray 2001; Gilbert 2002; Fagan 2012). There is a sense in which someone who testifies that *p* without believing that *p* is insincere. But if that person is a scientist who accepts that *p*, it may not be a sense that compromises the testimonial warrant that a recipient may acquire from the testimony. If so, some instances of warranted belief through scientific testimony differ from the ordinary case insofar as they are not defeated by this sense of insincerity of the testifier. This may be the case both in intra-scientific testimony and public scientific testimony. Given the highly collaborative nature of science, the belief of the spokesperson is less important than the fact that she conveys the result of a collective effort that may include distributed warrant.

In the case of intra-scientific testimony, consider a member of a research team who does not personally believe that an experiment may provide evidence for *p* but accepts so for the purpose of the collaboration. If this researcher testifies to a collaborator that the experiment provided evidence for *p*, it seems to me that the collaborator may acquire testimonial warrant for believing that this is so. In the case of public scientific testimony, consider a spokesperson for a research group who testifies that *p* in her role of spokesperson even though she does not believe that *p*. Assuming that the research group has found that *p*, the fact that the spokesperson does not personally believe it does not seem to undermine the testimonial warrant that laypersons may acquire through her testimony. There is definitely room for debate about these cases. But they serve to illustrate that the conditions for scientific testimony may deviate from cases of more mundane testimony. However, scientific testimony is hardly unique in featuring this type of testimonial warrant without testifier belief since it is partly explained by the collaborative and distributed origins of the testimony.

A further complication pertains to the role of social environment for scientific testimony. Specifically, it concerns *how* the reliability and sincerity of testifiers are secured and, relatedly, *what* explains recipients' entitlement to presuppose that testifiers are reliable and sincere. Again, it is far from clear that the accounts of testimony about mundane things in minimal background cases carry over to cases of scientific testimony. For example, I will argue that the cases of intra-scientific and public testimony call for different accounts.

Consider, for illustration, Burge's transcendental argument according to which the intelligibility of testimony is indicative of a reasonable and, therefore, truth-conducive source (Burge 1993). Even if such an argument succeeds as an explanation for testimonial entitlement in everyday cases of testimony, it is far less relevant in cases of scientific testimony, as Burge appears to recognize (Burge 2013). First, mere intelligibility is not sufficiently indicative of a testifier being a competent scientist or a reliable conveyer of science. Second, a layperson may not find that reliable but highly complex scientific testimony is intelligible. Similar limitations apply to very different accounts. Consider, for another example, a contractarian account according to which the existence of the social contract explains why testifiers tend to comply with social norms governing testimony (Simion forthcoming). Such an account is more plausible as a generic account of everyday testimony than as an account of scientific testimony. After all, both intra-disciplinary and public scientific testimony are structured by a mix of tacit disciplinary norms and explicit disciplinary conventions. Arguably, these are generated and sustained in ways that differ from how a general social contract governs testimony in minimal background cases.

However, the emphasis on the social norms that characterize the social environment in which the testifiers and recipients are embedded *is* important for understanding scientific testimony. Social norms have been argued to explain the default reliability. For example, Graham argues that testifiers have internalized a social norm that prescribes truthful and informative testimony and that testifiers tend to abide even when doing so does not further their immediate self-interest (Graham 2010, 2012b). This focus aligns well with the social externalist framework according to which systematic patterns of relations between a recipient of testimony and the general social environment partly explain the recipient's epistemic position. After all, such general patterns are held in place by the social norms that partly constitute the general social environment in question. Thus, an important route to understanding testimonial warrant consists in focusing on the social norms governing the relevant types of testimony (cf. Chapter 1.5). This includes the social norms that are distinctive to *science* in the case of intra-scientific testimony and the social norms that are distinctive of *science in society* in the case of public scientific testimony.

Given that an important aspect of the social environment of scientific testimony is the social norms that partly constitute it, it is worth considering such norms in a bit further detail.

2.4.c Norms and guidelines of scientific testimony: So far, I have argued that the general social environment of both intra-scientific and public scientific testimony is epistemically important. However, the general social environment is partly specified in terms of social norms, and such norms have independently been argued to govern testimony.[19] Furthermore, given that testimony is a species of assertion, the epistemic norms of assertion are relevant to understand scientific testimony (see Gerken and Petersen 2020 for an overview). Thus, social norms are important to understand scientific testimony. As noted (Chapter 1.5), viewing scientific practices through the lens of norms has been an influential approach since Merton proposed his four "norms of science" (Merton 1973): universalism (personal features and values of scientists are scientifically irrelevant), communism (scientific findings belong to the scientific community), disinterestedness (scientists should pursue a common scientific interest rather than their personal ones), and organized skepticism (the scientific community should challenge scientific theories and findings). Each of these norms has proven to be very controversial. But they are apt to illustrate the general idea that the scientific enterprise is partly but centrally characterized by social norms.[20]

Before turning to scientific norms, I will sketch a trichotomy of the nature of norms and normativity in general (following Gerken 2017a). In Chapter 1.5, I distinguished between *norms* (objective benchmarks of assessment that the agent need not have any cognitive access to) and *guidelines* (which are met only if they are, in some sense, *followed* by the agent). However, norms are constitutively characterized in terms of their *telos*, which I, following Thomson, will label "standards" (Thomson 2008). For example, the standard of epistemic norms is truth.

Standards, norms, and guidelines are easily conflated, and our folk epistemological practices are prone to such conflations. Consider, for example, truth-effects on epistemic assessment. These effects occur when two individuals meet a non-factive truth-conducive norm equally well but end up with a true and a false belief, respectively. Empirical findings indicate that laypersons provide different assessments of such individuals' norm-violation and epistemic competence (Turri 2016). However, there are both empirical and philosophical reasons to regard these findings as evidence of epistemic instances of *outcome bias* (Gerken 2020b). So, they exemplify how the standard of truth and a norm of truth-conduciveness may be mixed up.

Likewise, norms and guidelines may be run together, and this is particularly so because they may be similar or even identical in their formulation. This is particularly so because the norms governing scientific practices—including scientific testimony—are not natural but *social* norms (Longino 2002; Bicchieri 2006, 2017; Henderson and Graham 2017a, 2017b). That is, scientific norms are held in place by social conventions, such as sanction and reward structures. However, the

[19] Graham 2012b, 2015b; Gerken 2013b; Henderson and Graham 2017a, 2017b; Simion forthcoming.
[20] Hull 1988; Longino 2002; Strevens 2003, 2017; Fricker 2017.

fact that the norms are social does not entail that they are akin to guidelines that the scientists conceptualize and follow. Often, scientific norms are tacit insofar as they reflect a scientific practice that scientists are embedded in in a fairly unreflective manner. Normative commitments are often constituted by a practice in this way (Gerken 2012c). On reflection, scientists may sometimes articulate the norms—perhaps in an approximate manner. This may result in a more overtly prescriptive guideline which is a conceptualized approximation of the norm and which may figure in textbooks.

In some instances, a normative principle may serve double duty as a norm and a guideline. In fact, I will propose some candidates of principles that exemplify such a dual role. For example, science journalism has always been subject to more or less explicit norms and guidelines concerning the presentation of findings, theories, and hypotheses. Although my starting point for discussing science reporting is norms, I will also consider some principles of science communication *qua* guidelines. Of course, there is a significant step to implementable guidelines, such as those articulated by news organizations (e.g., BBC 2018a). So, the division of labor between philosophy and journalism raises some challenging methodological questions that I will begin to address in Chapter 6.

My general focus on epistemic standards is compatible with assuming that varieties of scientific testimony may be governed by social norms that reflect non-epistemic standards. For example, a standard for some science reporting may be that the audience comes to believe the scientific consensus (van der Linden et al. 2015, 2016 might be read in this manner). Guidelines may differ depending on the kind of scientific testimony in question. Guidelines for scientists may be articulated with esoteric technical terminology, whereas this would not be a good idea for guidelines for science journalists. So, examining scientific testimony through the normative lens involves a balancing act between considering norms that are sufficiently generic to be characteristic of science and sufficiently specific to be sensitive to the relevant social environments as well as the specific functions of the scientific testimony.

Finally, it is important to recognize that a comprehensive account of scientific testimony should consider norms that govern its *production* as well as the norms that govern its *consumption*. The former set of norms governing testifiers is perhaps more immediately recognized, and it may draw on the literature concerning the general norms of assertion. However, the latter set of norms governing the recipients' uptake of scientific testimony is equally important. In intra-scientific testimony, for example, it is important to recognize that a scientific collaboration involves norms for the uptake and use of testimony (Chapter 4.4). Even in the case of public scientific testimony, layperson recipients may be subject to norms governing the uptake of testimony (Chapter 7.4.d).

2.4.d Social and objective norms: It is important to distinguish between *operative social norms* (sometimes called *descriptive norms*), which reflect the actual scientific practice, and *objective norms*, which reflect the optimal scientific practice with regard to a standard such as the epistemic standard of truth.

One way to study science is to study the sanctions that operative social norms underwrite. Although such norms may be tacit, and hence not conceptualized by the scientists, they are regulative in virtue of authorizing sanctions and rewards.[21] For example, a scientist who is found to violate a scientific norm will typically face severe penalties from the scientific community. So, when the operative social norms are tacit, we can learn about them indirectly by observing the sanctions that they authorize. The operative social norms of scientific testimony are no different.

However, social norms and objective norms may come apart. Consider, for example, social psychology's *replication crisis* which stems from the fact that many established experimental findings have failed to replicate in subsequent work. The replication crisis may be partly diagnosed as a case in which the operative social norms concerning data analysis have drifted away from the objective epistemic norms of truth-conducive data analysis. However, social norms evolve over time and the pendulum may swing back toward closer proximity to objective norms. An example of social scientific norms that have evolved toward an objective ideal are found in the social sciences. Here methodological norms have evolved to favor *mixed methods*, and this is in large part for veritistic reasons (Creswell and Clark 2017; Zahle 2019).

In the case of intra-scientific testimony, I will primarily seek to articulate objective epistemic norms that are reflected imperfectly, but reasonably well, by the operative social norms. While there is room for wide social variation of scientific norms, it is not the case that anything goes within the scientific practice. The background scientific realist framework of the present book involves the assumption that the social norms of science are constrained by the epistemic functions of science. If social norms of a scientific discipline were to develop with the result that one could only meet them by maximizing laughter at the expense of the pursuit of truth, the norms would have ceased to be scientific norms. (Perhaps they would have become norms of comedy?) Given that the social norms of intra-scientific testimony are epistemically constrained, objective epistemic norms may be seen as idealizations of this epistemic dimension of the operative social norms. Consequently, my agenda is not to promote revisions of existing social norms of intra-scientific testimony. Rather, I take the scientific practices as a starting point from which objective norms may be extrapolated. For example, I consider how sanctions within scientific collaboration align with general epistemic norms of

[21] Elster 1989; Bicchieri 2006, 2017; Henderson and Graham 2017a, 2017b.

assertive speech to articulate an epistemic norm of providing intra-scientific testimony (Chapter 4.2). While this approach involves considerable idealization, the resulting norm is a candidate for an objective epistemic norm that reflects the epistemic dimensions of the operative social norms for intra-scientific testimony.

However, in some cases the differences between the objective and social norms are too considerable to make this approach viable. For example, the social norms of some types of public scientific testimony are at least as sensitive to non-epistemic aims as to the epistemic ones. Science reporting, in particular, must serve other masters than truth since it occurs in the attention economy that is the contemporary media landscape (Dunwoody 2014; Nisbet and Fahy 2015). So, while I will propose norms of science reporting that are congenial to the *best practice* in existing science reporting, these proposals do not purport to reflect the operative social norms but objective ideals. This ambition is ameliorative in nature. For example, the proposed norms may serve as inspiration for more concrete guidelines for public scientific testimony (Chapters 5 and 6).

Likewise, the operative social norms governing laypersons' uptake of public scientific testimony do not generally align well with objective epistemic norms. Consequently, my attempt to articulate an objective norm for laypersons' uptake of public scientific testimony has a revisionary and even ameliorative aim (Chapter 7.4.d). The proposed objective epistemic norm is one that the operative social norms should—from an epistemic point of view—approximate.

2.4.e Concluding remarks on individual and social aspects of testimonial uptake: Investigating testimony in general requires that social environments and the norms that partly constitute them, as well as the individual vigilance in the reception of testimony, be studied as parts of a complex whole. This is also true in accounting for scientific testimony. For example, I will argue that the epistemic norms governing scientific testimony and the vigilance with which they are enforced contribute to the high degree of reliability of scientific testimony. But since scientific testimony takes place in very different social contexts and is governed by importantly distinct norms, a piecemeal approach is required.

2.5 Concluding Remarks

Let us consider how the main conclusions of the chapter point forward. I began the chapter by providing an operational characterization of (linguistic) testimony as *an assertion that is offered as a ground for belief or acceptance on its basis*. I then moved on to characterizing testimonial belief as belief that is based on testimony and expanded the discussion to testimonial *acceptance*. This notion is particularly important for scientific testimony and for understanding group testimony. The focus on the collective aspects of scientific agency leads to one important interim conclusion: The roles of scientific testimony may not be understood if the analysis myopically focuses on the case of a single testifier and single recipient forming a testimonial belief. Of course, an analysis of such basic cases may inform more complex cases of testimony, such as the ones we find in science. But the more complex cases hardly reduce to the simple binary case. Thus, a desideratum for further discussions is the complex cases of collective scientific agency as central explananda in their own right.

The highly selective discussion of the foundational epistemological debates concerning testimony also yielded some important lessons for going forward. For example, reflection on the internalism-externalism debate gave rise to characterizations of species of warrant, such as *discursive justification*, which will be central in my account of scientific testimony.

While I sided with anti-reductionists' accounts of *basic* testimonial entitlements, I also argued that reductionists' focus on the recipients' capacities of assessing testifiers remains especially important for an account of scientific testimony. In particular, reflection on these capacities indicates that the *cognitive heuristics* that both testifiers and recipients rely on must be investigated. For example, attempts to articulate guidelines for public scientific testimony must consider the psychological biases of evolved heuristics for epistemic vigilance among laypersons. On the other hand, epistemically externalist arguments indicate the importance of the general social environment in which the testimony takes place. In particular, there is reason to consider scientific testimony alongside the social norms governing its use.

So, taken together, a central lesson from the foundational debates is that the entire nexus of individual vigilance and the social norms characterizing the social backdrop for testimonial exchanges must figure prominently in an account of scientific testimony. As I move on, these broad methodological conclusions will guide and constrain the investigation.

PART II

SCIENTIFIC TESTIMONY WITHIN SCIENCE

Part II of the book consists of two chapters which are concerned with the nature of scientific testimony and the roles it plays *within* the scientific practice. A central contention of the book is that a characterization of the scientific method is incomplete without a characterization of the nature, roles, and norms of scientific testimony. In this part, I will motivate this claim by focusing on scientific testimony's role in truth-conducive scientific collaboration.

Chapter 3 opens with a defense of the claim that what differentiates scientific testimony from other types of testimony is that it is properly based on scientific justification. Given the centrality of scientific justification in this characterization, the chapter also contains a characterization of some of scientific justification's central properties.

Chapter 4 consists of an examination of the nature and roles of scientific testimony within science and characterizations of some norms for it. For example, I articulate and defend an epistemic norm for *providing* intra-scientific testimony within a scientific collaboration. Moreover, I propose some norms and principles that pertain to the *recipients* of such intra-scientific testimony. In doing so, I argue that such norms partly, but centrally, explain the truth-conduciveness of scientific collaboration.

3
Scientific Justification as the Basis of Scientific Testimony

3.0 Scientific Testimony and Scientific Justification

I have two main aims in the chapter. The first is to argue that what distinguishes scientific testimony from other types of testimony is that it is properly based on scientific justification. The second is to partly characterize scientific justification by specifying some central assumptions about it that I will rely on in subsequent chapters.

In Section 3.1, I argue that what makes a piece of testimony a piece of *scientific* testimony is that it is based on scientific justification. In Section 3.2, I outline my methodology of characterizing scientific justification via some of its hallmarks. In Section 3.3, I argue that scientific justification is generally superior to other kinds of warrant. In Section 3.4, I argue that scientific justification is typically gradable. In Section 3.5, I motivate the assumption that scientific justification is typically a variety of discursive justification. In Section 3.6, I recapitulate the central conclusions.

3.1 Scientific Justification Distinguishes Scientific Testimony

What are the features of scientific testimony that make it a *scientific* testimony? In this section, I will argue that the distinctive feature is that it is properly based on scientific justification. I begin by clarifying my methodology.

3.1.a How to characterize scientific testimony?: I will not attempt to provide a reductive analysis of scientific testimony. Rather, my aim is to provide a working characterization of paradigmatic instances and focus the discussion on such instances. Thus, there may be counterexamples to the characterization that I will give. Although I think they will be fairly peripheral, they indicate some limitations of the working characterization.

I will approach my positive analysis by criticizing some alternative characterizations of scientific testimony. The flipside of pursuing a working characterization of paradigm cases, rather than a reductive analysis, is that criticism of

Scientific Testimony: Its roles in science and society. Mikkel Gerken, Oxford University Press. © Mikkel Gerken 2022.
DOI: 10.1093/oso/9780198857273.003.0004

alternatives cannot simply consist in peripheral counterexamples. Nevertheless, it is clarifying to state the various characterizations in biconditional form. Then counterexamples may be provided, and it may be debated whether they are peripheral or core cases.

A challenge for characterizing scientific testimony is that the species of it are rather different. Given the differences between, for example, intra-scientific testimony and science reporting, it is questionable whether they may be subsumed under a general characterization or whether a more piecemeal approach is required. I will try to solve this tension between overarching commonality and particular differences by providing a generic working characterization of scientific testimony that is consistent with recognizing central differences among its species.

3.1.b Scientific testimony is not distinctive in that the testifier is a scientist: It is natural to think about scientific testimony as testimony that is produced by a scientist. It is fairly standard to characterize phenomena in terms of their source. Moreover, adapting this approach to characterize scientific testimony captures a lot of cases. Intra-scientific testimony between collaborators is testimony by one scientist to another. Scientific expert testimony to the public is testimony by a scientist.

Of course, such a characterization calls for some specification of what it is to be a scientist. But this challenge is not a principled problem for the characterization. A characterization of X in terms of Y often calls for an explication of Y. But if the characterization is reasonable, the explication of Y will then teach us something about X. Another concern is that not every testimony from a scientist qualifies as a scientific testimony. Scientists are only human and tend to talk about all sorts of mundane things. Well, there are a few obsessive types who barely talk about anything but science. But we may set those individuals aside for the present purpose (and during dinner parties). Most scientists provide an abundance of testimony about the weather, the game, their rash, and trash TV. Such testimonies are not scientific testimonies but merely testimonies by scientists. It is natural to try to handle this concern by restricting the characterization to testimonies by scientists *qua* scientists. This restriction gives us the following characterization:

> ***Testifier Characterization***
>
> A testimony is a scientific testimony iff the testifier is a scientist testifying *qua* scientist.

While *Testifier Characterization* has a good deal going for it, it faces severe challenges. An overarching challenge is to provide a non-circular characterization of the idea of *testifying qua scientist.* To illustrate this overarching challenge by way of some more specific ones, I will consider each direction of the biconditional in turn.

Let us start with the right-to-left direction and reconsider whether the 'qua scientist' restriction is sufficient to salvage *Testifier Characterization*. The concern here is that scientists testify about all sorts of things *qua* scientists but that many of these testimonies are poor candidates for scientific testimony. For example, a scientist may testify that the dean makes changes merely for the sake of change, that the call for papers is now public, that there were 480 candidates for the postdoc position, or that a particular journal is triple-blind reviewed, etc. In short, scientists, *qua* scientists, provide heaps of testimony about the business of science and life of a scientist. These testimonies reflect important aspects of being a scientist, and the scientists are doing their work *qua* professional scientists in uttering them. However, the testimonies are not very good candidates for scientific testimony. Another phenomenon causes similar problems. This is *expert trespassing testimony*, which occurs when a scientific expert testifies in a domain of epistemic expertise different from her own (Gerken 2018b). I discuss this issue in Chapter 5.5, but egregious cases should not be regarded as scientific testimony.

A defender of *Testifier Characterization* might respond with a more fine-grained individuation of testifier roles. For example, it may be argued that a scientist who testifies that there were 480 candidates is not testifying *qua* scientist but *qua* university employee. However, this approach accentuates the need for a non-circular account of testifying *qua* scientist. It is too circular to note that the individuals are not testifying *qua* scientists because their testimonies are not scientific testimonies.

Against the left-to-right hand side, it should be noted that *Testifier Characterization* rules out what I take to be an important category of scientific testimony—namely, science reporting by non-scientists. When a non-scientist, such as a science journalist, reports a new scientific finding, this is an important example of public scientific testimony. I do not think that it is reasonable to exclude it as such simply due to the fact that the testifier is a non-scientist. To illustrate this point, compare two cases of a familiar type of science reporting: the TV weather forecast. In one case, the presenter is a meteorologist who is an accomplished researcher, and in the other case, the presenter is a journalist with no meteorological training. Both presenters base their forecast on an online resource and testify that it will rain in Kampala tomorrow. But assuming that the former testimony is scientific whereas the latter is unscientific just in virtue of the fact that the former presenter is a scientist strikes me as unpromising. Moreover, the case is instructive in that it suggests that it is the *basis* of the testimony that determines whether a testimony is scientific. This point will guide my positive proposal.

3.1.c Scientific testimony is not distinctive in that it concerns a scientific subject matter: Characterizing scientific testimony in terms of its content might seem to provide an alternative to *Testifier Characterization* that diagnoses why it fails.

According to a content-oriented characterization, a scientist's testimony that the dean makes changes for the sake of changes would not be a scientific testimony because the content concerns the business of science rather than science itself. On the other hand, the focus on content would have it that a medical scientist who testifies that mannose reduces the growth of tumor cells provides scientific testimony in virtue of the scientific content of her testimony. The considerations may motivate the following proposal:

Content Characterization

A testimony is a scientific testimony iff its content is scientific.

The *Content Characterization* calls for a characterization of what it is for a content of a testimony to be scientific. This is a serious limitation given that it is unclear whether a sufficiently robust notion of scientific content may be characterized. Moreover, even if a non-circular characterization may be provided, *Content Characterization* faces problems.

To see this, let's consider a couple of testimonial contents that a characterization of scientific content should include. Perhaps "flu shots cause the flu" and "coral reefs will adapt to warmer ocean temperatures" are reasonable candidates? A concern for *Content Characterization* is that laypersons may, for all sorts of random, and entirely unscientific, reasons, testify that flu shots cause the flu or that corals reefs will adapt to warmer ocean temperatures. Moreover, they may provide such testimonies without an inkling of understanding of the sciences relevant to justify these claims. But testimonies that are in no way based on the relevant science, and perhaps entirely out of sync with it, are poor candidates for scientific testimonies.

A possible defense of *Content Characterization* has it that my verdict runs together whether something is a scientific testimony and whether it is a good scientific testimony. But in response to this defense, it should be noted that the content of virtually every testimony may be subject to scientific investigation. Of course, it is hard to give a characterization of what makes an investigation scientific, but we may recognize paradigm cases well enough to make the point. Consider, for example, drunken banter about sports, such as the following testimony about a 1990s basketball star: "Jordan would have averaged 40 points a game with the current rules." By analyzing game data with sophisticated analytic tools, a scientist can provide scientific evidence for or against such a claim. Even in the absence of a demarcation criterion between science and non-science, it should be clear enough that this testimony is scientific testimony. And it is even clearer that a testimony with the same content made by a layperson on the basis of an alcohol-fueled Jordan fetishism is not scientific testimony of any sort. Since the content is the same in the two cases, the example illustrates both the difficulty of specifying the phrase 'scientific content' and the prospects of using it as the basis for an account of scientific testimony.

As above, a more constructive lesson from these cases is that the epistemic basis of a given testimony is central to whether it is a scientific testimony. This lesson is important since it indicates that *Content Characterization* may have trouble dissociating pseudo-scientific testimony from proper scientific testimony. So, the characterization would be not only inaccurate but also harmful in dealing with cases of anti-vaxxing, climate science denial, creationism, etc. that involve pseudo-scientific testimony. For example, if someone presents unrepresentative anecdotes as scientific evidence for the hypothesis that easier access to guns decreases gun violence, the person is providing pseudo-scientific testimony. This is because the relevant scientific justification about this correlation indicates that such anecdotes are unrepresentative. However, the content itself—a postulated correlation—is a good candidate for a scientific content. So, if the testimony were backed by scientific justification from analyses of demographic studies, it would have been a scientific testimony. However, according to *Content Characterization*, it does not matter whether the testimony is backed by scientific justification or not. So, in order to preserve the crucial distinction between scientific testimony and pseudo-scientific testimony, *Content Characterization* should be abandoned.

3.1.d Scientific testimony is properly based on scientific justification: Reflection on why the previous characterizations of scientific testimony fail is instructive. In particular, both the *Testifier Characterization* and the *Content Characterization* are prone to important classes of counterexamples that occur when the testimony is not properly based on any scientific justification. Taking the clue from these classes of counterexamples, I propose the following working characterization of scientific testimony:

> ***Justification Characterization***
>
> A testimony is a scientific testimony iff it is properly based on scientific justification.

For the purpose of a generic characterization, I use the phrase 'based on scientific justification' as a shorthand way of saying that the testimony is based on scientific research/evidence/procedures/investigations, etc. that has produced scientific justification for the content. *Justification Characterization* captures a large number of paradigmatic cases of scientific testimony—including those that cause problems for the characterizations criticized above. For example, *Justification Characterization* neatly keeps apart scientists' testimony about mundane matters from their scientific testimonies. For when scientists testify about the dean, or the number of applicants, their testimonies are not typically based on scientific justification. Furthermore, *Justification Characterization* classifies laypersons' uninformed testimonies with scientific contents (e.g., "flu shots cause the flu") as unscientific testimonies when they are merely based on emotions, social identity, drunkenness, etc. Moreover, it explains why some testimonies, such as "Michael Jordan would have averaged 40 points a game with the current rules" are unscientific when based on drunken Jordan fanaticism, and scientific when based on analysis of statistical data. More importantly, *Justification Characterization* helps classify pseudo-scientific testimony as such, insofar as it may be characterized as testimony which purports to be properly based on scientific justification although it is not. It would probably be too much to expect the characterization alone to pinpoint every case of pseudo-scientific testimony since they may be quite diverse. But *Justification Characterization* affords a clear diagnosis of central types of pseudo-scientific testimony.

Of course, there is a debate to be had about what *proper basing* amounts to. Indeed, the basing relation is a topic of debate in epistemology (Lord and Sylvan 2019; Neta 2019). A mere causal relation to scientific justification will result in an overly liberal characterization. Consider, for example, a four-year-old's testimony that the ice will melt if the freezer is not plugged in. There is a loose sense in which this testimony is based on scientific justification concerning refrigerators and electricity. But it is at most a highly derivative scientific testimony. On the other hand, proper basing does not entail meeting the relevant epistemic norms of scientific testimony since such a requirement would rule out epistemically inappropriate scientific testimony. The fact that proper basing falls between the extremes of a mere causal relation and complete fulfillment of epistemic norms provides some bearings and helps to account for further cases. One such case consists of a scientist who reports basic perceptually warranted observations to colleagues. For instance, a biologist might testify: "I've seen tadpoles on the lake's east shore. So, let's start the search for the yellow-legged frog there." This seems like a case of intra-scientific testimony but one that is based on basic perceptual warrant. However, in the context of the investigation, this warrant functions as scientific justification insofar as it may be recorded in observation protocols and figure in a scientific report as a rationale for initiating the search on the eastern shore. So, given such integration, a perceptually warranted testimony counts as a scientific testimony, and without it, it may be best classified as a testimony relevant for, or about, science.

So, generally, whether a warrant is a part of scientific justification partly depends on whether it serves an appropriate functional role in scientific inquiry. For example, if a simple perceptual observation is protocolled as a data point in a systematized set of evidence, it forms a part of the resulting scientific justification (see also Hoyningen-Huene 2016). This idea provides a further clue to the questions about proper basing. In the case of intra-scientific testimony, proper basing may require that the relevant justification for the content of the testimony is integrated in a scientific practice—e.g., figuring in observation protocols, being subjected to standard procedures, etc. In the case of science reporting, however, proper basing merely requires that the science reporter defers to the scientific community in the right manner—e.g., by interviews, surveying the literature, etc. So, proper basing of scientific justification does not require that one acquires it, and this assumption aligns with the principle *Non-Inheritance of Scientific Justification* (Chapter 2.2.b).

As with other philosophical distinctions, the *Justification Characterization* leaves ample room for gray zones between scientific and non-scientific testimony. Consider intra-scientific communication in the context of discovery where the relevant hypothesis has not yet been scientifically justified (Reichenbach 1938; Schickore 2018). Some such communication should be regarded as intra-scientific testimony. But this assumption is compatible with *Justification Characterization* given the noted point that being properly based on scientific justification is relative to the integration in scientific practice. Given this point, proper basing may be quite minimal in a context of discovery given that integration in this stage of scientific inquiry is subject to less demanding epistemic norms. The idea that weaker epistemic norms govern the context of discovery does not entail that there is a "logic of discovery" (Schickore and Steinle 2006; Schickore 2018). It merely entails denial of the idea that anything goes in the context of discovery. Typically, a hypothesis may be forwarded as true, plausible, or worthy of pursuit only if it is sufficiently compatible with available background scientific justification. Such minimal epistemic requirement *may* be regarded as a reason to be suspect of the idea of a context of discovery. But the key point is that one may accept the notion of a context of discovery which is partly characterized by laxer norms, such as more minimal epistemic basing requirements on intra-scientific testimony (cf. Chapter 4). That said, there may be cases where it is unclear whether a given testimony in an exploratory stage of scientific practice qualifies as a type of scientific testimony.

While *Justification Characterization* does not by itself settle every case, this should not be expected from a generic characterization that aims to capture different kinds of scientific testimony. Moreover, *Justification Characterization* provides resources for more fine-grained distinctions in terms of different specifications of the basing relation. As noted, intra-scientific testimony paradigmatically requires that the testifier's justification is integrated in a scientific practice whereas science reporting is often based on scientific justification by the right kind of deference to the scientific community. So, in contrast with *Testifier Characterization*, science reporting may be recognized as a derivative of nevertheless important species of scientific testimony. In this manner, *Justification Characterization* may address the noted challenge of providing a general working characterization of scientific testimony that allows recognition of the very distinct species of it.

3.2 Characterizing Scientific Justification

My proposal for a working characterization of scientific testimony, *Justification Characterization*, is articulated in terms of scientific justification. So, it behooves me to say a bit about the nature of scientific justification by highlighting some of its features that I will invoke in arguments or otherwise rely on. Some of these features contribute to a partial elucidation of scientific justification, and others are too generic. So, to clarify what I will (and will not) attempt, some background methodology is in order.

3.2.a Scientific justification and demarcation: The problem of characterizing scientific justification relates to the venerable demarcation problem of characterizing science well enough to draw the line between science and non-science, including, especially, pseudo-science (Hansson 2017). One prominent way to draw the line is to characterize science in terms of the distinctive type of justification that it enjoys. Alas, attempts to solve the demarcation problem have been unsuccessful. Some even think that the project of pursuing a demarcation criterion should be abandoned. For example, Laudan famously announced the demise of the demarcation question by proclaiming it to be "uninteresting and, judging by its checkered past, intractable" (Laudan 1983: 125). Similarly, scientific justification has not been successfully characterized in a manner that demarcates it from non-scientific (including pseudo-scientific) kinds of justification. However, failures to provide demarcation criteria do not entail that it is pointless to specify features of scientific justification which may, individually or in conjunction, constrain a more approximate characterization.

Once it is not a desideratum for a characterization of scientific justification to yield a demarcation criterion, more relaxed characterizations may be explored, and non-distinguishing but nevertheless important paradigmatic features of scientific justification may be noted. So, rather than adding to the list of failed demarcation criteria in terms of scientific justification, I will simply articulate some important properties of scientific justification.

3.2.b From necessary and sufficient conditions to hallmarks: It is generally hard to provide reductive analyses of substantive epistemic phenomena. The failure of sustained attempts to reductively analyze knowledge indicates this much. Yet, knowledge may be illuminated by a non-reductive elucidation that consists in considering its relations to other basic phenomena such as belief, justification, and so forth (Gerken 2018a). This approach exemplifies an *equilibristic methodology* because it pursues a wide reflective equilibrium among folk epistemological judgments, philosophical principles, empirical findings, and interrelations between different epistemic phenomena (Gerken 2017a). Whereas reductive analyses must not be circular, there is room for some circularity in an equilibristic elucidation (Gerken 2017a, 2018a).

Scientific justification may not be as basic a phenomenon as knowledge or belief. But an equilibristic methodology may nevertheless be apt to elucidate it. So, I will not pursue a characterization in terms of (jointly) sufficient or even necessary conditions. Rather, I will articulate some *hallmarks* of scientific justification. A hallmark of a phenomenon is a paradigmatic trait of it, although it need not be a necessary aspect of the phenomenon. Nor do hallmarks provide individually or jointly sufficient conditions for the presence of the phenomenon, although they may contribute to a non-reductive elucidation of it. For example, the ability to fly may be said to be a hallmark feature of birds. Although other animals also fly and a few bird species do not fly, a characterization of birds that left out flying would be wanting. That said, some hallmarks may be so generic that while they amount to *central* features of the phenomenon under elucidation, they do not amount to *distinguishing* features of it.

3.2.c Concluding remarks on characterizing scientific justification: My discussion of scientific justification will have far more modest aims than characterizations that are meant to serve as the basis for a demarcation criterion. I will merely seek to contribute to a principled but partial and non-reductive elucidation of scientific justification by articulating three hallmarks of it. Each of the hallmarks will be relied on in my subsequent investigation of scientific testimony.

3.3 Hallmark I: Scientific Justification Is Superior

The headline that "scientific justification is superior" is shorthand for the following more specific and highly qualified thesis:

> *Hallmark I*
>
> In many domains, scientific justification is generally epistemically superior to non-scientific types of warrant.

Hallmark I contributes to a principled motivation of the idea that scientific testimony deserves an authoritative role in public deliberation. For example, *Hallmark I* captures the idea that scientific justification regarding global warming, vaccine safety, and evolution is generally epistemically superior to non-scientific warrants. However, *Hallmark I* is not a claim about the absolute epistemic force of scientific justification but a comparable claim according to which scientific justification often is the *best available* warrant at a given moment in time. Moreover, the comparative claim is not fully general but restricted to domains. Given this domain restriction, epistemic superiority is not a necessary condition for scientific justification. Nor is epistemic superiority sufficient for scientific justification since non-scientific types of warrant are superior to available alternatives in certain domains.

But while *Hallmark I* does not provide a demarcation between scientific justification and other types of warrant, it nevertheless helps to characterize scientific justification as a type of warrant that is often the best available in a restricted domain of inquiry. Thus, *Hallmark I* helps to motivate the distinctive authoritative role of public scientific testimony. Moreover, its restricted nature indicates some limits of the authority of science. Consequently, it is important to gain some clarity on the sense in which scientific justification is superior to non-scientific alternatives. So, I will defend *Hallmark I* by explicating it through a number of qualifications. The defense touches on some grand debates in the philosophy of science that I can only give a highly partial and focused treatment for the purpose of defending *Hallmark I*. My main line of defense is that despite all its flaws scientific justification tends to be epistemically better than available non-scientific alternatives.

3.3.a Epistemic aims and the nature of science: Arguably, it is constitutive of science that it pursues the highest feasible degree of truth-conduciveness. Of course, practical factors may severely limit the pursuit of the epistemic aims of science in a given research context. Often the epistemically best methods are not feasible. Moreover, scientists may occasionally hold on to epistemically inferior methods, and they have done so throughout history. Likewise, non-epistemic aims—such as publishability, grantability, and narcissism—may compromise the epistemic aims of science. These factors contribute to science's epistemic fallibility. However, as a general rule, it would be contrary to the spirit of science if scientists held on to a method that they recognized as epistemically inferior to another feasibly adoptable method. *Hallmark I* is qualified in part because there are exceptions to this rule. There may be practical or ethical reasons to avoid an epistemically superior method. For example, the early (pre-clinical) tests of medical products are not performed on humans, although this might be an epistemically superior way to justify hypotheses about their effects on humans. But everything else being equal, scientists ought to prefer, and they do by and large prefer, the epistemically best feasible methods at their disposal.

This is evidenced by the fact that the scientific community allocates a great deal of its resources to refining and improving its existing methods and instruments to become more reliable (Kampourakis and McCain 2019). An important aspect of the scientific practice consists in inventing new instruments and methods that are epistemically superior to existing ones. A celebrated historical case is Tycho Brahe's improvements to both astronomical instruments and practices of observation and systematic record keeping (Blair 1990). A celebrated contemporary case is the building of the Large Hadron Collider, which enables tests of predictions of particle physics (Brüning et al. 2012). These cases are highlights of the general scientific practice of improving instruments and methods to become more accurate and reliable. This practice is a *part of science*, and this fact indicates the centrality of the scientific aim of pursuing the highest feasible degree of truth-conduciveness. Fortunately, the efforts are not entirely in vain. Science continually succeeds in producing new findings and novel theories that could not even be articulated by non-scientific approaches, and this supports *Hallmark I*.

As mentioned, *Hallmark I* is highly qualified. First, it is qualified in terms of domain (Chapter 3.3.c). Secondly, it is qualified in terms of feasibility, since science serves other masters than truth. These include practical factors, such as relevance or urgency, or professional ones, such as publishability or visibility. Third, since 'superior' is a relational term, *Hallmark I* is a comparative claim that does not suggest that science produces the best warrant humanly feasible. Fourth and finally, the qualification indicated by the term 'generally' allows for cases where scientific justification is inferior—perhaps because the scientists are in the grip of a misguided theory or questionable research practices (Chapter 3.3.e).

While these qualifications indicate important limits of the epistemic authority of science, they do not undermine it. To substantiate this assessment, I will consider some objections to *Hallmark I* and elaborate on the qualification in the process.

3.3.b The charge of scientism: Does the qualified claim that scientific justification is epistemically superior come with a commitment to an objectionable version of scientism? Scientism comes in many varieties (Ladyman et al. 2007; de Ridder et al. 2018). But since *Hallmark I* postulates epistemic superiority, I will focus on epistemological versions of scientism. However, these also come in different varieties. So, for simplicity's sake, I will lump views together in strong and weak versions of scientism.[1]

Formulations of strong epistemological scientism may be found in Cowburn, who characterizes scientism as "the belief that scientific knowledge is the only valid knowledge which we have" (Cowburn 2013: 47). Likewise, de Ridder characterizes epistemological scientism as the following view: "Science is the only source of justified belief or knowledge about ourselves and the world" (de Ridder 2014b: 25). *Hallmark I* does not include any commitment to such strong forms of scientism. For example, the view is compatible with assuming that perception,

[1] Cf. Gerken 2017a. Peels 2018 provides a more comprehensive taxonomy.

memory, and introspection are important and indispensable sources of warranted belief and knowledge. Mizrahi contrasts such strong scientism with the following characterization of weak scientism: "Of all the knowledge we have, scientific knowledge is the *best* knowledge" (Mizrahi 2017: 354). There are a couple of notable differences with *Hallmark I*. Mizrahi's weak scientism is articulated in terms of knowledge, and I prefer the formulation that scientific justification is generally superior to other types of warrant.[2] More pertinently, *Hallmark I* includes both a domain-restriction and the qualifier 'generally' that renders it compatible with exceptions even within the relevant domains. In contrast, Mizrahi's official formulation is general (although his presentation is compatible with restrictions). In sum, the present view is far weaker than strong scientisms and also a good deal weaker than weak scientisms, such as Mizrahi's (Mizrahi 2017).

Since I am not about to argue over labels, I leave it open whether *Hallmark I* is a version of scientism. The important upshot is that it does not constitute, entail, or commit me to an *objectionable* kind of scientism.

3.3.c Domain restrictions: As noted, *Hallmark I* is compatible with the assumption that particular propositions or general domains of inquiry are such that non-scientific warrant is superior to scientific justification. So, the scope of the superiority of scientific justification may be specified by noting some candidates of such domains.

Let me first consider the class of everyday beliefs which trivially enjoy warrant that is superior to scientific justification because the latter is non-existing. For example, I am warranted in believing that I had rye bread for lunch even though the matter has not been scientifically investigated. However, many everyday hypotheses may be subject to scientific scrutiny, which would produce superior warrant for them. A biochemical analysis would provide superior warrant that the bread is indeed based on rye rather than some other grain. This is not to suggest that my non-scientific warrant is deficient or that I lack non-scientific *knowledge* that I had rye bread for lunch. But these points do not entail that my non-scientific warrant for the belief cannot be improved by a scientific investigation. So, even for many everyday beliefs, the best warrant that we can feasibly achieve would come from scientific investigations that can rule out alternatives that our everyday cognitive resources (perception, memory, etc.) cannot.

This point is important to recognize in order to avoid delimiting *Hallmark I* too much. There are plenty of phenomena—perhaps social phenomena in particular—that are barely investigated scientifically. So, people's non-scientific warrant for beliefs about these phenomena is *de facto* superior to scientific justification. But this does not show that the non-scientific warrant that people

[2] For my case against knowledge-centric epistemology, see Gerken 2011, 2014a, 2015b, 2017a, 2018a, 2018d, 2020b, 2021.

may have for beliefs about such phenomena is a *type* of warrant that is superior to scientific justification. With this point in mind, let us consider some stronger cases for delimiting *Hallmark I*.

One candidate for non-scientific justification that is superior to scientific justification includes beliefs about our own mental states. A venerable philosophical tradition has it that introspection gives us privileged access to our mental states (Gertler 2015). According to this tradition, warrant for beliefs about our mental states formed by third-personal methods may be defeated in ways that the introspective warrant may not (Peels 2016). So, beliefs about one's own mind is a candidate for a domain in which at least some types of introspective, non-scientific warrant are superior to scientific justification. But this is far from uncontroversial since there are specific defeaters to introspective warrant (Carruthers 2011). These include motivational issues that may lead to self-deception, confabulation, and so forth. Nevertheless, introspection is an important candidate for a domain in which it is reasonable to restrict *Hallmark I*.

Another candidate for beliefs that are justified by a non-scientific source is philosophical doctrines that are justified by philosophical reflection. To vindicate this candidate case type, two assumptions must be defended: first, the assumption that philosophy is not a science, and second, that some philosophical hypotheses are better justified by philosophy than by science. I am partial to accepting both assumptions. But they are sources of heated debate, and it would be a digression to defend them here. Moreover, dialectically speaking, *Hallmark I* would be strengthened if I were wrong about one of these assumptions. Candidates of philosophical claims that may be justified better by philosophical reflection than by scientific justification include claims about general moral principles and other normative claims. Candidates from the non-normative realm include cogito-like thoughts, such as my thought that I am currently thinking (Gerken 2020a). In contrast, it is hard to see how such a claim might be justified better by scientific methods. Something similar may be true of Kripke's doctrine that some necessary truths are only knowable a posteriori (Kripke 1980; Gerken 2015c). So, the domain restriction of *Hallmark I* may go well beyond the normative realm.

Domain restrictions of *Hallmark I* reinforce the idea that it does not amount to an objectionable kind of scientism. As a matter of *focus*, I will tend to discuss the domains in which scientific justification *is* epistemically superior to other types of epistemic warrant. After all, these are most pertinent in debates about scientific testimony. As mentioned, I take it to be uncontroversial that scientific justification for hypotheses about climate change, vaccine safety, etc. is generally superior to any non-scientific type of warrant. However, the restrictions of *Hallmark I* are worth noting as they are relevant to the issue of scientists who testify about a domain of epistemic expertise other than their own (Chapter 5.5).

3.3.d The objection from the pessimistic meta-induction: Perhaps the most common philosophical argument invoked to cast doubt on scientific justification is the so-called *pessimistic meta-induction*. It is so-called because it proceeds from the premise that most scientific theories of the past have been false to the conclusion that current scientific theories are likely false (see Putnam 1978; Laudan 1981 for influential formulations). The argument is standardly invoked by anti-realists about science to cast doubt on the scientific realist assumption that approximating truth is an actual and reasonable aim of scientific theories (Psillos 1999; Godfrey-Smith 2003; Chakravartty 2011).

However, *Hallmark I* does not concern the reasonable aims of science or the absolute reliability of scientific theories. It is merely the comparative claim that in many domains, scientific justification is generally superior to non-scientific warrant. So, although my sympathies are with the scientific realists, my focus here is to consider whether the pessimistic meta-induction may compromise *Hallmark I*. My central line of defense will be that even if it is granted that most scientific hypotheses have turned out to be false, they have typically been closer to the truth than their contemporary non-scientific competitors.

However, it is worth noting that several criticisms of the pessimistic meta-induction are even less concessive. For example, the argument structure of the pessimistic meta-induction has been challenged (Lewis 2001; Mizrahi 2013). Moreover, the premise that most past scientific theories have been shown to be false may be criticized by noting that the theories have often contained grains of truths that allowed them to be *superseded*.[3]

The idea of superseding suggests that scientific justification is far more truth-conducive than an initial glance on the track record of science might suggest. This point strengthens the case for *Hallmark I* since it is merely the claim that scientific justification is generally superior to non-epistemic warrants in many domains. Scientifically justified theories may lack a great deal in truth-conduciveness and still be epistemic improvements on their non-scientific competitors. This is arguably a trait of many abandoned or superseded scientific theories of the past. For example, Dalton's model of the atom was superseded by Thompson's, Rutherford's, and Bohr's models, and revisions have continued. But Dalton's model is reasonably regarded as a better account of the basic components of matter than any non-scientific alternatives available at the time.

The latter point is important because the pessimistic meta-induction is *diachronic* insofar as it compares earlier theories to current ones. Exploring the consequences of this point vis-à-vis scientific realism would take us too far afield. However, *Hallmark I* is only committed to a *synchronic* comparison between hypotheses that are scientifically justified and their contemporary competitors

[3] See Fahrbach 2011; Park 2011; Kampourakis and McCain 2019 for varieties of this objection. For criticism, see Müller 2015. Dellsén 2018a surveys kinds of scientific progress.

that are not. Such a synchronic comparison bolsters *Hallmark I* against concerns from the history of science. One can surely find examples in which non-scientifically warranted assumptions turned out to be closer to the truth than the contemporary scientifically justified ones. But I venture to suggest that such cases are exceptions to the rule. This comparative point may be developed in response to a related concern that current science is unreliable.

3.3.e The objection from reliability and the replication crisis: Whereas the pessimistic meta-induction casts doubt on the justification for contemporary scientific theories by reflecting on the track record of science, some sources of skepticism derive from reflecting on current theories. Again, my central response to this source of skepticism will be a comparative argument that even flawed science tends to be epistemically superior to non-scientific alternatives.

Much contemporary skepticism of scientific justification has its source in various metrics. For example, in an influential paper, Ioannidis presented a formal model according to which more than half of published research is false (Ioannidis 2005). The paper has not gone without criticism. For example, its assumptions are far more plausible for biomedical, psychological, and social sciences than for many natural sciences. Furthermore, the details of the model and the methodology of employing it may be questioned. Nevertheless, the paper raises concerns about scientific fields for which the assumptions are plausible. Moreover, empirical research has featured failed replication attempts, which have led to a more indirect, but nevertheless serious, source of skepticism about scientific justification. This is the *replication crisis* for parts of disciplines such as psychology, biomedical sciences, and economics (Open Science Collaboration 2015; Gilbert et al. 2016). Since lack of replicability indicates a lack of reliability, the replication crisis casts doubt on the strength of scientific justification of entire fields.

In response, Bird has argued that the replicability failure is not best explained by questionable research practices, but by the fact that most tested hypotheses are false (Bird forthcoming). If so, well-conducted experiments are prone to produce false positives given a statistical significance threshold of $P < 0.05$ (Bird forthcoming). Moreover, there is a considerable step from the assumption that some disciplines are prone to replication problems to the conclusion that scientific justification is not superior to non-scientific warrant about the same domains (Oreskes 2019). Even if some fields feature unimpressive reliability and lack of replicability due to questionable research practices and problematic incentive structures, this does not compromise scientific justification as a *type* of epistemic warrant (Kampourakis and McCain 2019). It merely indicates that scientific justification is not immune to failures to live up to adequacy constraints of scientific methods.

Two different ways in which scientific justification is fallible may be captured by a distinction between *well-functioning* fallibility and *malfunctioning* fallibility: A cognitive competence is well-functioning fallible just in case optimal performance of the competence does not entail that it fulfills its cognitive function which is, in the epistemic case, that of generating true belief (Gerken 2013a: 79). Generalizing from individual competences to social systems of hypothesis-formation suggests the following diagnosis: Reflection on the history of science establishes well-functioning fallibility, whereas reflecting on select scientific disciplines indicates that science is also malfunctioning fallible (Martinson et al. 2005; John et al. 2012). Thus, scientific justification is the product of practices that are fallible in both of these ways. However, this recognition does not show that scientific justification is not superior to non-scientific types of epistemic warrant since these are also fallible in these regards (Douglas 2015; Oreskes 2019).

Moreover, many of the troubled disciplines have begun to take measures that address the trouble. For example, it has been proposed that false positives may be addressed by redefining statistical significance from $P < 0.05$—i.e., less than or equal to a 5 percent probability that the result occurred by chance. Specifically, it has been proposed to institutionalize a stricter threshold of $P < 0.005$ (Benjamin et al. 2018). Another measure involves systematic attempts to replicate and alter incentive structures that reward such efforts (Zwaan et al. 2018). Moreover, meta-analyses, which may be argued to serve scientific self-correction, are becoming increasingly important (Bruner and Holman 2019). Furthermore, there are movements in favor of recommending or requiring preregistration of study designs in order to address hindsight bias and confirmation bias. Such movements also seek to address questionable research practices such as *p-hacking*—the phenomenon that a dataset is manipulated or analyzed in a selective manner so as to obtain a statistically significant finding.[4] It is unlikely that such measures will provide a quick fix (Adam 2019). But the fact that structural problems are being recognized and that strategies for addressing them are being explored and implemented nevertheless indicates that self-correction is central to scientific practice (Kampourakis and McCain 2019). Of course, it must be asked whether this practice has epistemic payoffs. But at this point, it is important to recall that *Hallmark I* is a comparative claim and ask the question: What is the alternative?

Whereas science is generally imperfect and occasionally systematically flawed, it is doubtful that non-scientific warrants are any better in most domains investigated by a reasonably mature science (Douglas 2015). Even when the relevant science is subject to serious concerns, it is hard to identify a superior, or even

[4] Nosek et al. 2015, 2018; van't Veer and Giner-Sorolla 2016; Wagenmakers and Dutilh 2016.

equal, epistemic source. Despite the challenges facing social psychology, there are hardly any viable alternatives for warranting beliefs about, for example, social biases in intuitive judgment. Despite the challenges facing parts of the biomedical sciences, there are no viable alternatives for warranted beliefs about nutrition, vaccine safety, etc. In fact, it is often unlikely that any non-scientific alternative could even conceptualize and articulate the hypotheses in question.

This comparative argument may be strengthened once we raise the gaze from the particular cases of unreliable scientific postulates and consider the scientific justification as a communal source. Although the measures to self-correct and improve the methodology yielding scientific justification are also fallible and systematically flawed, alternative sources rarely have similar mechanisms (Kampourakis and McCain 2019). So, while scientific justification is far less perfect and far messier than enlightenment ideals might suggest, *Hallmark I* is not compromised by the challenges from unreliability and the replication crisis. Ioannidis, who has done as much as anyone to highlight unreliability, bias, and other problematic features of science, is clear on this point: "Our society will benefit from using the best available science for governmental regulation and policy" (Ioannidis 2018). Although scientific justification is more fallible than we would like to think, it is generally our best type of epistemic warrant in many domains.

3.3.f Concluding remarks on the epistemic superiority of scientific justification: *Hallmark I* is important because it motivates that scientific testimony should play a privileged role in public discourse and decision making. I have defended *Hallmark I* and, thereby, the role of scientific testimony by a comparative argument according to which the alternative sources of warrant are generally worse. However, a lesson from the bleaker views of science is that the epistemic strength of scientific hypotheses varies greatly from discipline to discipline as well as within a single discipline (Ioannidis 2018). This lesson is important to bear in mind when considering norms of both interdisciplinary and public scientific testimony. Epistemic norms of scientific testimony should reflect the fact that the nature and strength of scientific justification vary enormously among disciplines.

3.4 Hallmark II: Scientific Justification Is Gradable

The second feature of scientific justification that I will highlight is its gradability.

> *Hallmark II*
>
> Scientific justification generally comes in degrees of epistemic strength.

This hallmark is far less controversial than *Hallmark I.* Moreover, it is not a distinguishing property of scientific justification that it is gradable since this is true of just about any kind of warrant. However, I will rely heavily on *Hallmark II* given the view that differing degrees of scientific justification is something that should be reflected in the norms governing scientific testimony. So, I will briefly indicate the nature and source of the gradability of scientific justification.

In some cases, the gradability of scientific justification is explicitly marked or indicated by institutionalized practices. For example, meteorologists mark the epistemic strength of their predictions in terms of a probabilistic measure—the Brier score (Brier 1950). Likewise, philosophers of science have emphasized that International Panel on Climate Change's (IPCC) reports standardly indicate the degree of epistemic strength via coarse-grained confidence ratings (Steele 2012; Betz 2013; John 2015a). It is controversial whether practice vindicates or compromises the value-free ideal of science in part because IPCC reports are science-policy hybrids. However, it is common ground in these debates that the practice of IPCC reflects the gradability of the scientific justification for the relevant hypotheses. Similarly, the standardized stages of testing pharmaceutical products also reflect degrees of strength of the scientific justification that a given pharmaceutical drug is effective or that it does not have severe side effects. Finally, it is a general practice to explicitly indicate the strength of the statistical evidence for a given hypothesis (see, e.g., American Psychological Association 2009). This practice also indicates the gradability of the relevant type of scientific justification.

Reflection on the more general *kind* of warrant that is characteristic of, at least, the natural sciences also supports *Hallmark II.* Empirical scientists rarely, if ever, possess justification that amounts to a proof. Philosophers of science have emphasized that empirical scientific hypotheses and theories are typically justified by induction (Hall and Hájek 2002) or abduction (Lipton 2003). However, both of these types of warrant are gradable. Indeed, it is a desideratum for a theory of induction that inductive justification comes in degrees, and even Popper, who was a skeptic about induction, made sure to reflect the idea of gradability in his account of corroboration (Popper 1934).

Of course, some scientific claims are justified by direct observation. One may observe a herd of reindeer over time and on this basis form the hypothesis that reindeer grow a new set of antlers every year. However, some observations are more reliable than others. So, even scientific justification from direct observation may give rise to questions concerning the degree of scientific justification.

So, scientific justification is typically *gradable.* Arguably, it is the least controversial of the three hallmarks of scientific justification that I propose. But it is one that is important to recognize in order to articulate norms for scientific testimony and to understand problems that concern the uptake of it.

3.5 Hallmark III: Scientific Justification Is Articulable

The final property of scientific justification that I will highlight here is that is generally articulable.

> ***Hallmark III***
>
> Scientific justification generally involves a high degree of *discursive* justification.

Hallmark III is cast in terms of *discursive justification*, which I characterized as follows: S's warrant for believing that *p* is a discursive justification iff S is able to articulate some epistemic reasons for believing that *p* (Chapter 2.3.a). Given this characterization, *Hallmark III* has it that scientific justification for *p* generally involves the ability to articulate a good deal of it as epistemic reasons for believing that *p*. (Note that this does not require that S believes *p* herself, as she may merely accept it.)

While many other types of justification are discursive, *Hallmark III* helps distinguish scientific *justification* from basic entitlements and from justifications that are minimally discursive. The latter include mere references to the source, such as "she said so" or "I seem to recall." It also helps explain why someone who forms a hypothesis on the basis of a hyper-reliable oracle is not scientifically justified despite meeting *Hallmark I*. So, although *Hallmark III* does not form part of a demarcation or reductive characterization of scientific justification, it does, alongside other hallmarks, contribute to some principled elucidation of it.

Hallmark III explains why I have mainly used the phrase 'scientific justification' instead of the more generic phrase 'scientific warrant.' I do not want to argue that there are no externalist types of scientific warrant—i.e., scientific entitlements. But according to *Hallmark III*, the central type of scientific warrant is epistemically internalist in the sense that it is articulable. I will first motivate *Hallmark III* and then consider candidate exceptions to it.

3.5.a Motivation from scientific practice: A central strand of motivation for *Hallmark III* comes from reflection on scientific practice. The core of scientific publishing consists in articulating a particular justification for a particular hypothesis or theory. Since scientific justification is gradable (as *Hallmark II* has it), the scientific justification may differ in strength, and disciplinary conventions determine how strong the justification must be to be worthy of publication. In mathematics, the scientific justification for a theorem tends to consist of a proof, and, consequently, articulating a proof is central to mathematical publications. In the medical sciences, scientific justification tends to consist of randomized controlled studies, and, consequently, articulating the data and methodology of such studies is central to publication in medical science publications. Often, there are fairly fixed disciplinary conventions outlining how the justification must be articulated—e.g., in terms of a section outlining methodology, one specifying the data found, one analyzing the data, etc. (see American Psychological Association 2009 for an example). The high degree of explicit conventionalization of how scientific justification must be articulated in publication indicates that the articulability characteristic of scientific justification is not a peripheral aspect of it. Rather, it is part of what makes it the type of warrant that counts as *scientific* justification in the relevant science. Thus, the role of a scientist is partly that of being able to explicate scientific justification (see Khalifa and Millson 2020).[5]

Empirical research on contemporary scientific practice provides another line of support for *Hallmark III*. For example, Wagenknecht reports on a series of semi-structured interviews: "I learnt that from the scientists' perspective an ethical imperative of "dialoguing" is a fundamental feature of good team research: When statements seem dubious or incomprehensible, a scientist should ask questions, and when asked a question, a scientist should provide a good explanation" (Wagenknecht 2015: 170). Specifically, Wagenknecht concludes that scientific testifiers should be explanatorily responsive in the sense that they should possess "the willingness and the skill to address the listener's epistemic needs" (Wagenknecht 2015: 172). While this formulation focuses on character traits of individual scientists, the study supports the present weaker claim about the nature of scientific justification—viz., that it is discursive.

So, both reflection and empirical research on the practice of philosophy provide some initial support for *Hallmark III*. In the subsequent sections, I will augment this motivation.

3.5.b The motivation from replicability and communal scrutiny: Philosophers of science have argued that the requirement of being able to provide epistemic reasons is connected to the scientific ideals of objectivity and replicability, which have long been recognized as central to science (Longino 2002). A clear early statement may be found in Popper: "the objectivity of scientific statements lies in the fact that they can be inter-subjectively tested" (Popper 1934/2002: 22).

A central idea underlying this stance is that scientific hypotheses are supposed to be objective in the sense that their epistemic basis is not idiosyncratic to particular scientists. The ideals of scientific objectivity and intersubjectivity are concretely manifested in the ideal of *replicability* (Kahneman 2012). However, replicability requires that both the data and the methodology used in justifying a

[5] Interestingly, this approach has important historical roots. For example, in attempting to lay the methodological foundation for the new sciences, Descartes famously distinguished between *cognitio* and *scientia*. While I will not let this point carry argumentative weight, it is interesting to observe that the word 'science' derives from the term 'scientia,' which is characterized as distinctively secure and harder to achieve than *cognitio* (Descartes, Replies 2, AT 7:141).

scientific hypothesis are articulated in such a manner that other scientists may attempt a replication (Winsberg et al. 2014). But this amounts to articulating the warrant for it, and this, in turn, amounts to providing discursive justification.

More generally, scientific hypotheses must be explicitly motivated to be scrutinized not just in terms of replication but also from different viewpoints. Hoyningen-Huene puts the point as follows: "the scientific community must be organized in such a way that all knowledge claims are scrutinized by its members from as many possible different points of view. We are thus looking for the social reflection of something epistemological: the highly systematic defense of knowledge claims" (Hoyningen-Huene 2016: 109). Hoyningen-Huene takes systematicity to be the main characteristic feature of science, but one need not accept this view to appreciate the point that scientists are expected to explicitly justify their claims. The scientific community partly consists in institutionalized conventions and platforms for not merely replication but also critical scrutiny in general. Longino articulates the point in arguing that scientific practice involves "recognised avenues for the criticism of evidence, of methods, and of assumptions and reasoning" (Longino 1990: 76–81, 2002: 129ff.). More generally, Longino's account of a truth-conducive scientific community highlights the importance of explicating scientific justification given that she argues that the scientific process largely consists in "activities involving discursive interactions among different voices" (Longino 2002: 99; see also 134). Such criticism of evidence, methods, and assumptions is feasible only if they are articulated. But that, in turn, requires that articulability is central to scientific justification. If the justification is not articulated, it is hard for the scientific community to scrutinize it.

How does *Hallmark III* relate to radically collaborative research? In such collaborations, no single individual may comprehend all or even most of the justification for a hypothesis set forth by a large research group (Winsberg et al. 2014; Huebner et al. 2018). However, this fact does not compromise *Hallmark III*. First of all, discursive justification does not require that the individual is able to articulate every aspect of the justification, but only that she is capable of articulating some of it as epistemic reasons. Second, the fact that no single individual is capable of articulating every aspect of the scientific justification does not mean that it is not articulable by someone in the group. In fact, it would be a fault of the collaborative work if the scientific justification for the relevant hypothesis could not be articulated by anyone in the group or by the group as a cognitive unit. Indeed, the *group*'s ability to articulate the justification for a hypothesis may be central to an account of scientific group knowledge (de Ridder 2014). Finally, the research communications produced by a large group are by no means exempt from the general requirement of providing discursive justification for scientific hypotheses and theories. However, cases of massive collaboration raise important questions about the norms of the production and consumption of intra-scientific testimony (Chapter 4).

3.5.c Motivation from the aims of science: A different type of rationale for *Hallmark III* may also shed light on the relationship between the articulability and the epistemic superiority of scientific justification (i.e., *Hallmark I*). The type of rationale that I have in mind concerns the aims of science.

Here, I assume that an important aim of science is a veritistic one—i.e., the pursuit of true theories and hypotheses. While the assumption is a controversial one in the philosophy of science, it is part of the broad realist framework that I take for granted in this book (Psillos 1999; Godfrey-Smith 2003). Given the assumption that the pursuit of truth is an aim of science, I will argue that since discursive justification is generally truth-conducive for the scientific community, discursive justification is characteristic of scientific warrant.

Although I will give a specific argument for this conclusion, the general idea is not novel. It is widely recognized that critical scrutiny from a number of perspectives is a truth-conducive practice that requires that scientific justification be explicated. Thus Kitcher: "scientific debates are resolved through the public articulation and acceptance of a line of reasoning that takes considerable time to emerge" (Kitcher 1993: 344). Indeed, Kitcher highlights that scientific justification must be explicated in order for this process to take place: "the working out of this line of reasoning depends crucially on the presence in the community of people who are prepared to work on and defend rival positions" (Kitcher 1993: 344). So, from a communal perspective, the scientific justification that a single scientist or research group has acquired must be made explicit in order for truth-conducive criticism to take place. This partly explains the institutionalized practice of articulating scientific justification.

The value of explicating justification is sometimes met with criticism from epistemic externalists who take the traditionally internalist property of articulability to be *epistemically* inert. According to such criticisms, scientific justification is at most articulated for practical purposes—i.e., to convince others. In response to this criticism, I will rehearse an argument for the truth-conduciveness of discursive justification that I have given on several occasions (Gerken 2013b, 2015a, 2020a).

Here is a structural basis for the argument: Assume that we have three collaborating scientists S1, S2, and S3. Each of these scientists provides testimony in favor of a hypothesis that bears on their joint research. However, the three hypotheses are jointly inconsistent. Assume, for example, that S1 testifies that the hypothesis that *p* is true, that S2 testifies that the hypothesis that *p* entails *q* is true, and that S3 testifies that the hypothesis that *not-q* is true. If the collaborators realize that they hold jointly inconsistent hypotheses, they ought to engage in communal revision of the hypothesis set if doing so is feasible. However, if all the scientists are merely entitled and lack any access to the grounds of their respective hypothesis, they lack resources for rationally deciding which hypotheses should stay and which should go.

In contrast, if each of the collaborators is *discursively* justified, they are able to articulate their epistemic reasons for their respective hypothesis. Doing so can enable them to improve the group's epistemic basis for making a truth-conducive communal revision. What enables a truth-conducive group revision of the hypothesis set in a truth-conducive manner is that the epistemic reasons for and against the hypotheses in question may be weighed against each other. An interesting example of this occurs when there are asymmetric defeasibility relations among the epistemic reasons for the various hypotheses. Assume, for illustration, that some of the reasons for the hypothesis that *p* also defeat the epistemic reasons for the hypothesis that *not-q*. Assume, moreover, that the epistemic reasons for the hypotheses that *not-q* do not defeat any of the reasons for the hypothesis that *p*. *Ceteris paribus*, the hypothesis that *not-q* should go in such a case. But unless the reasons for both hypotheses may be put on the table, the group will not be in a good position to reach that conclusion.

The schematic example may be made a bit more concrete. Assume that the scientists are economists who rely on three different models as justifications for their respective hypothesis. It may be that the assumption *p*, which is justified by S1's model, M1, provides strong evidence that a central idealization in S3's model, M3, is misguided. Everything else being equal, the economists will have a defeasible reason to abandon S3's hypothesis that *not-q*. But, again, they will only be in a position to acquire this reason if S3 is able to articulate the justifying model, M3, well enough to make the problematic idealization evident to her collaborators. Such cases exemplify why articulability is a beneficial *epistemic* property and not merely a conventional or pragmatic one. This provides us with another reason to accept *Hallmark III*.

Of course, it must be recognized that deliberation is not *invariably* truth-conducive. The scientists may have a bias in favor of their own hypothesis and Solomon has argued that pressure to reach a consensus can "lead to phenomena such as groupthink and to suppression of relevant data" (Solomon 2006b: 28). However, the case above represents a truth-conducive mode of collective belief-revision that I take to be fairly prevalent in science (Longino 2002).

Given that discursively justified testimony is generally truth-conducive from a communal standpoint, *Hallmark I* (epistemic superiority) and *Hallmark III* (articulability) are closely related. This is in part because collaborating scientists' ability to articulate scientific justification contributes to the epistemic force of scientific collaboration. The ability to articulate scientific justification is truth-conducive in the positive sense of helping to organize scientific research in an effective manner. Moreover, it is truth-conducive in a negative sense insofar as it allows for critical scrutiny that helps to minimize individual mistakes, biases, and idiosyncrasies (Longino 1990, 2002). In a reasonably well-functioning scientific community, the articulability of scientific justification contributes to its comparative epistemic superiority. So, *Hallmark I* and *III* offer perspectives that reinforce the thesis *Collaboration's Contribution* according to which scientific collaboration contributes immensely to the epistemic force of science (Chapter 1.3.b). Moreover, since the epistemically beneficial articulation of epistemic reasons consists in large part in intra-scientific testimony, the argument also reinforces the thesis *Testimony's Contribution* according to which intra-scientific testimony is an epistemically vital part of scientific collaboration (Chapter 1.3.c).

An alternative teleological rationale for *Hallmark III* may appeal more to those who hold dear the idea that science has the aim of providing understanding and explanation rather than mere true belief.[6] To articulate the justification for a hypothesis is to provide a candidate explanation of why it is true: "[R]easons in general are explanations, and . . . a justification is a particular kind of explaining, namely, an explanation of why something is right or just" (Hyman 2015: 135–6). Moreover, McCain has argued that such explanations are central to the generation of scientific knowledge due to their role in inference to the best explanation (McCain 2015; see also McCain and Poston 2014). This reinforces the idea that discursive justification is central to scientific justification and knowledge.

Likewise, articulating the justification for a hypothesis contributes to understanding why the hypothesis is true. So, given that providing explanations or understanding is a central aim of science, it is plausible that articulating epistemic justification is also an aim of science. If so, there is another sense in which scientific justification is characteristically discursive. Note that this rationale does not require that explanation or understanding be the sole, primary, or ultimate aim of science. The rationale only requires that it is an important aim of science to provide explanations or understanding.

So, given the commonly proposed aims of science, there are a range of teleological motivations for *Hallmark III* insofar as articulating scientific justification may contribute to these aims.

3.5.d Concluding remarks on the articulability of scientific justification: I have motivated the view that articulability is central to scientific justification from a number of angles. If a scientist cannot articulate any reasons for the results and hypotheses that she communicates to her collaborators or the wider scientific community, she is unlikely to have much of a career as a scientist. This point will be important in considering the norms of intra-scientific testimony (Chapter 4). Moreover, I will argue that the articulability of scientific justification is central to the privileged status of public scientific testimony in public debates (Chapters 5 and 6).

[6] For perspectives on this issue, see Grimm 2006; Khalifa 2013, 2017; Strevens 2013.

3.6 Concluding Remarks on Scientific Justification and Scientific Testimony

The central ambition of this chapter has been to provide a working characterization of scientific testimony that distinguishes it from other types of testimony. According to my proposed characterization, what makes testimony scientific is that it is properly based on scientific justification. Given this working characterization, scientific testimony may be specified in terms of a specification of scientific justification. However, I have not sought to add to the long list of failed attempts to provide a specification that can distinguish scientific justification from non-scientific warrant. Rather, I have provided a partial specification by highlighting three *hallmarks* of scientific justification that I will assume throughout:

> *Hallmark I*
>
> In many domains, scientific justification is generally epistemically superior to non-epistemic types of warrant.
>
> *Hallmark II*
>
> Scientific justification generally comes in degrees of epistemic strength.
>
> *Hallmark III*
>
> Scientific justification generally involves a high degree of *discursive* justification.

Each of these hallmarks provides an important bearing on the accounts of intra-scientific and public testimony that I will develop. Moreover, the explanation and defense of the three hallmarks have also given rise to some important points in their own right. For example, I have highlighted that providing discursive justification is part of the aim of science. Likewise, I provided some arguments that, from a communal standpoint, discursively justified testimony is truth-conducive, and noted how this reinforces the theses *Collaboration's Contribution* and *Testimony's Contribution* (set forth in Chapter 1.3).

Finally, I have argued that epistemic norms of scientific testimony should reflect the fact that the nature and strength of scientific justification vary greatly between disciplines and between hypotheses within a given discipline. These are substantive conclusions in their own right that will also play important roles in what is to come.

4
Intra-Scientific Testimony

4.0 The Roles of Intra-Scientific Testimony

In Chapter 1.1.c, I gave a broad and approximate characterization of intra-scientific testimony as *scientific testimony from a scientist that has collaborating scientists as its primary audience and which aims to further future scientific research*. As will soon become evident, there are many types of intra-scientific testimony and they play a great variety of roles in scientific work. The aim of this chapter is to partly characterize the roles that intra-scientific testimony plays in scientific collaboration by articulating some of the norms governing it.

Here is the plan: In Section 4.1, I consider the characterization of intra-scientific testimony and draw some distinctions that will inform the substantive discussion. In Section 4.2, I articulate and develop an epistemic norm of intra-scientific testimony that applies to the testifiers. In Section 4.3, I consider the recipient's uptake of intra-scientific testimony. In Section 4.4, I develop a norm of uptake of intra-scientific testimony. In Section 4.5, I consider the significance of intra-scientific testimony in scientific practice. In Section 4.6, I summarize the main conclusions.

4.1 The Characterization of Intra-Scientific Testimony

Consider the characterization of intra-scientific testimony as *scientific testimony from a scientist that has collaborating scientists as its primary audience and which aims to further future scientific research*. This characterization has two related components. The first one concerns the primary audience of collaborating scientists, whereas the second concerns the aim of furthering future research. As noted in Chapter 1.1.c, the two components are neither individually necessary nor jointly sufficient for intra-scientific testimony. But the characterization captures paradigmatic examples of intra-scientific testimony and helps distinguish it from central cases of public scientific testimony that are aimed at non-scientists. It also helps distinguish it from central cases of inter-scientific testimony that are aimed primarily at non-collaborating scientists. But, as emphasized, intra-scientific testimony may overlap with both inter-scientific and public scientific testimony. However, my focus in this chapter will be *science in the making* and the roles that testimony plays within collaborative scientific processes.

Scientific Testimony: Its roles in science and society. Mikkel Gerken, Oxford University Press. © Mikkel Gerken 2022.
DOI: 10.1093/oso/9780198857273.003.0005

Note that this characterization of intra-scientific testimony gains further specification due to the general *Justification Characterization* of scientific testimony as testimony that is properly based on scientific justification (Chapter 3.1.d). Consider, for example, a scientist's testimony to a colleague: "There is freshly brewed coffee in the lab kitchenette." This testimony is directed at another collaborating scientist with the aim of furthering future research—indeed, it may be *crucial* in this regard. But since it is not based on scientific justification, it is not scientific testimony, and *a fortiori* it is not intra-scientific testimony. So, combining the characterization of intra-scientific testimony with the general *Justification Characterization* of scientific testimony renders the former more specific and avoids misclassification of important cases. Still, the characterization does not amount to a reductive definition of intra-scientific testimony. (There are exotic cases in which both conjuncts may be met by a testimony that is not intra-scientific testimony.) So, the characterization provides an approximation that distinguishes central cases of intra-scientific testimony from central cases of inter-scientific and public scientific testimony. However, that is a good enough operative characterization for the present discussion, which focuses on such central cases.

4.1.a Species of intra-scientific testimony and their commonalities: The characterization of intra-scientific testimony remains a broad one that admits of many species. Just think about the differences among testimonies to a graduate student, a departmental colleague, a departmental collaborator working in another subfield, a collaborating colleague from another department, a collaborator in an interdisciplinary research group, etc. Testimonies in these different contexts do not even begin to exhaust the varieties of informal intra-scientific testimony.

But while this variety of intra-scientific testimony must be recognized, there are enough important common denominators to develop some general and principled assumptions. Moreover, not every variety in communicative context constitutes a distinct *type* of intra-scientific testimony. For example, some variances in the testifier-recipient relationship or the specific aim of the testimony are best thought of as mere varieties in conversational context. So, it is an important desideratum for a general norm of intra-scientific testimony that it may account for such contextual variations. Consequently, I will pursue fairly general principles and norms of intra-scientific testimony that reflect what its varieties have in common.

I will not attempt to provide a taxonomy of species of intra-scientific testimony, although it is worth recalling the distinction between *natural* and *formal* testimony (Chapter 1.1.a). Many cases of intra-scientific testimony are on a continuum between formal and natural testimony. But although natural intra-scientific testimony is ubiquitous, some intra-scientific testimony has a formal character. A written annual progress report from a project member to a principal investigator may be a species of formal intra-scientific testimony. Such a report could include the following intra-scientific testimony:

> "Preliminary analysis of postdoc Y's data indicates that further data must be collected to control for an unforeseen confounder."

However, a closely related natural intra-scientific testimony would occur if the principal investigator asserted the following to postdoc Y in the hallway after a group meeting:

> "You need to re-run the study in order to control for this confounder that we discussed."

The philosophical challenge in providing norms for such testimonies is to articulate them at a level of generality that subsumes both the formal and the informal versions while also accounting for the differences between them.

4.1.b Disciplinary, interdisciplinary, and multidisciplinary collaboration: Both intra- and inter-scientific testimony may be characterized as disciplinary or interdisciplinary. The disciplinary-interdisciplinary distinction admits of gray zones given that disciplinary borders may be porous. They may be porous in terms of substance. For example, it may be unclear where psychology ends and sociology begins. Likewise, it may be unclear whether a research question falls within philosophy or linguistics. Furthermore, disciplinary borders may be *methodologically* porous. Distinct scientific fields may require very similar methods and competences. For example, statistics is used in data analysis in multiple fields. But even less generic competences may cut across disciplines. For example, interview methodology is important in fields such as communication studies, psychology, and anthropology. These topical and methodological overlaps partly explain why there are substantive gray zones between disciplines.

Yet, the distinction between disciplinary collaboration and interdisciplinary collaboration has its place. It is typically clear enough that someone excavating a Viking settlement and classifying its artifacts is doing archeology and that someone analyzing the acidity of a water sample is engaged in chemistry. The vague and porous borders of some disciplines do not entail that there are no distinct disciplines. Hence, some scientific collaborations are accurately described as interdisciplinary ones.

Recall the distinction between interdisciplinary and *multi*disciplinary collaborations (Chapter 1.2.c). Interdisciplinary research involves some degree of integration of at least two disciplines (Klein 2005; Andersen and Wagenknecht 2013). In contrast, multidisciplinary research may, following Holbrook, be characterized as research in which at least two disciplines contribute to work on a given research problem in a fairly isolated, perhaps sequential manner (Oskam 2009; Holbrook 2013). So, whereas interdisciplinary collaboration is characterized by a significant degree of integration in terms of methods and terminology, multidisciplinary collaboration is not. For example, anthropologists, sociologists, legal scholars, and historians may form a collaborating research group to make predictions about future migration patterns. Insofar as each of them adopts some measure of perspectives, methods, or terminologies from the other disciplines, they are working interdisciplinarily. In contrast, a group of biologists may systematically gather wild salmon to get a representative sample and hand over the sample to a group of chemists who analyze it to determine the level of mercury that it contains. In this case, the division of labor is so clearcut that the collaboration is multidisciplinary research. Since the notion of *integration* admits of degrees, there are cases that are not clearly classifiable as interdisciplinary or multidisciplinary research (Klein 2010). However examples such as the ones mentioned above provide reasonably clear cases of interdisciplinary and multidisciplinary research.

The cases of interdisciplinary and multidisciplinary scientific collaboration raise important problems for intra-scientific testimony. These problems concern the risk of misunderstanding due to equivocation, which arises because different fields use terminology differently. Moreover, epistemic norms of scientific testimony should reflect that the nature and strength of scientific justification vary enormously between disciplines. This may lead to communicative problems insofar as a collaborator in discipline A may expect a higher degree of scientific justification backing intra-scientific testimony than is required in discipline B. On a theoretical level, the point indicates a challenge for articulating general norms governing the production and consumption of intra-scientific testimony in such scientific collaborations.

4.1.c Concluding remarks on the characterization of intra-scientific testimony: The fact that there are a number of distinct species of inter-scientific testimony provides a challenge for giving a general account of it. And the fact that intra-scientific testimony occurs in a wide variety of communicative contexts adds a further layer of complexity. But this does not mean that it is futile to articulate a general account. The different species of intra-scientific testimony have features in common that make them subsumable under general norms. Consequently, I will proceed by first articulating and defending an epistemic norm of providing intra-scientific testimony.

4.2 The Epistemic Norms of Intra-Scientific Testimony

In this section, I turn to a central theme of the book—namely, the norms of scientific testimony. More specifically, I will articulate and defend an epistemic norm of providing intra-scientific testimony that is based on a more general epistemic norm of assertion. So, I will set the stage by considering epistemic norms of assertive communication in general.

4.2.a The nature of intra-scientific testimony and its norms: Recall that I regard norms as objective benchmarks of good performance relative to some *standard*—e.g., truth in the epistemic case (Chapter 2.4.c). This is also true of social norms, such as those governing intra-scientific testimony, which are held in place by the scientific community, broadly construed, through incentives and sanctions (Frost-Arnold 2013; Rolin 2020). The relevant norms are largely tacit to the scientists themselves. But they may be articulated via reflection on systematic features of scientific practice and the standard of truth. I take the operative social norms of intra-scientific testimony to approximate the objective epistemic norms reasonably well but imperfectly since science is subject to a range of constraints other than the pursuit of truth. So, my attempt to articulate epistemic norms of intra-scientific testimony will appeal to the epistemic aspects of scientific research alongside broader considerations concerning the epistemic aims of science. Given this focus on the epistemic functions of intra-scientific testimony, my proposed epistemic norms involve considerable idealization.

Being able to take part in a norm-governed scientific practice reflects aspects of scientific expertise that may be categorized as both interactional and contributory expertise. But since the scientists may lack reflective access to the norms, they are distinct from *guidelines*, which are only met if they are, in some sense, followed by the scientists. I will not develop such guidelines for intra-scientific testimony. But the norms that I will articulate may constrain such guidelines, which are often simplified approximations of the norms.

Given that testimony and, hence, intra-scientific testimony is characterized as an assertive expression which centrally functions to convey its content to an audience, norms of assertion apply to testimony (Gerken and Petersen 2020). However, given that intra-scientific testimony is a distinctive assertive expression, it is not plausible to merely subsume it under a general epistemic norm of assertion. It is important that the proposed norm is distinctive in that it reflects what differentiates intra-scientific testimony from other assertive expressions. So, once I have put some concrete norms of intra-scientific testimony on the table, I will conclude by defending the thesis *Distinctive Norms*, according to which *the epistemic contribution of scientific collaboration depends on distinctive norms of intra-scientific testimony* (Chapter 1.5.b).

Generally, I will try to exemplify how epistemic norms concerning the production and consumption of intra-scientific testimony may reveal important facets of it. A better understanding of the general norms of intra-scientific testimony contributes to an account of scientific collaboration and its epistemic force.

4.2.b An epistemic norm of intra-scientific testimony: The preceding discussion indicates that an epistemic norm governing the production of intra-scientific testimony must be a reasonably general norm of assertion that reflects the noted variety in intra-scientific testimony. In Chapter 3.1.d, I noted that the epistemic requirement in the context of discovery might be weaker than at other stages of scientific investigation. But I have also noted that the epistemic norms of scientific testimony should reflect that the nature and strength of scientific justification varies enormously among disciplines. Thus, there is a bit of tension between the need for a general account and the need for specificity. My proposal seeks to strike the balance between generality and specificity in two ways. First, it marks a species of assertion that is distinct in standardly requiring scientific justification. Second, the norm requires that the degree of scientific justification is determined by the particular communicative context.

To illustrate the relationship between the general epistemic norm of assertion and the epistemic norm of intra-scientific testimony, I will start by presenting the former by drawing on previous work. The widely shared idea in these debates is that assertion is, perhaps constitutively, governed by an epistemic norm (for an overview, see Gerken and Petersen 2020). The dispute is primarily over the specification of the epistemic conditions that an asserter must meet in order to meet this epistemic norm. Here, I just present my favorite epistemic norm without much attention to the dialectics of arguing that it is superior to competitors.[1] I adopt the general *Warrant-Assertive Speech Act* norm of assertion ("WASA", for short):

> WASA
>
> In a conversational context, CC, in which S's assertion that *p* conveys that *p*, S meets the epistemic conditions on appropriate assertion that *p* (if and) only if S's assertion is properly based on a degree of warrant for believing that *p* that is adequate relative to CC.

The basic idea underlying WASA is that conversational context determines the degree of warrant that the asserter must possess. (The sufficiency claim is parenthetical due to complications that need not concern us here.) So, in some conversational contexts, a very high degree of warrant is required for assertion, and in others a lesser degree of warrant will do. Thus, WASA is a *sliding-threshold* epistemic norm of assertion. The conversational context determines how high a threshold of warrant an asserter must meet in order to be in an epistemic position to assert. The generally relevant determiners of the conversational context include the following features:

[1] For such arguments, see Gerken 2011, 2012a, 2014a, 2015a, 2015b, 2017a, 2018a, 2020b.

(i) alternative assertions (including qualified assertions),
(ii) the availability of evidence for the asserted content (or what is conveyed by it),
(iii) the urgency of conveying the asserted content (or what is conveyed by it),
(iv) the relevant stakes,
(v) social roles and conventions.

For example, it is reasonable to criticize an asserter who asserts something of extreme importance on the basis of very poor warrant in a non-urgent context where she could have easily obtained further evidence. According to WASA, this is because she violates the epistemic norm of assertion. Assume, on the other hand, that the content of the assertion is not particularly important, that the speaker has extremely strong warrant, and that it would be extremely difficult to obtain further evidence. If it turns out that she asserted a falsehood, the speaker may reasonably defend herself by noting these contextual features. Given the distinction between norms and standards (Chapter 2.4.c), WASA offers the diagnosis that the speaker met the epistemic *norm* of assertion although she did not meet the *standard* of truthful assertion.

Because WASA is a general norm, it is cast in terms of epistemic warrant since epistemic entitlement may be all that is required in some conversational contexts (see Gerken 2017a: 159). But in *discursive* conversational contexts, the speaker is expected to be able to articulate a justification for the contents of her assertions (Gerken 2012a). This assumption is a key to deal with the challenge of articulating a distinctive epistemic norm of intra-scientific testimony that exemplifies a general norm of assertion. Given the assumption that scientific testimony is testimony that is properly based on scientific justification which is typically discursive (cf. *Hallmark III*), my suggestion is that conversational contexts characteristic of intra-scientific testimony are *discursive contexts*. Typically, a scientist who testifies that *p* to a collaborator or to a broader scientific community is required to be able to articulate reasons for believing or accepting that *p*.[2]

Since intra-scientific testimony is characteristically based on scientific justification, it is natural to take it to require scientific justification. Indeed, this requirement helps distinguish intra-scientific testimony from assertion generally as well as from other types of testimony. In previous work, I argued that intra-scientific testimony is standardly governed by a Discursive Justification-Assertion account ("DJA" for short), which governs discursive conversational contexts in general (Gerken 2012, 2015a). Discursive contexts are communicative contexts that require discursive justification. So, DJA is structurally similar to WASA, but it requires a degree of contextually determined discursive justification. In applying

[2] Kitcher 1993; Longino 2002; Gerken 2015a; Khalifa and Millson 2020.

DJA to intra-scientific testimony, I noted that it remained a fairly generic epistemic norm: "a further specification of the distinctively *scientific* context may ultimately be required. If this turns out to be the case, a yet more specific subspecies of the epistemic norm may be required. But I will adopt DJA as a starting point" (Gerken 2015a: 576). In consequence, I will here provide a slightly more specific epistemic *Norm of Intra-Scientific Testimony* ("NIST" for short):

> NIST
>
> In a context of intra-scientific communication, CISC, in which S's intra-scientific testimony that *p* conveys that *p*, S meets the epistemic conditions on appropriate intra-scientific testimony that *p* only if S's intra-scientific testimony is properly based on a degree of scientific justification for believing or accepting that *p* that is adequate relative to CISC.

While NIST is a mouthful with all its qualifications, the basic idea is, in a simplified nutshell formulation, that *context determines the degree of scientific justification required for intra-scientific testimony.* NIST develops the more general DJA in several ways. First, it is restricted to contexts of intra-scientific communication rather than to discursive contexts in general. Second, it requires *scientific* justification whereas DJA only requires *discursive* justification and WASA only requires some kind of warrant (entitlement or justification). Third, NIST does not require justification for belief but justification for *belief or acceptance.* Finally, although there are only peripheral cases in which proper basing on contextually adequate scientific justification is not sufficient for epistemically appropriate intra-scientific testimony, I have eliminated the parenthetical sufficiency condition which I included in DJA. Thus, NIST only sets forth a necessary condition.[3]

Given that communicative contexts in which intra-scientific testimony takes place are diverse, NIST is a general norm under which a variety of distinct types of intra-scientific testimony may be subsumed. Some kinds of intra-scientific testimony tend to demand a high degree of scientific justification. Examples include cases of non-urgent formal intra-scientific testimony where the stakes

[3] The two former restrictions simply tailor NIST more specifically to intra-scientific testimony. But the third development—the introduction of the disjunct 'or accepting'—corrects my earlier proposal according to which S needed discursive justification for *believing* that *p*. I have always been explicit that S did not in fact need to believe that *p* (Gerken 2012a: 378). But this was to account for cases involving asserters who do not believe what they assert (Lackey 1999, 2006a; Graham 2000, 2006). For intra-scientific testimony, however, it is more relevant to recall that collaborating scientists often merely accept that *p* (Chapter 2.1.b–c). Hence, scientists may, in some contexts, only be required to provide scientific justification for accepting that *p*. Of course, it is a big question whether scientific justification for acceptance differs from scientific justification for belief. But given the importance of acceptance in accounts of scientific collaboration, a norm cast in terms of scientific justification for belief is likely too restrictive. So, to make NIST fit intra-scientific testimony in all types of scientific collaboration, it requires scientific justification for the disjunction of belief or acceptance.

are extremely high (see the case *WIND SPEED* a few paragraphs below). Other kinds of intra-scientific testimony only demand a lesser degree of scientific justification.

Another contextually determined feature is proper basing. If the testifier's role in the collaboration is to produce the relevant bit of scientific justification, she may be required to be able to articulate it on the spot. But in other contexts, the testifier may only be required to be aware that the relevant scientific justification is to be found in the larger epistemic network. Note that the contextually determined requirements on justificatory strength and basing relation are independent variables: In principle, a lax constraint on proper basing is compatible with a high degree of scientific justification. Likewise, a demanding constraint on proper basing is compatible with requiring a more modest degree of scientific justification. So, although the contextual requirements on *proper basing* and *adequate degree* of scientific justification tend to go hand in hand, they may vary independently.

NIST appeals to the hallmark properties of scientific justification, such as the assumption that the justification is gradable (*Hallmark II*) and that it be discursive (*Hallmark III*). The (qualified) epistemic superiority of scientific justification (*Hallmark I*) does not figure explicitly in NIST, although it figures implicitly given that NIST requires scientific justification. Given the nature of scientific collaboration, it is reasonable to suppose that, as a general rule, the degree of scientific justification required for intra-scientific testimony is standardly comparatively high. However, there may be exceptions to this rule. S's intra-scientific testimony that there are eighty participants in each condition of the study that is based on S's fallible memory of the study design or data set may be adequate in some context whereas other contexts require a systematic check of the data set. Recall that although memory provides the immediate warrant, it is part of a scientific justification insofar as it is integrated with the scientific investigation—after all, it is memory of a scientific study design or data set (Chapter 3.1.d).

In contrast, S may know that *p* on the basis of mere entitlement without being in an epistemic position to provide intra-scientific testimony. For example, S may be sufficiently reliable to know that a finding is statistically significant by eyeballing the dataset. But this epistemic basis might be insufficient for providing formal intra-scientific testimony that the finding is statistically significant in a non-urgent context where stakes are high. In such a context, a proper statistical analysis would be required (Gerken 2015a gives an example).

Given that NIST requires scientific justification and given that scientific justification is paradigmatically discursive, NIST captures an important aspect of what Rescorla labels a dialectical model of assertion which is based on the following idea: "by asserting a proposition, I commit myself to defending the proposition when faced with challenges and counter-arguments" (Rescorla 2009: 100). However, NIST does not *merely* consist of a dialectical requirement, but also involves an epistemic requirement. After all, NIST requires scientific *justification*. A scientist would often derail effective collaboration if they provided intra-scientific testimony on a very poor epistemic basis. Moreover, NIST is *restrictive* in Rescorla's sense of involving restrictions on what a scientist may testify in the first place (Rescorla 2009). The following case illustrates these points:

> *WIND SPEED*
>
> A meteorologist, Anita, provides, on the basis of a hunch, the following intra-scientific testimony to her collaborator, Andy: "The storm will have wind gusts with a maximum speed of 100 km/h when it makes landfall." Andy believes her and plugs this assumption into a model in order to predict the resulting flooding. Hours later, incoming evidence indicates that Anita's testimony is highly inaccurate. So, Andy challenges Anita's testimony: "The latest measurements indicate that the maximum wind speed upon landfall will exceed 200 km/h. What was the evidence for thinking that it would max out at 100 km/h?"

It is natural to assume that Anita violates the norm of intra-scientific testimony as soon as she provides it independently of whether she is willing to retract it or defend it against challenges. For example, it is reasonable to criticize Anita for providing the initial testimony whether or not she retracts it or defends it. Note also that even if Anita's defense is dialectically successful for non-epistemic reasons (e.g., because Andy is smitten with her), her initial assertion remains at odds with proper scientific conduct. NIST explains this well given that her testimony was based on no warrant—much less scientific justification—whatsoever. So, *WIND SPEED* favors both NIST and a restrictive conception of it.

A further difference between NIST and dialectical models, as characterized by Rescorla's gloss, is that the latter is individualistic given that the *testifier* is committing to being able to defend the assertion herself. This may be an overly strong requirement for a norm that governs intra-scientific testimony which occurs in scientific collaborations that may distribute the relevant scientific justification across multiple individuals (Chapter 1.3–4). In cases of radically collaborative research, the scientific justification is distributed to the point where no single scientist is able to articulate or even comprehend every aspect of it. In such cases, the defensibility may—at least to some extent—be deferred to collaborators or other scientists within the testifier's broader network (Winsberg et al. 2014; Klausen 2017). But although the testifier may not be required to articulate the scientific justification herself, she may be at fault. For example, she might be reasonably faulted if she cannot cite any higher-order reason that there is scientific justification of the contextually required degree in the network.

NIST addresses the tension between providing a general norm and respecting disciplinary and contextual differences among various species of intra-scientific testimony. It does so in terms of the idea that context determines the required degree of scientific justification. This makes for a structurally general norm that allows for contextual variance in the degree of scientific justification that intra-scientific testimony must be based on. The relevant contextual factors include the general ones such as (i)–(v) mentioned above. Additional factors may be more specific to scientific collaboration. For example, the stage of inquiry may determine the degree of scientific justification required. For example, more lax standards may be in place in the early stages of explorative inquiry akin to contexts of discovery. As noted, in early "brainstorming" contexts of intra-scientific communication, a tacit hedge may be presupposed such that intra-scientific testimony may be brought forward on the basis of non-scientific warrants or very meager scientific justification. Sometimes context demands that this hedge is made explicit ("I don't have any evidence for this, but . . ."). Overall, such cases exemplify a central aspect of NIST—namely, that the demand for scientific justification may vary considerably with context.

Moreover, a variety of (v)—namely, *disciplinary* conventions—is an important contextual parameter. Disciplines are partly individuated in terms of the types of scientific justification that are demanded. For example, a sociologist and an anthropologist may provide intra-scientific testimony about the very same hypothesis, *p*. But they are required to provide different kinds of scientific justification for an intra-scientific testimony that *p* to collaborators within their respective fields. In consequence, it is *an aspect of scientific expertise* to be (implicitly) familiar with the kind and degree of scientific justification expected in various contexts in one's discipline. Specifically, it is a dimension of interactional expertise. However, it is also an important aspect of contributory expertise insofar as the ability to collaborate effectively is a central prerequisite for contribution in the discipline in question. Likewise, familiarity with the contextually determined justification requirements of adjacent disciplines is critical for interdisciplinary and multidisciplinary collaborations and gives rise to distinct pitfalls. Miscommunication may occur when epistemic requirements differ between disciplines.

Disciplinary conventions are, to a large degree, determined by the desired end result of the collaboration. The desired end result of a collaboration is often a scientific publication and, hence, a form of inter-scientific testimony. Consequently, the norms of intra-scientific testimony must be sensitive to the norms of inter-scientific testimony. So, it is informative to note that various disciplines have very different epistemic standards for inter-scientific testimony. In fact, the epistemic thresholds for publication may vary in a single discipline. For example, in high particle physics, "The threshold for "evidence of a particle," corresponds to $p = 0.003$, and the standard for "discovery" is $p = 0.0000003$" (Lamb 2012). However, the history of physics suggests that inter-scientific testimony may be made on the basis of a far lesser degree of scientific justification (Dang and Bright forthcoming; see also Fleisher 2021). Thus, the strength of required scientific justification for publication varies within a broad discipline, physics, and within sub-disciplines thereof. Although such cases concern inter-scientific testimony, they indicate how variations in communicative context may in general affect the required degree of scientific justification. Given that the norms of intra-scientific testimony must be sensitive to the norms of inter-scientific testimony, the cases provide some indirect motivation for NIST.

In sum, NIST allows for considerable disciplinary and contextual variation on the requirements of scientific justification while remaining a general norm at the structural level (see also Gerken 2014a).

4.2.c Norms, guidelines, and sanctions: Often, the idea of constitutive norms is motivated by an analogy with explicit sets of rules. While such analogies with rules of board games or traffic laws bring out the idea that norms mark how the practice is performed correctly or acceptably, they may encourage the idea that the norms directly guide the practitioners. In a board game, for example, the explanation of why players play correctly is almost invariably that they have explicit knowledge or understanding of the rules.

In contrast, I conceive of norms as objective benchmarks of assessment that the agent need not be able to conceive and which, therefore, differ from guidelines which are only met if the agent follows them (Chapters 1.5, 2.4.c). So, a norm, such as NIST, is not typically conceptualized by the scientists who are concerned with science rather than with philosophy of science. Nevertheless, the norms help to structure scientific practice. This may be because the scientists are tacitly adapting to the social structure of sanctions and rewards in a fairly irreflective manner. But it may also be because they are familiar with broad methodological guidelines that reflect them. For example, commonplace slogans such as *do not conclude anything stronger than what the evidence supports* reflect NIST. Likewise, scientists are subject to discipline-specific methodological requirements which may often play the role of providing more or less explicit guidelines in virtue of which they are sensitive to the norms.

NIST and other norms of intra-scientific testimony are idealizations of social norms (Graham 2015b; Bicchieri 2017). They are characteristic of the epistemic dimensions of the social practice of communicating within a scientific collaboration. As mentioned, I take the epistemic dimensions of the operative social norms of intra-scientific testimony to align imperfectly but reasonably well with objective epistemic norms. This assumption contributes to an account of the comparable epistemic force of science. But it is important to bear in mind that NIST represents a considerable idealization insofar as it represents an *epistemic* norm in abstraction from the aspects of intra-scientific practice that reflect competing non-epistemic aims of science. However, this type idealization is not particular to the norms of intra-scientific testimony. For example, the epistemic dimensions of the operative social norms of data collection are also practically constrained. Consequently, they are also imperfectly aligning with the epistemic norms of data collection that are concerned with truth-conduciveness. Following the norms of data collection reasonably well is part of being a scientific expert. Similarly, it is a part of *scientific expertise* to conduct one's inter-testimonial practice in accordance with the norms that govern this practice.

The question as to *how* social norms are sustained through social practice is a complex and controversial one.[4] So, here I will only focus on its aspect that concerns sanctions. It is widely agreed that social norms are closely related to social sanctions for norm violation. When the norm in question is a social norm—and it seems safe to assume that this is the case when it comes to norms of communication—the norm is upheld, at least in part, by sanctions and rewards in the relevant community (Graham 2015b; Fricker 2017). In the *WIND SPEED* case outlined above, Anita would likely be on the receiving end of sanctions for her blatant violation of NIST. Given that she was testifying *qua* scientist in a high-stakes context on the basis of a hunch, her position in the scientific community may be implicitly or explicitly compromised. The implicit ways might include a general lack of uptake of her testimony and lack of opportunity to enter future collaborations. The explicit ways might include loss of privileges or even loss of her job given the egregious nature of her norm violation. Such social sanctions help ensure that scientists generally testify in accordance with the epistemic norms of intra-scientific testimony.

Like the norms of intra-scientific testimony, the norms of sanctions are also contextually determined, and it is plausible that some of the same contextual factors are relevant in determining the nature of the sanctions. For example, sanctions will often be harsher in cases where the stakes are high. However, the sanctions for violating norms of intra-scientific testimony tend to be very strong in general. Cases of norm violation so egregious that they constitute fraud or dubious research practices often end the relevant culprit's academic career. Consider, for instance, the highly publicized cases of scientific misconduct by Marc Hauser in the US, Milena Penkowa in Denmark, and Diedrik Stapel ("The Lying Dutchman") in the Netherlands. In these cases, the scientists were not only fired from their position but entirely ostracized from the scientific community. There are no second acts in scientific fraudsters' academic careers.

There are several reasons for the hard sanctions against scientific misconduct, and one of them is closely related to the nature of scientific collaboration. Given that scientific collaboration is based on a highly specialized division of cognitive labor, it may be infeasible for collaborators to examine the scientific justification for the testimonial input of their collaborators. This is particularly so in cases of

[4] Elster 1989; Bicchieri 2006, 2017; Henderson and Graham 2017a 2017b; Fricker 2017.

radically collaborative research and in cases of multidisciplinary research. Recall that in multidisciplinary research, scientific collaboration among researchers in various disciplines takes place without any methodological integration. Consequently, the collaborators are particularly ill-equipped to check the epistemic basis of each other's intra-scientific testimony. This leaves considerable room for violating the epistemic norms of intra-scientific testimony (Frost-Arnold 2013). In consequence, sanctions must be fairly strong. In slogan: *Sanctions for violating norms of intra-scientific testimony are very strong because opportunities for detection are fairly poor.*

While the slogan requires qualifications, it indicates that scientific collaboration consists of a complex interplay between norms governing the production of intra-scientific testimony and the norms governing the consumption of intra-scientific testimony to be discussed shortly.

4.2.d Concluding remarks on the epistemic norm of intra-scientific testimony: The epistemic norm of intra-scientific testimony, NIST, inherits its structure from a general epistemic norm of assertion. However, the fact that NIST requires scientific justification for belief or acceptance helps to distinguish intra-scientific testimony from other assertive speech acts. Moreover, NIST may reflect the variety of intra-scientific testimony given that it requires a contextually determined degree of scientific justification. Thus, NIST does a reasonable job of striking the balance between generality and specificity. Moreover, I have argued that it aligns well with the practices of collaborating scientists and the sanctions related to norm violation.

4.3 Uptake of Intra-Scientific Testimony

The epistemic norm of intra-scientific testimony set forth in the previous section is one that applies to the *producers* of intra-scientific testimony. In this section, I turn to the norms that concern the *consumers* of intra-scientific testimony. In particular, I will consider collaborating scientists who receive and make use of intra-scientific testimony in a scientific collaboration.

4.3.a Consuming intra-scientific testimony: As emphasized in Chapter 1.3.b, scientific collaboration is a truth-conducive social structure that enables the scope and quality of research by enabling hyper-specialization and by maximizing different perspectives and creative input (Longino 2002; Hackett 2005). However, given that scientific collaborators are specialists in separate areas, they are often incapable of checking the epistemic basis of each other's claims. Moreover, even if they were capable of doing so, it would often be counterproductive from a resource standpoint. For example, Hackett has argued that continuous checking of collaborators may negatively affect their independence and creativity (Hackett 2005). Everybody who has been micromanaged will add that it may also impede their motivation.

Given that checking and controlling scientific collaborators are often infeasible and sometimes counterproductive, collaborating scientists are epistemically dependent on each other.[5] Reflection on the epistemic dependence in scientific collaboration led Hardwig to set forth a thesis that I will label *Hardwig's Dictum*:

Hardwig's Dictum

"a scientific community has no alternative to trust" (Hardwig 1991: 706)

There are grains of truth to *Hardwig's Dictum*, but it calls for qualifications (some of which are provided in Hardwig's own work (Hardwig 1985, 1991, 1994)). Hardwig considers "blind trust" a fairly automatic formation of belief in the content of testimony in the absence of countervailing evidence. However, I will prefer Wagenknecht's terminology of *complete* trust, which is, roughly, trust in the absence of epistemic reasons beyond the testimony (Wagenknecht 2015). However, the notion (or notions) of trust stands in need of specification (Hawley 2012, 2019; Carter 2020). Insofar as I discuss trust, I restrict the discussion to epistemic trust in the content of the testimony—i.e., trust that the content is true. However, I will often simply discuss testimonial *uptake*, which may be characterized as testimonial belief or acceptance. This is because my focus is the function of intra-scientific testimony in collaboration rather than the nature of the attitude that scientists have towards each other. While these issues are, of course, related, they may come apart. Perhaps the norms of scientific collaboration may, in some contexts, require one to accept the intra-scientific testimony from someone that one does not trust. Likewise, a scientist may trust someone but not be in an epistemic position to accept his intra-scientific testimony.

Hardwig's Dictum marks the insight that scientific collaboration often requires a default uptake of intra-scientific testimony since scientists collaborate in large part because they lack the competencies of their collaborators. However, the idea that scientific collaboration requires complete trust is too strong. Yet, the motivating assumption—that hyper-specialization that characterizes scientific collaboration demands automatic and fairly uncritical uptake—is widely shared. Milgram argues that one must be an expert in order to reliably identify other experts and argues from this assumption that the cost of hyper-specialization is "a great endarkenment" (Milgram 2015: 30, 44–8).

But while the costs of hyper-specialization are well worth recognizing, the assumption that only experts in domain D1 are able to reliably identify other

[5] Hardwig 1985, 1991; Wilholt 2013; Andersen and Wagenknecht 2013; Gerken 2015a.

experts in domain D1 is questionable. In many cases, laypersons may reliably identify the experts in, for example, medicine. We may not be able to distinguish a genuine medical expert from a quack in terms medical evidence. However, social structures often allow us to reliably identify medical experts. Our warrant about the social environment is sometimes a mere entitlement, but often we have some degree of discursive justification as well. If someone asks me why I accept the doctor's medical testimony, I can respond that she is employed in a medical establishment that only hires qualified doctors. So, I do not draw on warranted beliefs about medicine but about the social environment. Scientists are often in an even better position to do so. This is an important clue to the articulation of the norms governing the uptake of intra-scientific testimony. It is also an important clue to diagnose problems that arise when the social structures do *not* permit scientists in one domain to reliably identify the relevant scientific experts in another domain. This may be the case when scientific issues are politicized or if the methods of a discipline are corrupted.

In sum, *Hardwig's Dictum* sloganizes the important insight that the central benefits of scientific collaboration—namely, fine-grained epistemic specialization—render collaborating scientists epistemically dependent on each other. However, scientists do not entirely lack resources to assess the credibility of actual and potential collaborators, and they should not be naïve. The norms of intra-scientific uptake should strike a balance between these requirements. In order to approach such norms, I will consider some of the resources for assessing intra-scientific testifiers that collaborating scientists have at their disposal.

4.3.b Evidence of testifier reliability: A number of philosophers of science have argued that scientists have considerable resources for assessing scientists with expertise in areas other than their own (Fricker 2002; Wagenknecht 2015). Others have argued that even laypersons have some means to assess the testimony of scientific experts.[6] I will first consider the general resources since they are also available to collaborating scientists.

Goldman considers how a layperson who is not an epistemic expert in D may deal with apparent expert disagreement about a proposition in D without acquiring epistemic expertise in D herself (Goldman 2001). However, Goldman distinguishes between esoteric and exoteric reasons and statements (where 'N' denotes a layperson): "Esoteric statements belong to the relevant sphere of expertise, and their truth-values are inaccessible to N in terms of his personal knowledge, at any rate" (Goldman 2001: 94). In contrast, "Exoteric statements are outside the domain of expertise; their truth-values may be accessible to N" (Goldman 2001: 94). The esoteric/exoteric distinction extends from *statements* to reasons, arguments,

[6] Goldman 2001; Collins and Evans 2007; Sperber et al. 2010; Turner 2014; Anderson 2011; Grundmann forthcoming.

etc. Put in these terms, the problem for the layperson is that many of the statements, reasons, and arguments that are required to adjudicate between diverging views concerning a domain of epistemic expertise, D, are esoteric.

However, Goldman argues that the relevant statements, reasons, and arguments are not *wholly* esoteric. Among the partly exoteric considerations, he enlists the following sources of evidence (Goldman 2001):

(A) Arguments presented by the contending experts to support their own views and critique their rivals' views.
(B) Agreement from additional putative experts on one side or other of the subject in question.
(C) Appraisals by "meta-experts" of the experts' expertise (including appraisals reflected in formal credentials earned by the experts).
(D) Evidence of the experts' interests and biases vis-à-vis the question at issue.
(E) Evidence of the experts' past "track records."

In each case, Goldman argues that these conceptual resources are of genuine help to the laypersons although they are highly fallible individually as well as in conjunction. For example, appraisals from meta-experts and past track record do not guarantee epistemic expertise. Such appraisals are even less likely to establish superior epistemic expertise with regard to D in cases of conflicting testimonies. More generally, the sources are highly dependent on a social environment that is reasonably hospitable. The indicators of scientific expertise may be manipulated in order to convey the appearance of scientific expertise (Goldman 2001; Guerrero 2016). If this occurs too often in the recipient's social environment, the value of the indicators is greatly diminished. Another concern is that each resource is only partly exoteric. For example, the arguments that disagreeing scientists set forth tend to be esoteric at least in part. Similarly, an epistemic assessment of a scientist's track record is rarely an entirely exoteric affair. So, despite some serious limitations of the sources, they remain valuable and accessible to laypersons in varying degrees.

As mentioned, Goldman is concerned with how laypersons may assess apparently expert disagreement, and I will return to this issue in Chapters 5 and 6. However, the core problem carries over to some cases of scientific collaboration—especially multidisciplinary collaboration. Here, a collaborating scientist who is an epistemic expert in D1 may lack epistemic expertise in the domain of expertise, D2, of actual and potential collaborators. In such cases, the scientist is akin to a layperson in terms of his ability to assess intra-scientific testimony from experts in D2.

However, in most cases, a collaborating scientist who lacks epistemic and contributory expertise with regard to D2 is nevertheless in a better position than a non-scientist layperson. First of all, as noted in the discussion of epistemic expertise in Chapter 1.2.a, the domains D1 and D2 may overlap, and this is often the case in scientific collaboration. Scientists often collaborate with *adjacent* disciplines (Andersen 2012). Secondly, scientific expertise typically involves generic competencies that allow for some assessment of individuals, hypotheses, and evidence in fields other than one's own. For example, scientists may more readily understand standard technical terminology such as 'control condition' or 'confounder,' and this puts them in a better position than laypersons when assessing (potentially) collaborating scientists.

The idea that scientists deploy a number of resources for assessing intra-scientific testimony from collaborators from very different disciplines has not gone unrecognized (Fricker 2002; Wagenknecht 2015). For example, Fricker argues that "in many cases—and certainly in communication between co-researchers in the sciences—she will have a body of evidence bearing on the issue of the speaker's trustworthiness" (Fricker 2002: 382).[7] Some of the sources of evidence that Fricker notes align with Goldman's general sources considered above. In addition, she notes the background social knowledge about the scientific community and scientists' roles in it: "basis for trusting each other lies in their knowledge of each other's commitment to, and embedding within, the norms and institutions of their profession" (Fricker 2002: 382). Likewise, Fricker reasonably notes that a collaborating scientist standardly has information about the testifier and her access to evidence about the content of her testimony. More generally, most contexts of intra-scientific communication are hardly ever *minimal background cases* but rather cases in which the recipient has some general and context-specific background information that underwrites her trust in intra-scientific testimony.

Before proceeding, let me note another relevant property of intra-scientific testimony. Compare the following testimonies:

Ozone 1

"The Montreal Protocol has led to a reduction in the emissions of chlorofluorocarbons which has, in turn, reversed the depletion of ozone in the stratosphere"

Ozone 2

"The Montreal Protocol is healing the ozone layer"

I take it that in a minimal background case, a (potentially) collaborating scientist who testifies *Ozone 1* is more likely to be a scientific expert in atmospheric composition research than someone who testifies *Ozone 2*. This is due to the former testimony's technical terminology, lack of metaphor, and more precise causal claim. These features of the *content* of the intra-scientific testimony are (fallibly) indicative of scientific competence. In contrast, the crude and slightly

[7] For criticism, see Douven and Cuypers 2009; Burge 2013; Turner 2014.

metaphorical causal testimony of *Ozone 2* does not indicate scientific competence. Thus, the comparison illustrates that the very content of intra-scientific testimony may be an indicator of scientific expertise, although it is, of course, a fallible indicator.

As is generally the case, context is important for evaluating content in this manner. In some communicative contexts, *Ozone 2* would be quite appropriate, and *Ozone 1* would be overly pedantic. So, the recipient must have some sensitivity to the characteristic contexts of intra-scientific communication. However, the collaborating scientist need not understand every aspect of the scientific justification that is indicated by the intra-scientific testimony. This is so even when it is, as in the case of *Ozone 1*, only gestured at rather than fully articulated. The use of technical terminology, causal explanations, sketches of methodology, etc. may be sufficiently reliable indicators that the testimony is in accordance with NIST. Hence, the *content* itself may—at least to a recipient who is a scientist—indicate that the testimony is based on an adequate degree of scientific justification or it may raise a red flag. Thus, recipients of intra-scientific testimony are not generally in minimal background cases. Rather, they have multiple resources available for assessing the epistemic trustworthiness of intra-scientific testimony. Some resources pertain to the context of collaboration. Other resources are also available to laypersons, although scientists are generally better equipped to exploit them.

Empirical evidence suggests that scientists do exploit some of these resources. For example, Wagenknecht argues, on the basis of semi-structured interviews with a research group, that intra-scientific trust requires personal acquaintance and is developed through iterations of reliance on intra-scientific testimony (Wagenknecht 2015; see also Kusch 2004). On the basis of her study, she concludes that "[p]racticing scientists do not make do with attitudes of all-encompassing trust and, I argue, their judgment should be taken seriously, since it reflects their professional experience and mundane strategies in actual collaborative scientific practice" (Wagenknecht 2015: 164).

Thus, Wagenknecht's interviews provide an important antidote to the idea that intra-scientific collaboration is characterized by complete trust or uncritical uptake. But, although Wagenknecht does not highlight this, they also illustrate the point that although the trust is rarely complete, a fairly unquestioning uptake is often required for effective scientific collaboration. Consider, for example, this excerpt from an interview with a senior biologist: "of course I can question your expertise, but if we want to work together, I have to consider you as an expert in your field. Otherwise it's not worthwhile. If I question the other person's expertise every time we did an experiment, we would never get started with an experiment" (Wagenknecht 2015: 168).

So, Wagenknecht provides evidence for two assumptions. She emphasizes that intra-scientific testimony takes place against an initial and continual assessment of collaborators. But her evidence also suggests that an effective ongoing scientific collaboration often requires a fairly unquestioning uptake of intra-scientific testimony. A theoretical account of the uptake of intra-scientific testimony must capture both assumptions. Moreover, an account should capture that context helps determine the appropriate degree of intra-scientific trust. Reflection on the various stages of collaboration supports this assumption. Early stages of collaboration may involve a good deal of vetting, and in some contexts (e.g., hiring postdocs) the vetting is often formalized. Once the collaboration is ongoing, however, some degree of default uptake is often called for. Plausibly, the parameters that determine the degree of scientific justification required for *producers* of intra-scientific testimony (i.e., stakes, urgency, etc.) are also parameters of the context that partly determines when it is reasonable and expected that the recipient forms the relevant testimonial belief.

The upshot that the context of intra-scientific communication partly determines the appropriate degree of trust bears on Fricker's conclusion that default entitlement "shrinks to irrelevance in the explanation of the basis on which scientists take each other's word in the scientific community" (Fricker 2002: 373). Fricker is right that focusing exclusively on testimonial entitlement may lead to underestimation of the significance of these strands of evidence. On the other hand, an exclusive focus on the strands of evidence available to the scientists may lead to underestimation of the importance of the social environment of scientific collaborations. Likewise, a myopic focus on the individual scientist's assessment of collaborators may lead to a failure to recognize the work that the norms governing intra-scientific testimony do in securing its general reliability. Scientists are embedded in a social structure which partly consists of epistemic norms of intra-scientific testimony, such as NIST. Therefore, the social environment that scientists are embedded in helps to ensure that intra-scientific testimony is generally trustworthy. This environmental fact explains an important basis of rational uptake of intra-scientific testimony that does not require that the recipient appreciates or assesses individual testifiers. Hence, general facts about the social environment help explain an important foundational aspect of the warrant for belief in intra-scientific testimony (see also Burge 2013).

More specifically, the fact that contexts of intra-scientific communication are rarely minimal background cases does not compromise the general idea of testimonial entitlement. Nor does it compromise the idea that an effective division of cognitive labor requires that intra-scientific testimony must often be accepted without excessive assessment. However, it does suggest that the warrant for intra-scientific testimony is often more cognitively demanding for the recipient than warrant for more mundane forms of testimony (see also Wilholt 2013; Miller and Freiman 2020).

Finally, it bears repeating that the social environment may be polluted insofar as the indicators of scientific expertise may be systematically rigged to create the appearance of scientific expertise (Guerrero 2016; Weatherall et al. 2020). The fact that it is not only individual vigilance but also the social environment that bears on warranted uptake of intra-scientific testimony indicates an important source of fallibility: An epistemically inhospitable social environment is a significant obstacle even for scientists who are fairly sophisticated evaluators of scientific expertise. In some scientific fields, the social environment is polluted because non-epistemic incentives such as visibility, grantability, publishability, employability, etc. become too dominant. This may compromise effective intra-scientific communication and, thereby, scientific collaboration and, thereby, the epistemic force of science. But while it is important to take this threat seriously, it is also important to recall that the scientific environment has some features which help to curb these problems (cf. Chapter 3.3).

4.3.c Concluding remarks on the uptake of intra-scientific testimony: Reflection on the resources which scientists may use to assess intra-scientific testifiers suggests scientists must strike a balance between a default uptake and an appropriate assessment of testifier. One lesson from considering such resources is that recipients are heavily dependent on the relevant social environment. For example, intra-scientific testimony is situated within a social structure that includes implicit and explicit marks of epistemic competence. In consequence, recipients of intra-scientific testimony are vulnerable to "pollution" in the relevant social environment. I will seek to articulate norms of intra-scientific uptake that respect these points.

4.4 A Norm of Uptake of Intra-Scientific Testimony

On the basis of the considerations above, I will articulate a norm of intra-scientific uptake. Likewise, I will articulate an epistemic principle concerning the warrant that recipient scientists acquire in cases of default uptake of intra-scientific testimony.

4.4.a The need for norms governing the uptake of intra-scientific testimony: As in the case of the norms governing the production of intra-scientific testimony, there are also norms governing its consumption—i.e., the uptake and use of intra-scientific testimony. If there were no such norms, considerable resources would have to be allocated to ensure that the recipients believe, or at least accept, a given instance of intra-scientific testimony and make appropriate use of it.

For example, if a collaborator provides intra-scientific testimony that part of the data set is unreliable because an instrument was not calibrated, effective collaboration standardly requires that she can count on the recipients to accept that this is so. If so, the relevant part of the data set may be excluded, or the experiment may be rerun after calibration. The key point is that effective collaboration requires that appropriate intra-scientific testimony is not simply ignored. Hence, recipients of intra-scientific testimony are under some normative obligation with regard to the consumption—i.e., the uptake and use—of it. Of course, it is reasonable to reject an intra-scientific testimony if one has good reasons for doubting it. However, this will typically involve articulating these reasons for the testifier or other members of the collaborating group (Longino 2002; Wagenknecht 2015). Thus, the norms of uptake of intra-scientific testimony also reflect *Hallmark III* of scientific testimony, that it is based on discursive justification. This augments the arguments from Chapter 3.5.c that the articulability of scientific justification contributes to the truth-conducive structure of scientific collaboration because it enables critical scrutiny from various perspectives. This point should be reflected not only in the norms governing the production of intra-scientific testimony but also in the norms governing its consumption.

The idea that the uptake of intra-scientific testimony is also governed by norms has not gone unrecognized, although theorists have put the point slightly differently. For example, Mayo-Wilson provides the following (conflicting) norms as examples: "believe others in the absence of conflicting information" and "only believe those you know to be reliable" (Mayo-Wilson 2014: 56). Longino sets forth a number of norms that are characteristic for scientific collaborators and members of the wider scientific community (Longino 2002). These include the following social—in some cases institutional—requirements of effective intra-scientific communication: *Venues*: "There must be publicly recognized forums for the criticism of evidence, of methods, and of assumptions and reasoning" (2002: 129). *Uptake*: "There must be uptake of criticism" (2002: 129). *Public Standards*: "There must be publicly recognized standards by reference to which theories, hypotheses, and observational practices are evaluated and by appeal to which criticism is made relevant to the goals of the inquiring community" (2002: 130). More controversially, Longino argues for *Tempered Equality*: "communities must be characterized by equality of intellectual authority" (2002: 131).

Longino qualifies each of these norms in various ways. But the general point that scientific collaboration requires social norms governing "discursive interactions" is not dependent on her specific norms. Moreover, it is noteworthy that Longino's proposed norms concern the scientific community at large rather than merely those who provide intra-scientific testimony. So, I see the attempt to articulate social norms of uptake of intra-scientific testimony as part of the larger project of characterizing the scientific process as a social process that is governed by distinctive norms.

4.4.b A norm of uptake of intra-scientific testimony: Norms governing the uptake of intra-scientific testimony must reflect that scientific collaborations are based on a fine-grained division of labor that often requires a fairly unquestioning uptake of intra-scientific testimony (Frost-Arnold 2013). But this point should be juxtaposed with the point that scientific competence involves a meta-competence that concerns the ability to recognize competent scientific collaborators and some sensitivity to defeaters of intra-scientific testimony (Rolin 2020). So, the normative expectation of default uptake that is central to intra-scientific testimony is triggered when recipient has strong and undefeated warrant for believing that the testimony is properly based on adequate scientific justification. So, I cooked up a *Norm of Intra-Scientific Uptake* (NISU) that I think reconciles these points:

> **NISU**
>
> In a context of intra-scientific communication, CISC, in which S's intra-scientific testimony that *p* conveys that *p*, the default attitude of a collaborating scientist, H, should be to believe or accept that *p* if H has strong and undefeated warrant for believing that S's testimony that *p* is properly based on adequate scientific justification.

In a simplified nutshell formulation, NISU has it that *if H has strong and undefeated warrant for believing that S's intra-scientific testimony that p is epistemically adequate, then H should, as a default, believe or accept that p.*

NISU is a defeasible sufficient condition for intra-scientific uptake, but not a necessary one given that H may have independent reasons to believe or accept that *p*. If H has no further evidence concerning *p*, the principle may also provide a necessary condition.

NISU concerns the recipients' *default* attitude, and this default is easily defeated. For example, it is defeated in cases where the recipients have some warrant for doubting that *p* or for doubting the testifiers. Importantly, NISU merely requires that H's *propositional* warrant for believing S's testimony is properly based on adequate scientific justification. It does not require that H forms the corresponding belief. Moreover, the required recipient warrant may be an entitlement that is in large part explained by aspects of the social environment that H has not reflected upon (*pace* Fricker 2002: 373).

The requirement of strong and undefeated "meta-warrant" sets forth a high threshold for an epistemic requirement to believe or accept on the basis of intra-scientific testimony. I leave open whether there is some room for contextual variation of the epistemic threshold. But, if so, it will at most vary with epistemic factors such as availability of alternative evidence. In contrast, practical factors such as stakes do not raise or lower the threshold since NISU is an *epistemic* norm of uptake rather than an all-things-considered norm. Moreover, even if the threshold varies with epistemic factors, the requirement of strong and undefeated warrant sets the lowest threshold high.

Reflection on cases of norm violation may motivate NISU. To wit: Assume that Rosa is an expert in D1 but not in D2. For this very reason, Rosa collaborates with Ali, who she has excellent warrant for believing to be a scientific expert in D2 but not in D1. During their collaboration, Ali testifies that a proposition, *p*, which belongs to D2 (and not to D1), is true. But rather than accepting Ali's testimony, Rosa begins to investigate whether *p* herself, and in doing so she delays the project for weeks because she lacks the prerequisite scientific competences for investigating whether *p*. Assume, finally, that *p* is relevant to their investigation although not crucial to it. In this case, Rosa's endeavor is at odds with the norms of scientific collaboration. On a single occasion, her behavior may merely be seen as odd and counterproductive. But if generalized, the collaboration would break down and sanctions would be likely. Sanctions for violations of NISU may tend to be less harsh than sanctions for violating NIST. But if Rosa's behavior persisted, she might well face career challenges. For example, it could be difficult for her principal investigator to recommend her as a team member. So, the general patterns of sanctions for violations of social norms provide some support for NISU.

A variation of the case exemplifies how the epistemic norm, NISU, may be overridden by practical factors. Assume that *p* is crucial to the investigation and that Rosa is aware of this. However, she is also aware that it is hard to realize that *p* is crucial unless one has epistemic expertise in D1. Assume, moreover, that Rosa is aware that there is no particular urgency. Assume, finally, that Rosa is aware that there is a set of *p*-relevant evidence, E, that would not be consulted as a manner of routine. If Rosa lacks warrant that Ali is aware of the high stakes, the lack of urgency, and the further available evidence, she would be practically or even overall reasonable to not accept Ali's testimony right away. Depending on the nature of the collaboration, it might still be inappropriate to double-check *p* herself. But Rosa would be reasonable to indicate the high stakes to Ali or to question whether he had consulted the set of evidence, E. NISU explains this verdict since Rosa does not possess sufficiently strong warrant for believing that Ali's testimony is properly based on adequate scientific justification. Thus, the epistemic requirement to believe or accept set forth by NISU may be overridden by practical factors or factors that pertain to social norms such as politeness. This is a general feature of epistemic norms which only capture one aspect of all-things-considered norms (Gerken 2011: 531).

4.4.c An epistemic corollary regarding the uptake of intra-scientific testimony: In this section, I will argue that NIST and NISU fit together in a manner that motivates an important epistemic corollary. Roughly, this is the thesis that recipients whose uptake of intra-scientific testimony is in accordance with NISU acquire a default warrant for their testimonial belief.

Importantly, NISU does not require that the recipient understand the scientific justification for *p* that NIST requires the intra-scientific testimony to be based on. This point indicates that the recipient does not simply inherit the testifier's scientific justification (cf. the *Non-Inheritance of Warrant* principle from Chapter 2.2.b). Unless the testifier explicates her justification for the content of her testimony, the recipient's warrant is importantly distinct from the testifier's warrant, and frequently it is merely an epistemic entitlement.

The fact that NIST ensures a contextually adequate degree of scientific justification of intra-scientific testimony helps explain why belief-formation in accordance with NISU is epistemically rational. There are two components to this explanation. First, the content of the intra-scientific testimony is backed by a contextually adequate degree of scientific justification, which is a comparatively strong type of warrant (cf. *Hallmark I* from Chapter 3.3). Second, the testifier typically has an interest in being truthful since she may be severely sanctioned if she is caught in violating NIST.

These two components are fairly close analogs of the key aspects of testimonial entitlement in general—namely, reliability and sincerity (Gerken 2013b; Gelfert 2014). So, they help motivate an epistemic principle of *Warranted Intra-Scientific Uptake* ("WISU" for short):

WISU

In a context of intra-scientific communication, CISC, in which S's intra-scientific testimony that *p* conveys that *p*, a collaborating scientist, H, is, as a default, warranted in believing or accepting that *p* if H has strong and undefeated warrant for believing that S's testimony that *p* is properly based on adequate scientific justification.

In a nutshell, WISU has it that *if H has strong and undefeated warrant for believing that S's intra-scientific testimony that p is epistemically adequate, then H is, as a default, warranted in believing or accepting that p.*

WISU only sets forth a default sufficient condition since warranted belief that S manifests scientific competence is hardly a necessary condition on being warranted. For example, H may be warranted in believing or accepting that S knows that *p* but not on the basis of her scientific competence. The default warrant for believing the content of a collaborator's intra-scientific testimony is easily defeated. For example, it may be defeated if the recipient has strong warrant for believing that *p* is false.

While the recipient often acquires a testimonial entitlement, there are plenty of cases in which she may acquire more than that. She might understand the basic nature of the scientific justification that the testimony is based on. For example, a sociologist might collaborate with a neuroscientist and understand that her intra-scientific testimony that *p* rests on fMRI studies. Even though the sociologist does not understand much else about fMRI techniques, he will have some small degree of discursive justification for believing that *p*. The discursive justification is partly substantive, depending on the sociologist's familiarity with fMRI studies, and partly social, depending on the sociologist's familiarity with his collaborator and her competences.

Let me address a potential misunderstanding: WISU does *not* motivate reductionism about testimonial warrant from intra-scientific testimony. For example, the fact that the principle is restricted to cases in which the recipient, H, has warrant for believing that S's testimony is properly based on adequate scientific justification does not show that the testimonial entitlement reduces to such higher-order warrant. Generally, none of the norms I have developed entails that any warrant for intra-scientific testimony reduces to the recipient's non-testimonially warranted beliefs about the testifier, the background context, etc. Rather, WISU is compatible with the idea that the recipient may acquire default testimonial entitlement in minimal background cases. This idea is part of the view, although it is not captured by the principle. Likewise, both NISU and WISU are compatible with a broadly social externalist framework (Gerken 2013b, 2020a). According to such a framework, the social environment characteristic of scientific collaboration provides a central aspect of the explanation why the recipient is warranted in beliefs based on the uptake of intra-scientific testimony. Likewise, the framework has it that general features of the social environment may defeat or diminish testimonial warrant (Gerken 2013b; Guerrero 2016).

Together NISU and WISU reflect the idea that part of scientific competence consists in a meta-competence that includes at least two aspects. The first aspect consists in some ability to recognize a competent scientific collaborator. The second aspect consists in some sensitivity to defeaters of intra-scientific testimony. The meta-competence required to obtain entitled testimonial belief is fairly undemanding since it only requires the recipient to reliably identify the testifier as possessing relevant scientific competence and comprehend the content of the testimony. Yet, the meta-competence required by the norms of scientific uptake is related to interactional expertise. That said, it does not primarily concern the linguistic aspects of interactional expertise that Collins and Evans highlight when they characterize it as "expertise in the language of a specialism" (Collins and Evans 2007: 28). However, since meeting the norms of intra-scientific uptake contributes to rendering scientific collaboration effective and truth-conducive, the competences that they require are also aspects of contributory scientific competences.

4.4.d Concluding remarks on intra-scientific uptake: The ambition of articulating norms of intra-scientific uptake faces two independently plausible but seemingly opposing assumptions. The first assumption is that scientists collaborate in large part because they lack expertise in their epistemic collaborators' field and that they, therefore, have to accept their intra-scientific testimony (cf. *Hardwig's Dictum*). The second assumption is that scientists have, and often use, some exoteric resources for assessing collaborators (cf. Wagenknecht 2015; Fricker 2017). In response, I have articulated NISU as a default norm that respects both of these assumptions and argued that forming testimonial belief or acceptance in accordance with this norm usually results in warranted testimonial belief. This is in part due to the social environment that helps secure that collaborators testify in accordance with NIST. However, NISU and WISU are default principles that may be defeated by specific contextual factors as well as general environmental ones. A recipient should not always accept an intra-scientific testimony, and her testimonial belief is not always warranted.

4.5 Collaboration and Norms of Intra-Scientific Testimony

In this section, I will put the preceding treatment of intra-scientific testimony and the norms governing it to use in defending the thesis *Distinctive Norms*, that the epistemic contribution of scientific collaboration depends on distinctive norms of intra-scientific testimony. This idea is important in its own right. But it will also serve as a premise in my final argument that distinctive norms of intra-scientific testimony are vital to the scientific methods of collaborative science (Chapter 7.2). So, I will conclude the chapter by briefly explicating the idea and motivating it in the light of the more concrete norms of intra-scientific testimony that I have developed.

4.5.a The need for norms: Recall the thesis *Distinctive Norms*, which I set forth in Chapter 1.5.b:

> ***Distinctive Norms***
> The epistemic contribution of scientific collaboration depends on distinctive norms of intra-scientific testimony.

This thesis reflects the broad idea that hyper-specialized division of cognitive labor which characterizes scientific collaboration calls for intra-scientific testimony which, in turn, can serve its functions only if it is governed by distinctive norms. The norms of intra-scientific testimony outlined in this chapter exemplify such norms insofar as they are distinctive of it. That is, they go beyond generic norms of testimony. Moreover, NIST and NISU are cast in terms of scientific justification, and this reflects the characteristic aspect of scientific testimony, namely that it is based on scientific justification (Chapter 3.1.d). A hallmark (*Hallmark III*) of scientific justification is that it is *discursive*, and this is also part of what makes the proposed epistemic norms distinctive. For example, NIST requires that the intra-scientific testifier articulate the scientific justification for the content of the testimony, and this is important to the epistemic force of scientific collaboration (Chapter 3.5 and below). Thus, the norms are distinctive of intra-scientific testimony in that they reflect one of its central functions in scientific collaboration—roughly, the function of transferring information from highly specialized collaborators in a manner that retains a high degree of reliability.

As noted in Chapter 1.5.b, *Distinctive Norms* may be motivated in analogy with norms of scientific collaboration that do not directly involve testimony. For example, it may be a tacit distinctive norm of scientific collaboration that a collaborating scientist should perform the type of research that one is the most qualified for and not the type of research where collaborators are better qualified. Such norms and guidelines typically play an important epistemic role. A proper distribution of cognitive labor tends to yield a more truth-conducive investigation. Norms of scientific collaboration may also be more specific and explicit. As noted, there are distinctive scientific norms governing double blinding procedures in many types of experimentation and there are norms governing the choice of statistical test. Some such norms are so explicit that they are reflected in concrete guidelines in textbooks (Aschengrau and Seage 2020; UCLA Statistical Consulting Group 2021). These scientific norms and guidelines also contribute to truth-conduciveness of the scientific practices in question. For example, double blinding may help to minimize a number of epistemic problems such as placebo effects, confirmation bias, data mining, etc.

So, many of the distinctive norms of collaborative science contribute to its epistemic force. *Distinctive Norms* articulates an instance of this general phenomenon. Just as there are distinctive scientific norms governing how to double blind a study or chose a suitable statistical model, there are distinctive norms governing inter-scientific testimony. Both types of norm make an epistemic contribution. NIST and NISU are idealizations of the operative social epistemic norms of intra-scientific testimony. The idealization consists in abstracting away from disciplinary variations and the fact that the operative social norms of science are shaped by many other standards than truth—e.g., publishability, visibility, and so forth. But while the operative social norms in science are complex and messy, the force of scientific collaboration depends in part on the fact that they involve epistemic norms.

Importantly, norms proposed here do not exhaust norms of scientific collaboration or even the methodological norms of intra-scientific testimony. For example, it is reasonable to suppose that there are norms governing when it is acceptable and expected to provide testimony in the first place. A scientist may be required to testify the results of a pilot study to a collaborator and simultaneously required to not testify the results to a member of a competing research group. So, such norms may help clarify the distinction between intra- and inter-scientific testimony. Further unexplored norms of intra-scientific (and inter-scientific) testimony concern its content and its articulation. For example, it may be that social norms govern when simplifications or uses of generics are permissible (DeJesus et al. 2019). I will not develop such further norms of intra-scientific testimony here, but they are worth noting to augment the motivation for *Distinctive Norms* and to call attention to an avenue of future research.

But while the norms that I have motivated, NIST and NISU, are by no means exhaustive of the relevant norms, they are central to the practice of intra-scientific testimony.

4.5.b The rationale from the norm governing the production of intra-scientific testimony: This line of reasoning for *Distinctive Norms* begins with the assumption that the epistemic force of science is partly explained because it involves a collaboration that is characterized by a hyper-specialized division of labor (Chapters 1.3 and 3.3). Since hyper-specialization distributes epistemic work, scientific collaboration depends centrally on reliable intra-scientific testimony (Chapters 1 and 4.1.b, 4.3.a). However, the reliability of intra-scientific testimony is secured in part by social norms that are associated with communal sanctions for violation and incentives for abidance (Chapters 2.4.c–d and 4.2.a, 4.4.a).

The last step in this line of motivation is an abductive one. The presence of distinctive social norms of intra-scientific testimony is postulated as a central part of the explanation of why it is generally reliable. A fairly standard account of why a particular practice is maintained in a community involves an appeal to social norms governing the practice (Bicchieri 2006, 2017; Graham 2015b; Henderson and Graham 2017a, 2017b). In particular, social norms are often postulated to explain stable social practices that benefit the community, but not necessarily the individual, such as the testifier who has to devote resources to meet the (often demanding) norm. In the present case, the epistemic demands on intra-scientific testimony, articulated by NIST, are determined by a context that is generally fairly demanding. This is in large part due to the fact that the consequences of false intra-scientific testimony are high for both the immediate collaborators and the wider scientific community.

I have proposed norms of production and uptake of intra-scientific testimony that can help secure the reliability of intra-scientific testimony in scientific collaboration. The present rationale for *Distinctive Norms* is an abductive one according to which the best explanation for why truth-conducive scientific collaboration can function is that truth-conducive norms govern the process. Note that this rationale motivates both components of *Distinctive Norms*. That is, it motivates both that the norms of intra-scientific testimony are distinctive and that they are important contributors to the epistemic force of scientific collaboration.

The rationale from NIST has the flavor of a transcendental argument. It assumes the truth-conduciveness of scientific collaboration as a premise from which it is argued that, since distinctive norms of intra-scientific testimony are required by the best explanation of this assumption, such norms are in fact governing intra-scientific testimony. (A notable difference with Kantian transcendental arguments, however, is that the present motivation is abductive and does *not* assume that these norms are conditions of possibility for truth-conducive collaboration.)

The rationale is general insofar as it does not require a commitment to any particular epistemic norm, but only to an epistemic norm of intra-scientific testimony that is sensitive to the standard of truth. Moreover, *Distinctive Norms* may also be motivated, at least in part, by arguing that epistemically effective scientific collaboration requires non-epistemic norms, such as norms of relevance, that are distinctive of intra-scientific testimony. Thus, the motivation for *Distinctive Norms* exemplifies a commonplace motivation for assuming that a particular social practice is norm-governed.

4.5.c The rationale from norms governing the consumption of intra-scientific testimony: The motivation of *Distinctive Norms* from the reception of intra-scientific testimony begins in a manner similar to the previous line of reasoning. Recall the thesis *Collaboration's Contribution* according to which scientific collaboration contributes immensely to the epistemic force of science (Chapter 1.3). Given that scientific collaborators are often specialized in non-overlapping and esoteric domains, it can be infeasible for them to check each other's intra-scientific testimony (Chapter 4.3.a). Even when recipients can double-check an intra-scientific testimony, this would often be such a waste of cognitive resources that it would be contrary to the point of collaboration. Finally, providers of intra-scientific testimony must be able to assume that what they communicate to collaborators is generally accepted and put to use unless the recipient appropriately challenges them. Consequently, effective scientific collaboration frequently requires an incomplete but fairly high default degree of uptake of intra-scientific testimony (Chapter 4.4). Social norms associated with sanctions for violations provide an important part of the explanation of why scientists tend to accept their collaborators' intra-scientific testimony when they have contextually adequate warrant for believing that that their intra-scientific testimony is properly based on adequate scientific justification (Chapters 2.4.c–d, 4.2.a, and 4.4.a). Given that NISU is distinctive of intra-scientific testimony, the assumption that it helps to explain the epistemic force of scientific collaboration motivates *Distinctive Norms*. Of course, NISU does not provide the full story. Scientists' default unquestioning acceptance of intra-scientific testimony is also partly explained by the cold, hard fact that it can be infeasible to check the intra-scientific testimony.

As above, the present rationale for *Distinctive Norms* does not hinge on the specific norms that I have developed in this chapter. Likewise, the rationale is an abductive one and, as such, may be criticized by postulating alternative explanations of the trusting behavior characteristic of scientific collaboration. But in conjunction with the considerations above, the rationale from NISU provides a compelling reason to accept *Distinctive Norms*.

4.5.d Concluding remarks on norms of intra-scientific testimony: *Distinctive Norms* is an important thesis in its own right given that it specifies an important aspect of the link between scientific collaboration and the significance of scientific testimony. Tacit appreciation of norms of intra-scientific testimony, such as NIST and NISU, is a part of scientific expertise. In particular, it is part of the aspects of interactional expertise and T-shaped expertise (discussed in Chapter 1.2.c) in virtue of which a scientist may be an effective scientific collaborator. In addition to this self-standing point, *Distinctive Norms* also contributes to my overarching aim of characterizing the significance of intra-scientific testimony in scientific practice. In particular, it will do so in Chapter 7.2.a, where it will serve as a premise in an argument that intra-scientific testimony is a vital part of science.

4.6 Concluding Remarks on Intra-Scientific Testimony

This chapter has dealt with the role of scientific testimony in science in the making. In particular, I have considered the relationship between scientific collaboration and intra-scientific testimony. For example, I developed an epistemic norm of providing intra-scientific testimony, NIST. According to NIST, the degree of scientific justification that intra-scientific testimony requires depends on the context in which it is provided. I then turned to the recipients of intra-scientific testimony and developed a norm of intra-scientific uptake, NISU, along with an epistemological principle, WISU. Finally, I returned to the larger issue of the role of and need for norms governing intra-scientific testimony. Here, I appealed to the previous considerations in arguing for the thesis *Distinctive Norms*, that scientific collaboration based on the division of cognitive labor requires distinctive norms governing intra-scientific testimony.

The thesis that scientific collaboration requires norms of intra-scientific testimony, and the development of central examples of such norms, help to motivate the idea that intra-scientific testimony is a vital part of scientific practice. Although I will not argue for this thesis until Chapter 7, the discussion of specific norms of the production and consumption of intra-scientific testimony makes the significance of intra-scientific testimony in scientific collaboration more tangible. Given the centrality of collaboration to scientific practice, the discussion illuminates an important, and sometimes overlooked, aspect of the scientific enterprise and the methodology that characterize it.

PART III

SCIENTIFIC TESTIMONY IN SOCIETY

Whereas Part II of the book primarily concerned the nature of scientific testimony and its roles within science, Part III primarily concerns the roles and significance of scientific testimony in the larger society. In particular, I will consider norms and guidelines for two types of public scientific testimony: scientific expert testimony and science reporting. In doing so, I consider the uptake of testimony and some of the principled obstacles for it. I conclude by stepping back to consider the significance of scientific testimony within the scientific enterprise and in society.

In Chapter 5, I consider some general challenges for public scientific testimony that have to do with psychological biases and the social environment. I then turn my attention to scientific expert testimony and consider some norms and guidelines for it. Chapter 6 concerns science reporting. I outline and criticize some common science communication strategies. On this basis, I articulate a novel alternative in terms of a general norm for science reporting. I conclude the chapter by applying this norm to the widely discussed question of balanced reporting.

5
Public Scientific Testimony I
Scientific Expert Testimony

5.0 Scientific Testimony in the Wild

The primary audience of scientific expert testimony is laypersons who lack epistemic or contributory expertise in the relevant domain (Chapter 1.2). Consequently, scientific expert testifiers face distinctive challenges. This chapter identifies some of these challenges and discusses how they may be addressed.

In Section 5.1, I consider the roles that scientific experts play in public deliberation. In Section 5.2, I consider some challenges associated with communicating science to laypersons. In Section 5.3, I consider some folk epistemological obstacles to the uptake of public scientific testimony. In Section 5.4, I develop and defend an *epistemic* norm as well as a *presentational* norm of scientific expert testimony. In Section, 5.5, I consider the problem of scientific expert testimony in a domain of epistemic expertise other than the scientist's own. In Section 5.6, I summarize some of the central conclusions.

5.1 The Roles and Aims of Scientific Expert Testimony

Scientific expert testimony is similar to intra-scientific testimony and differs from science reporting in that the testifier is a scientific expert. However, it is similar to science reporting and differs from intra-scientific testimony in that the intended primary recipients are laypersons rather than scientists. Thus, most of the recipients will be unlikely to possess relevant epistemic expertise and may be unfamiliar with even basic aspects of scientific practice. Consequently, the aims of expert scientific testimony are very different from those of intra- or inter-scientific testimony. So, I will begin by considering the roles that scientific experts play in society.

5.1.a The roles of scientific experts in the public domain: Scientific experts play many different roles in society. To narrow the discussion somewhat, I will focus on scientific expert testimony in contemporary societies that pursue ideals of deliberative democracy. But despite this focus, there is a wide variety of types

Scientific Testimony: Its roles in science and society. Mikkel Gerken, Oxford University Press. © Mikkel Gerken 2022.
DOI: 10.1093/oso/9780198857273.003.0006

and functions of scientific expert testimony.[1] For example, scientists do not only testify about factual matters. Sometimes their testimonies implicitly recommend courses of action or involve other normative claims (Peters forthcoming b). As a further restriction, I will focus on factual testimony as far as possible. I use the phrase 'as far as possible' since factual testimony may have directive implicatures (Gerken 2017a). Indeed, some argue that the distinction between the roles *scientist* and *scientific advisor* is spurious.[2] This discussion is situated in the debate about values in science (Steele 2012; Winsberg 2018). As mentioned, I will largely sidestep the grand debate by relying only on a fairly minimal assumption—namely, that the reasonably expected practical consequences of scientific expert testimony partly determine when and how it may be offered (Steele 2012; Wilholt 2013; John 2015a, 2015b). This assumption is compatible with a wide range of views regarding the value-free ideal and the advisory roles of scientists.

Scientists convey factual hypotheses to the general public for several reasons. We should not discount that scientists may simply be excited by their findings and wish to share them. Moreover, they may be motivated by enlightenment ideals or ideals of deliberative democracy that value a well-informed public. However, as often highlighted, it is also important to recognize social incentives for science communication, such as influence, prestige, visibility, and such (Whyte and Crease 2010).

Different types of scientific expert testimony occur on different platforms, which come with their own sets of constraints. In publications dedicated to popularizing science, it may be appropriate to dwell on the process of discovering esoteric findings with no obvious relevance. In contrast, many media platforms tend to require simplified and sparse descriptions of the results with a focus on their practical ramifications. However, these crude generalizations only represent a fraction of the relevant platforms for science communication. For example, social media provide an array of platforms with their own associated challenges such as algorithmic filter bobbles (Lazer et al. 2018). The issue is complicated even further once we consider cases in which scientific expert testimony is not initiated by the scientist. Frequently, scientists are *called upon* to provide scientific expert testimony that serves as input to policymaking or public deliberation. Other scientists serve this role as part of their job. For example, the European Commission has a Group of Chief Scientific Advisors who are to provide "high quality, timely and independent scientific advice on specific policy issues" (European Commission 2019). Likewise, institutions such as the Environmental Protection Agency and many others employ scientists for advice on policy matters in the US.

[1] Jasanoff 1990; Kappel and Zahle 2019; Oreskes 2019; Baghramian and Croce 2021.
[2] Douglas 2009: 82. See also Jasanoff 1990: 76; Gundersen 2018.

Many cases of solicited scientific expert testimony are cases of formal testimony, which takes the form of a report which may or may not include policy advice. However, solicited scientific expert testimony may be far more informal, as when scientists are called upon to be interviewed. So, there is also a great variety of solicited scientific expert testimony. Moreover, there are gray zones between *scientist-initiated* and *solicited* scientific expert testimony. So, although this distinction is good to have on hand, the science communication context matters for the appropriate form and content of scientific expert testimony.

5.1.b Truth as fundamental to scientific expert testimony: As noted, scientific expert testimony serves a plurality of functions (Kappel and Holmen 2019). However, I assume that veritistic functions are fundamental among them. This assumption derives from two background assumptions. The first is the broadly realist assumption that approximating truth is an actual and reasonable aim of scientific theories (Psillos 1999; Godfrey-Smith 2003; Chakravartty 2011). The second assumption is that science is epistemically superior to alternative sources in most domains in which it inquires (cf. Chapter 3.3).

Given these assumptions, it is natural to see truth as central to the aims of scientific expert testimony. Goldman articulates the idea as follows: "The standard goal of people seeking informational experts is precisely to learn the *true* answers to their questions" (Goldman 2018: 6; original italics).

However, it is tricky to further articulate this idea precisely. For example, it may be challenged by the recognition that "science unabashedly relies on models, idealizations, and thought experiments that are known not to be true" (Elgin 2017: 1). Perhaps, it may be said that this reliance is in the service of truth-approximation (Khalifa 2020; Frigg and Nguyen 2021). Moreover, it would be misguided to articulate it simply in terms of disseminating truth, since the best scientifically justified testimony may be false. A better candidate is the view that scientific expert testimony should facilitate that the public's beliefs or credences in scientific hypotheses align with the strength of the scientific justification for them. Since strong justification is by its nature conducive to truth, this articulation comes closer to the mark. As in the case of norms generally, the distinction between the *standard* of truth and the *norm* of truth-conduciveness in terms of cognitive competence is of help (Chapter 2.4.c–d; Gerken 2017a). Disseminating truth may be said to be the standard of correctness of scientific expert testimony, whereas its norms should be cast in terms of strength of scientific justification.

Furthermore, a *basing constraint* must be met. For example, it would be problematic if people's belief aligned with the strength of the scientific justification out of fear or because of manipulation. On the other hand, it is too demanding to require that recipients' beliefs reflect the strength of the scientific justification in virtue of understanding it.[3] Rather, the basing constraint involves the laypersons' uptake of public scientific testimony reflecting some general appreciation of the epistemic force of science. Thus, the public should optimally defer to public scientific testimony in virtue of some appreciation of its comparative epistemic force. I call this type of uptake *appreciative deference* and elaborate on it in in Chapter 7.4.c.

Important prudential roles of science may be partly explained by its veritistic roles. For example, scientific expert testimony that serves certain practical ends, such as policymaking, does so in virtue of the desideratum of basing political decisions on hypotheses that are most likely to be true. Similarly, scientific expert testimony that primarily serves to justify public investments in science should do so in light of the desideratum of science providing an epistemically superior source. In general, the fact that scientific expert testimony serves multiple roles does not compromise the idea that the pursuit and dissemination of truth is fundamental to scientific expert testimony. Thus, the central roles of scientific expert testifiers are veritistic (Goldman 1999).

5.1.c Explanation, understanding, and science literacy as desiderata of science reporting: Some of the best scientific expert testimony takes the form of an explanation. Consider a scientific expert who is interviewed about the correlation between pesticide use and the dwindling bee population. She not only may testify *that* pesticides negatively impact the bee population but may furthermore provide an explanation *why* this is so. As already noted, providing a scientific explanation is closely related to explicating a scientific justification (Chapter 3.5.c).

Insofar as the scientific explanation is articulated in a way that the audience may appreciate, it may increase their *understanding* of why scientists regard the proposition in question as true or plausible (Lombrozo et al. 2008; Huxster et al. 2018). This assumption does not require a strong view about the debated relationship between explanation and understanding. The assumption required is the fairly modest one that explanation appreciable by an audience may increase that audience's degree of understanding of the nature of the hypothesis and the justification for it.[4] So, laypersons may not merely acquire testimonial entitlement for believing the hypothesis in question but moreover some modest discursive justification for believing it.

In Chapter 3.5.c, I argued that possessing discursive justification for an inconsistent set of hypotheses epistemically improves scientists' communal belief revision. Although mere entitlement is also of value in public deliberation, a scientifically literate public is typically better positioned than an illiterate one to engage in truth-conducive public deliberation. While optimistic, the assumption is

[3] Keren 2018; de Melo-Martín and Intemann 2018; Slater et al. 2019.
[4] Grimm 2006; Khalifa 2013, 2017; Strevens 2013; Boyd 2017.

considerably weaker than the much-discussed idea of epistemic democracy—roughly, the contention that public deliberative processes among non-experts are at least as epistemically good as other types of deliberation.[5] It is far more modest to assume that a population that is scientifically literate and well-informed is generally epistemically superior to one that is not. But even this more modest assumption is complicated by various psychological mechanisms and social factors that have been found to hamper people's ability to engage in truth-conducive deliberation (Taber and Lodge 2006; Kahan et al. 2017). These obstacles to the positive enlightenment picture will take center stage from Section 5.2 onwards. However, I will argue that while they need to be taken very seriously, they do not ultimately compromise the idea that the central role of scientific experts in society is to provide epistemically authoritative testimony. So, I take it to be a reasonable desideratum of science to provide scientific expert testimony that contributes to a generally informed and scientifically literate public.

5.1.d Concluding remarks on the roles of scientific experts: Scientific expert testimony is critical in the societal division of cognitive labor. In particular, it plays a key role in ensuring that the public is well-informed and that political decisions are made on a strong epistemic basis. The public's beliefs should accord with the best scientific justification and do so out of appreciation of the epistemic authority of science. However, recent empirical research has identified some principled obstacles for approximating this ideal through scientific expert testimony. I now turn to this issue.

5.2 Laypersons' Uptake of Public Scientific Testimony

How easy things would be if scientists could simply provide scientific expert testimony on matters of public interest with the result that the public formed beliefs accordingly. However, empirical research indicates considerable complications for this broad enlightenment picture. In this section, I outline some general background assumptions concerning human cognition and articulate how they raise challenges for public scientific testimony. This section and the next one serve as background for the present chapter on scientific expert testimony as well as for the next chapter on science reporting. Hence, I will discuss the empirical work in relation to the broader category of public scientific testimony (cf. the taxonomy in Chapter 1.1.b).

[5] Peter 2008; Kitcher 2011; Landemore 2012; Holst and Molander 2019.

5.2.a Folk heuristics and biases in laypersons' uptake of public scientific testimony: Humans are bounded cognizers who, due to limited cognitive capacities, rely on cognitive heuristics in judgment, decision, and action. Following Kahneman, a cognitive heuristic may be characterized as "a simple procedure that helps find adequate, although often imperfect, answers to difficult questions" (Kahneman 2011: 98). Since heuristics are cognitive cost-minimizing mental shortcuts, they are associated with biases, which may be characterized as systematic deviations from what I call the "gold standard" response—i.e., the optimal response that the agent is capable of providing.[6]

These abstract characterizations may be made more concrete by considering an experimental paradigm, such as the influential Wason four-card selection task (Wason 1966, 1968). Participants are presented with four cards and informed that each of them has a letter on one side and a number on the other side. The visible sides of the cards read 'A,' 'K,' '4,' and '7.' The participants are then asked to select all and only the cards that must be turned over to determine the truth of the following rule:

If a card has a vowel on one side, then it has an even number on the other side.

The choice of 'A' and '7' is what I call the *gold standard* selection since this may reveal the rule to be false. Yet, most participants select either 'A' alone or 'A' and '4.' But the card '4' need not be checked to determine the truth of the rule given that it is consistent with *any* letter being on the other side of an even number. Hardly any participants pick the card 'K' but virtually all pick the card 'A' (Wason 1966, 1968; Wason and Johnson-Laird 1972). So, the mistakes are highly systematic and predictable. Hence, they do not indicate a mere *performance error* according to which haphazard mistakes are to be explained by distraction, fatigue, etc. Rather, the systematicity of the mistake indicates a *bias* that reflects some signature cognitive limitations of the agent and the operative heuristics.

The bias underlying the selection task is standardly taken to be a strand of evidence for confirmation bias, which is, roughly, the tendency to seek evidence that may confirm a hypothesis and fail to seek, or even ignore, evidence against it (for different strands of evidence, see Hahn and Harris 2014). Notably, performance on the selection task improves considerably if it is cast in terms of some thematic contents involving deception rather than the abstract number/letter contents (Griggs and Cox 1982; Fiske et al. 2010). Likewise, the content of public scientific testimony may bear on whether its uptake is affected by confirmation bias.

[6] Gerken 2017a. For broader frameworks, see Evans 2010; Kahneman 2011; Stanovich 2011.

Confirmation bias has played an important role in the philosophy of science. As noted in Chapter 2.4.a, Popper cast his critical rationalism as an antidote to scientists' natural inclination to pursue evidence confirming their hypotheses (Popper 1934, 1963). While contemporary philosophers of science are skeptical of Popper's general framework, they have suggested that confirmation bias is a challenge that scientific methods are required to overcome (Hergovich et al. 2010; Bird 2019). Others have argued that confirmation bias may contribute to collective scientific rationality (Solomon 1992; Smart 2018; Peters forthcoming a).

Confirmation bias is also found among laypersons (Nickerson 1998). Moreover, when the hypothesis in question is central to laypersons' values and social identity, confirmation bias may be amplified, resulting in a potent form of *motivated cognition*.[7] But whereas science may impose procedural regulations to curb confirmation bias, it is less clear how to address it when communicating science to the lay public.

Given laypersons' cognitive limitations and lack of knowledge about science, it is reasonable to suppose that their assessments of scientific testifiers often rely on folk epistemological heuristics. These heuristics often serve bounded cognizers well in ordinary cases of testimony, but when applied to scientific testimony, they may lead recipients astray. This is a general pattern that often explains the occurrence of biases. For example, this idea is also taken to explain outcome bias—roughly, the tendency to let information about the outcome of a procedure unduly affect an assessment of the procedure. Outcome bias is important for public scientific testimony since it also impacts epistemic assessments (Gerken 2020b). But the present point is that outcome bias exemplifies a structural feature of many biases: "the generally useful heuristic of evaluating decisions according to their outcome may be overgeneralized to situations in which it is inappropriate" (Baron and Hershey 1988: 571).

Due to our cognitive limitations, we rely on cognitive heuristics, and overreliance on generally useful heuristics may result in biased judgments. Given that this is a general source of bias, it is plausible that the folk epistemological heuristics used to assess ordinary testimony may be problematic when applied to assess public scientific testimony. Consequently, it is paramount to investigate the heuristics responsible for the laypersons' uptake of public scientific testimony.

5.2.b The challenge of selective uptake: An overarching challenge for public scientific testimony is one that I will call the *Challenge of Selective Uptake*. It arises from the fact that even laypersons who generally accept public scientific testimony nevertheless reject it with regard to specific domains or hypotheses. Moreover, they do so despite very strong scientific justification in favor of, and scientific consensus about, the hypotheses in question.

[7] Taber and Lodge 2006; Kahan et al. 2007; Kahan 2015b; Levy and Ross 2021.

For example, there is extremely strong and growing scientific justification that there is no correlation between MMR vaccines and autism (Hviid et al. 2019). Nevertheless, increasing public concern about a vaccine-autism link has led to ongoing measles outbreaks (de Melo-Martín and Intemann 2018; Papachrisanthou and Davis 2019). In consequence, WHO in 2019 named vaccine hesitancy as a top ten threat to global health (WHO 2019). This claim occurred even before the COVID-19 pandemic, which has made it abundantly clear how serious a problem vaccine skepticism is. Developing vaccines is a remarkable scientific feat, but it turns out that it is an even harder feat to convince skeptical laypersons to get vaccinated.

Likewise, studies of laypersons' denial of the scientific hypothesis of anthropogenic global warming (henceforth "AGW") indicate tenacious climate science skepticism despite decades of scientific consensus on the issue. Studies in the US have found that conservative Republicans exhibit low trust in climate scientists and are more likely to doubt the existence of AGW (Funk and Kennedy 2016). Notably, both liberals and conservatives dismiss ideology-inconsistent science (see Nisbet et al. 2015). However, the US population is not in general skeptical of science, and general trust in science has remained fairly stable for decades (Smith and Sons 2013; Funk and Kennedy 2016). Despite a modest decline among conservatives, trust in the general scientific community remains high across the political spectrum (Gauchat 2012; Kovaka 2019).

Yet, laypersons' uptake of scientific testimony is selective. Some people who trust a meteorologist's report that they are in the path of a storm do not trust a climate scientist's report that we are on a path to global warming. Some people who trust their doctor who tells them that the antibiotics prescribed for their pneumonia are safe do not trust her when she tells them that the MMR vaccine recommended for their children is safe.

This asymmetry of trust is remarkable because it may obtain even though the scientific expert testimonies, such as the ones above, are based on equally strong scientific justification of the same type. Hence, the cases of selective skepticism about climate science and selective skepticism about medical science exemplify a more general challenge, which I will articulate as follows:

> ***The Challenge of Selective Uptake***
>
> Laypersons who generally accept public scientific testimony fail to accept public scientific testimony concerning select, equally well-warranted, scientific hypotheses.

The reason why the *Challenge of Selective Uptake* is a challenge for public scientific testimony is that laypersons' trust in it does not track the epistemic strength of the scientific justification (de Melo-Martín and Intemann 2018). Laypersons such as climate science deniers and anti-vaxxers fail to proportion their doxastic attitudes towards scientific hypotheses to the strength of scientific justification for them. As WHO's warning and vaccine skepticism during the COVID-19 pandemic indicates, selective scientific uptake is a major concern for public health. Moreover, it is a concern for public deliberation. Consequently, much effort has gone into diagnosing the social and psychological mechanisms that explain the *Challenge of Selective Uptake*.[8]

The first thing to note is that there is not a mono-causal explanation for why laypersons exhibit selective uptake. Rather, there is a complex set of causes, which range from the social information environment to psychological mechanisms (O'Connor and Weatherall 2019). With regard to the information environment, much effort has been directed at understanding whether social media and its expansion of platforms have contributed to the *Challenge of Selective Uptake*. Indeed, notions such as *fake news*, *echo chambers*, *information bubbles*, etc. have become commonplace explanations.[9] Some laypersons are in epistemically inhospitable information environments in virtue of being confronted with misleading evidence about scientific hypotheses and misleading higher-order evidence about the evidential status of these hypotheses (Lazer et al. 2018). Further, problematic *strategic expertise contexts*, which involve non-alignment between the experts' interests and laypersons' interests, may give the layperson some reason to avoid complete trust in the experts (Guerrero 2016). All of these features of the social environment constitute serious obstacles to laypersons' warranted uptake of scientific expert testimony and may contribute to the *Challenge of Selective Uptake*. One dimension of the problem is that inhospitable social environments may corrupt the indicators of scientific expertise that Anderson, Goldman, and others have considered (Goldman 2001; Anderson 2011; cf. Chapter 4.3.b). For example, Guerrero has argued that such indicators are unreliable in strategic expertise contexts (Guerrero 2016). Moreover, indicators of scientific expertise may be systematically manipulated, as witnessed by case studies of, for example, Big Tobacco.[10]

On the other hand, it bears mentioning that the lack of uptake of some public scientific testimony may be overestimated because people may profess a non-scientific belief to express support in a political debate rather than to express a corresponding belief (Bullock and Lenz 2019; Levy and Ross 2021). While such *expressive responding* may indicate that some apparent selective uptake is merely apparent, it hardly accounts for all cases, since the professed beliefs have been found to affect behavior. For example, Republicans expressing a negative attitude towards the Affordable Care Act were found to be less likely to enroll than those expressing a more positive attitude (Lerman et al. 2017). More generally, some studies suggest that expressive reporting's inflation of partisan gaps is only modest (Peterson and

[8] Weber and Stern 2011; Figdor 2013, 2018; Fischhoff 2013; Jamieson et al. 2017.
[9] Frost-Arnold 2014; Rini 2017; Boyd 2018; Iyengar and Massey 2019; Nguyen 2020.
[10] WHO 2000; Tong and Glantz 2007; Oreskes and Conway 2010; Weatherall et al. 2020.

Iyengar 2019). So, recognizing expressive responding is compatible with considering genuine cases of selective uptake, and this will be my primary focus.

In addition to the noted explanations, researchers have investigated psychological factors, such as folk epistemological biases and social biases, and found considerable evidence that they contribute to the selective uptake of public scientific testimony. Moreover, there are likely interaction effects between the recipients' information environment and the heuristics that their uptake relies on. Hence, simplistic mono-causal explanations of selective uptake should be resisted. Rather, laypersons' information environment and their general psychological features should be considered in relation to each other.

5.2.c Concluding remarks on laypersons' uptake of public scientific testimony: While a comprehensive diagnosis of the challenges that science communicators face is multi-dimensional, empirical research on the psychological dimensions has resulted in some partial, but important, explanatory progress. On the basis of such research, science communicators have been preoccupied with articulating strategies for science reporting, and I will contribute to this effort. To do so in an empirically responsible manner, I will first consider some of the psychological explanations of the *Challenge of Selective Uptake* and related challenges for public scientific testimony.

5.3 The Folk Epistemology of Public Scientific Testimony

The multidisciplinary field *science of science communication* has provided a good deal of empirical evidence concerning laypersons' selective uptake of public scientific testimony (Fischhoff 2013; Dunwoody 2014; Kahan 2015a; Jamieson et al. 2017). In this section, I outline some central strands of this body of empirical research. I will suggest some folk epistemological biases that require more empirical investigation than they have received.

5.3.a Motivated cognition and identity-protective cognition: Converging empirical evidence suggests that varieties of motivated cognition yield important challenges for public scientific testimony. Roughly, motivated cognition may be characterized as the, often unconscious, tendency to privilege or discard information in ways that favor those of one's antecedent views that one is motivated to preserve.[11] Thus, motivated cognition may be seen as a variety of confirmation bias that concerns those beliefs that participants are motivated to preserve. On this view, confirmation bias may be amplified by motivational factors.

[11] Kunda 1987, 1990; Hart et al. 2009; Hart and Nisbet 2012; Corner et al. 2012; Kahan et al. 2012; Sinatra et al. 2014.

The range of motivational factors that may impact people's uptake of public scientific testimony is wide. For example, people's desire to maintain a positive self-image may lead to motivated cognition with regard to public scientific testimony. In a classic study, participants read an article which stated that caffeine consumption causes health problems in women but not in men. It was found that women who had a generally high caffeine consumption found the article to be less convincing than women who consumed little or no caffeine (Kunda 1987).

But motivated reasoning is not restricted to individual motivations, since *social* identity is also an important factor. Consequently, the idea of *identity-protective cognition* is central to empirical investigation of public scientific testimony. Roughly, identity-protective cognition is a variety of motivated cognition that concerns issues that are central to the subject's social identity, and these include political views.[12]

Recent empirical work under the heading *cultural cognition* has focused on how social motivations, such as those tied to cultural belonging and political viewpoints, impact the uptake of public scientific testimony. For example, evidence indicates that participants' assessment of whether scientists with strong credentials are trustworthy experts correlates with how the scientists' testimonies aligned with the cultural commitments of the participants (Kahan et al. 2011). In one condition, a scientist testified that it is premature to conclude that CO_2 emissions cause global warming. Participants holding hierarchical and individualistic political views were more inclined to regard the scientist as knowledgeable and trustworthy than participants holding egalitarian and communitarian outlooks (Kahan et al. 2011). This picture reversed in a condition where the scientist testified that AGW is beyond reasonable scientific doubt. These cultural commitments also correlated with acceptance of the existence of a scientific consensus about AGW (Kahan et al. 2011; Kahan 2017). Moreover, evidence suggests that motivated cognition is a general phenomenon, since liberals and conservatives appear to be as likely to interpret evidence in accordance with their political convictions (Kahan 2013; Nisbet et al. 2015; Frimer et al. 2017).

In sum, recent empirical research provides converging evidence for thinking that motivated cognition and identity-protective cognition, which is related to values, self-conception, and social belonging, strongly impact laypersons' uptake of public scientific testimony. This broad conclusion is bolstered by a number of related studies.

5.3.b Polarization, backfire effects, and science literacy: Several studies closely related to the research on motivated cognition concern group polarization. Polarization comes in many varieties and may be measured in various ways

[12] Sherman and Cohen 2006; Kahan et al. 2012; Kahan 2016; Lewandowsky et al. 2018; van Bavel and Pereira 2018; McKenna 2019.

(Bramson et al. 2017). However, I will primarily consider cases in which the belief becomes more consolidated or becomes stronger due to some stimuli: "members of a deliberating group predictably move toward a more extreme point in the direction indicated by the members' predeliberation tendencies" (Sunstein 2002: 176; see also Sunstein 2009; Boyd 2018). The phenomenon tends to be found in cases where antecedent polarization on factual matters aligns with differences in political or ideological views (Kahan 2016).

Group polarization effects might be seen as partly explained by motivated cognition. After all, group deliberation involves sustained assessment of evidence (Taber and Lodge 2006; Kahan and Corbin 2016). However, group members do not merely become more confident in their antecedent beliefs, decisions, or choices but alter the content of those beliefs in a more extreme direction (Isenberg 1986; Sunstein 2009). According to social comparison theory, group members' formation and revision of opinion are influenced by opinions of their group members and the desire for social approval within the group. This may lead to adopting more extreme views, and when this is common among group members, the process will reiterate (Jellison and Riskind 1970). According to self-categorization theory, group members seek to consolidate the group as distinct from other groups and to preserve their membership in it. They do so by seeking to conform strongly to their in-group's views when confronted with out-groups' views and by signaling group membership to the other group members (Turner 1982; Turner et al. 1989). Some hypothesize that social media contributes to the phenomenon (Del Vicario et al. 2016). Other studies highlight the nature of the deliberating context, since free unstructured discussion has been found to be more polarizing than discussion structured by explicit norms (Strandberg et al. 2019).

Group polarization has been studied in connection with public scientific testimony, since a better understanding of the latter would contribute to a diagnosis of the *Challenge of Selective Uptake*. If laypersons' response to public scientific testimony is more concerned with social factors, such as consolidating group membership, than with truth, it is less surprising that their uptake is selective in a manner that is epistemically problematic.

Moreover, dramatic instances of group polarization have been found in the wake of science communication. These are so-called backfire effects or boomerang effects of public scientific testimony. Roughly, this is the phenomenon that lay recipients gravitate to believing the *opposite* of the scientific testimony that they receive.[13] For example, a study found a backfire effect of information about climate change among Republicans when it was described as affecting socially distant groups (Hart and Nisbet 2012). Likewise, a study of Republicans' reactions to message framings of climate change provided evidence of backfire effects (Zhou

[13] Nyhan and Reifler 2010; Kahan et al. 2011; Hart and Nisbet 2012; Hallsson 2019.

2016). Again, motivated and identity-protective cognition are thought to be central to an explanation of these findings (Sherman and Cohen 2006; Kahan 2016, 2017).

Perhaps the most disturbing empirical evidence is a set of findings suggesting that increased science literacy does not lead to improved acceptance of scientific testimony but to increased polarization. Here is how Kahan et al. put it: "Members of the public with the highest degrees of science literacy and technical reasoning capacity were not the most concerned about climate change. Rather, they were the ones among whom cultural polarization was greatest" (Kahan et al. 2012: 732; see also Kahan 2013; Kraft et al. 2015; Kahan et al. 2017).

Although these findings should be taken seriously, several studies found positive correlations between increased knowledge about the subject matter and acceptance of public scientific testimony.[14] Moreover, other studies indicate that backfire effects are rare—even for polarized issues (Wood and Porter 2019). The methodological lesson here is that the effects of public scientific testimony are highly contextual and should be investigated in relation to specific communication types and strategies.

The present section has merely introduced some of the key findings concerning phenomena such as confirmation bias, motivated and identity-protective cognition, group polarization, backfire effects, etc. However, these phenomena interact with folk misconceptions about science as well as with some less recognized folk epistemological biases.

5.3.c Folk misconceptions about science: Laypersons' assessment and uptake of public scientific testimony takes place against a backdrop of more or less articulate presuppositions about the nature of science, scientific practice, and scientific justification. Occasionally, these presuppositions are mistaken and such folk misconceptions about science may be a contributing factor to the *Challenge of Selective Uptake* and even to outright science skepticism.[15]

A central contention of this book is that the *science-before-testimony* picture, according to which reliance on testimony is orthogonal to the spirit of science, is misguided. Rather, scientific testimony is central to both the epistemic force of science and a reasonable societal division of cognitive labor. However, the *Nullius in verba* sentiment that one should avoid trust in authority in favor of personal examination is widely echoed in the contemporary slogan "do your own research." This idea is commonly promoted by individuals who are selectively skeptical about issues such as vaccine safety or climate science as well as by more organized conspiracy-mongers such as QAnon. This suggests that a more or less explicit folk conception of science remains in the grip of an anti-testimonial picture. It

[14] Shi et al. 2015; Shi et al. 2016; Guy et al. 2014; Bedford 2015; Weisberg et al. 2018.

[15] de Melo-Martín and Intemann 2018; Kampourakis and McCain 2019; Kovaka 2019; Gerken 2020c.

moreover indicates that this picture contributes to various types of science skepticism, ranging from highly selective lack of uptake to full-blown conspiracy. The appeal of this misguided anti-testimonial picture of science may be especially potent in conjunction with motivated cognition. After all, motivated cognizers who self-identify as pro-science may seek to rationalize their selective uptake by embedding it in a congenial conception of science and scientific rationality. The appeal to intellectual autonomy that lies in the simplified *Nullius in verba* tradition offers such a conception.

A related concern is that laypersons may associate scientific justification with *crucial experiments* that provide decisive evidence for or against a hypothesis. An example which is so paradigmatic that it has entered the vernacular is that of a *litmus test*. The idea that scientific justification is generally based on such crucial experiments is misguided given that much scientific justification is inconclusive since it is based on inductive or abductive evidence (Douglas 2015). Lakatos even questions whether crucial experiments may serve to falsify a hypothesis (Lakatos 1974). If crucial experiments are seen as characteristic of scientific justification, laypersons may see hypotheses based on different sources of scientific justification as falling short of the required epistemic standard. For example, the model-based justification that plays a central role in climate science may appear to constitute subpar scientific justification if it is juxtaposed with crucial experiments.[16]

More generally, our folk epistemology often exhibits *source bias* insofar as we are prone to trust some type of evidence-sources, such as vision, more than other equally reliable sources, such as inference (Zamir et al. 2014; Turri 2015b; Gerken 2017a). There is limited evidence pertaining to source bias with regard to public scientific testimony. But the available evidence is compatible with the hypothesis that justification by (crucial) experiments is seen as more compelling than justification by models or other sources of scientific justification that laypersons are unfamiliar with.

The misconception that scientific justification depends on crucial experiments may go hand in hand with an individualistic and often gendered misconception that I will call a *great white man fetish*. According to this misconception, scientific progress owes to an individual (white male) genius who comes up with an entirely new theory and who carries out the crucial experiment that proves it and, thereby, proves the scientific community wrong (Storage et al. 2016; Slater et al. 2019). This mythological picture of solitary genius scientists (Galileo, Newton, Darwin, Einstein) as central to scientific progress may still be perpetuated in science education (Allchin 2003). The focus on individual genius is another dimension of the misguided *science-before-testimony* picture that I am seeking to replace with a *testimony-within-science* alternative. However, the damaging anti-testimonial

[16] See Kovaka 2019. For further perspectives, see Winsberg 2001, 2012; Bailer-Jones 2009; Oreskes 2019.

folk conception of science and its role in society may remain a serious obstacle to laypersons' reasonable uptake of public scientific testimony.

5.3.d Further folk epistemological heuristics and biases: Much work on science communication has focused on motivated cognition. While this focus is reasonable, it should not become myopic. A wide range of further folk epistemological biases may impact laypersons' uptake of public scientific testimony. In this section, I suggest some candidates of factors that bear on laypersons' uptake of public scientific testimony. Although these suggestions are empirically grounded, they have not been as carefully investigated vis-à-vis public scientific testimony as motivated cognition.

A recent body of empirical literature suggests a tight folk epistemological connection between judgments about knowledge and judgments about being in a good enough epistemic position to act (Turri 2015a, 2016; Turri and Buckwalter 2017). While some epistemologists think this folk epistemological connection reflects proper epistemological principles, I have argued that it is best understood as a useful folk epistemological heuristic.[17] In normal circumstances, the degree of epistemic warrant involved in knowledge is more or less the degree that is required for action. However, there are circumstances in which action requires stronger warrant than knowledge requires. In yet other circumstances, it is epistemically rational to act on warrant that is weaker than what is required for knowledge (Gerken 2011, 2015b, 2017a). This is important because it may be too coarse-grained to characterize scientific findings in terms of knowledge or absence thereof. It is true that public scientific testimony must use coarser epistemic categories than those that scientists use within the scientific process (Steele 2012; John 2015a). But it would be a mistake to reduce the epistemic complexity to the categories of knowledge and non-knowledge. At some point in history, the AGW hypothesis was warranted to a degree that fell short of what is required for knowledge. Likewise, the evidence that smoking causes cancer was at some point in history strong enough to provide the epistemic basis for health advice and legislation, although it was not strong enough to underwrite scientific knowledge. Generally, there are cases in which it is urgent to act on the best available scientific evidence even though it is not strong enough to underwrite scientific knowledge. However, due to the folk epistemological action-knowledge link and the default use of 'knows' in epistemic assessment, this may be troublesome to communicate.[18]

The problem with communicating science in terms of knowledge is increased due to *salient alternative effects* on knowledge ascriptions. Roughly, this is people's disinclination to accept ascriptions of knowledge in the face of contextually salient

[17] Gerken 2011, 2015b, 2017a, 2017b, 2020b.

[18] Gerken 2017a; Pinillos 2018; Worsnip 2021. See also Winsberg 2018.

error-possibilities.[19] I have argued that salient alternative effects are often best explained by an epistemic focal bias that is characteristic of our folk epistemology (Gerken 2013c, 2017a). Due to such a focal bias, laypersons may in some circumstances be increasingly skeptical of claims that the scientific community knows some hypothesis when it is also mentioned that epistemically irrelevant confounders have not been controlled for. Big Tobacco's attempts to sow doubt about scientific justification concerning the hazards of smoking are illustrative (Oreskes and Conway 2010). Likewise, scientists' claims to knowledge may too easily be denied in a convincing manner by calling attention to confounders that are epistemically irrelevant in the sense that knowledge does not require them to be ruled out (see Gerken 2017a for the underlying theory of knowledge). To make the matter worse, the salient alternative effects may interact with the folk association between knowledge and action. If it is fairly easy to raise doubts about scientific knowledge by making error-possibilities salient, and knowledge is regarded as epistemically necessary for action, the action-guiding role of public scientific testimony may be hampered.

Extrapolating from the salient alternative effect on knowledge ascriptions, we may consider a more general putative phenomenon of *epistemic qualification effects* on public scientific testimony. Roughly, this is the idea that qualifications which indicate uncertainty about a scientific hypothesis may affect laypersons' uptake of it (van der Bles et al. 2019; Gustafson and Rice 2020). In the worst case, laypersons may regard such honest qualifications about epistemic uncertainty of the relevant hypotheses as a sign of unreliability.[20] Likewise, scientific experts' claim to knowledge may be undermined by noting uncertainty in its evidential basis. A general concern underlying both salient alternative effects and qualification effects is that laypersons rely on heuristics that have evolved to detect deception or unreliability in ordinary cases of testimony. For example, if a testifier highlights uncertainty or error-possibilities in ordinary testimony, laypersons may, in some contexts, take it to indicate some measure of unreliability. One study found that recipients' trust in both the communicated content and the source decreased when the uncertainty is communicated verbally but not when it is communicated numerically (van der Bles et al. 2020; Gustafson and Rice 2020). The matter is important because scientific expert testifiers often provide such qualifications, and some scientific institutions, such as IPCC, have standardized them in verbal formats.[21] A study of IPCC probabilistic testimony found that recipients interpreted verbal qualifications in a way that was systematically

[19] Schaffer and Knobe 2012; Nagel et al. 2013; Alexander et al. 2014; Buckwalter 2014; Buckwalter and Schaffer 2015; Gerken and Beebe 2016; Turri 2015a, 2017; Waterman et al. 2018; Grindrod et al. 2019; Gerken et al. 2020.

[20] Fischhoff 2012, 2019; Jensen et al. 2017; Osman et al. 2018; Kampourakis and McCain 2019; Gustafson and Rice 2020.

[21] For discussions, see Steele 2012; Gelfert 2013; Betz 2013; John 2015a.

inconsistent with IPCC's guidelines. Specifically, recipients underestimated high probabilities and overestimated low probabilities when interpreting verbal qualifications (Budescu et al. 2014).

It bears mentioning that, in some contexts, laypersons may take indications of source uncertainty as a sign of trustworthiness (Jensen 2008; Karmarkar and Tormala 2010). But further evidence suggests that this effect depends on the content of the public scientific testimony (Jensen and Hurley 2012). In general, the effect of epistemic qualifications is complex and highly dependent on context, content, recipients, and mode of communication.[22] Ultimately, I will recommend epistemic qualifications in many science communicating contexts. But it is important that they may, *in some contexts*, contribute to skepticism about public scientific testimony.

Furthermore, epistemic qualification effects may interact with a discursive expectation that one is supposed to be capable of defending one's beliefs against challenges. Such a discursive expectation is characteristic of many ordinary conversations. But laypersons who have just accepted public scientific testimony are often unable to rebut specific challenges and may, therefore, come to regard themselves as unwarranted or they may even come to regard the hypothesis as unwarranted. I will call this phenomenon *discursive deception*. In epistemological terms, discursive deception may occur when laypersons equivocate *discursive justification* with *warranted belief*. When unable to respond to a challenge, one may feel epistemically deficient and "overreact" by retracting claims to warrant or knowledge.[23] This makes life easier for climate science deniers, etc. They just need to provide a challenge to the relevant hypothesis in order to defeat existing testimonial entitlement or generate doubt about whether the hypothesis is warranted in the first place. Thus, discursive deception is among the serious obstacles to a reasonable uptake of public scientific testimony.

In conclusion, there is a range of deep-seated folk epistemological biases which may interact with motivated cognition and may lead to potent forms of science skepticism among the lay public.

5.3.e Considerations from social psychology: Further problems of applying folk epistemological heuristics in the uptake of public scientific testimony pertain to laypersons' in-groups and out-groups. Empirical research suggests that laypersons' epistemic assessments of in-group members are different, and generally more favorable, than epistemic assessments of out-group members. For example, laypersons are generally more inclined to trust and cooperate with in-group than

[22] Jensen et al. 2017; Osman et al. 2018; van der Bles et al. 2019, 2020; Gustafson and Rice 2020.

[23] Elsewhere, I have argued that this diagnosis explains Pyrrhonian skepticism in the form of Agrippa's Trilemma (Gerken 2012a). I think this is a nice example of how foundational philosophical research may become relevant in diagnosing societally relevant problems such as science skepticism.

out-group members (see Balliet et al. 2014 for a meta-analysis). Moreover, laypersons are prone to rely on crude stereotypes in their assessment of out-group individuals and extrapolate their own perspective to members of in-groups (Ames 2004; Robbins and Krueger 2005; Ames et al. 2012).

Generally, the in-group/out-group dynamics may lead us to overestimate members of one's in-group and underestimate members of one's out-group.[24] So, whereas identity-protective reasoning primarily concerns laypersons' inclination to form judgments in a manner that reflects the core values of their in-group, the social dynamics go beyond this tendency. For example, a scientific expert testifier may be regarded as an out-group member by recipients who are not well-educated or uphold social values that are incongruous with those values that scientists are perceived to uphold. If scientist experts are regarded as out-group members, it may affect judgments of their trustworthiness independently of how the *content* of the testimony relates to the recipients' core values. Conversely, if a scientific expert testifier is regarded as an in-group member, her trustworthiness may be overestimated even though the content of the testimony does not align with the relevant group values. These over- and underestimation effects provide important challenges for public scientific testimony insofar as some social groups may regard scientists or even science journalists as belonging to an elitist out-group.

Research on social stereotypes suggests a similar picture. It indicates that laypersons rely on cost-effective but systematically fallible heuristics and imperfect social stereotypes in categorizing each other.[25] With regard to science communication, an important dimension of this is that social categorization in terms of gender, race, and age is interwoven with *cognitive* traits such as competence or trustworthiness (Porter et al. 2008; Rule et al. 2013; Todorov et al. 2015).

Spaulding sums up this research as follows: "Simply in virtue of being part of particular social category we may upgrade or downgrade a person's knowledge or competence" (Spaulding 2016: 436; see also Gerken 2017a: 104 on knowledge-stereotypical properties). Thus, a broad lesson from the research on social cognition is that laypersons may overestimate someone's epistemic position if she is categorized as a member of their in-group or as belonging to a social category associated with epistemic competence. Likewise, laypersons may underestimate someone's epistemic position if she is categorized as a member of an out-group or as member of a social group associated with epistemic incompetence.

[24] Brewer 2001; Balliet et al. 2014; Spaulding 2018; O'Connor and Weatherall 2019.
[25] Hassin et al. 2005; Bargh 2007; Uleman et al. 2008; Ames et al. 2012; Carter and Phillips 2017.

I articulate these assumptions as a pair of principles (following Gerken 2022):

> ***Epistemic Overestimation***
>
> Both accurate and inaccurate social stereotypes may lead evaluators to overestimate a subject's epistemic position.
>
> ***Epistemic Underestimation***
>
> Both accurate and inaccurate social stereotypes may lead evaluators to underestimate a subject's epistemic position.

Epistemic Overestimation and *Epistemic Underestimation* concern laypersons' assessment of agents' epistemic position. Hence, they bear on laypersons' uptake of public expert testimony and they may have problematic ramifications in terms of epistemic injustice (Fricker 2007; Gerken 2021).

It bears mentioning that social psychology is one of the fields that face methodological challenges insofar as a number of signature studies have not been replicated and experimental paradigms—notably the *Implicit Association Test*—have been criticized as falling short of proper scientific standards. Consequently, there is currently reason to be cautious about specific hypotheses and findings in social psychology (but see Bird forthcoming). However, the mentioned over- and underestimation effects of social stereotypes and in-group/out-group dynamics are based on converging strands of evidence. So, although future research may require qualifications to *Epistemic Overestimation* and *Epistemic Underestimation*, I take these principles to be reasonable working hypotheses.

5.3.f Concluding remarks on laypersons' uptake of public scientific testimony: In contrast to the at times myopic focus on varieties of motivated cognition, it is important to consider a broad spectrum of folk epistemological biases relevant for the uptake of public scientific testimony. Consequently, I have highlighted the putative ramifications of salient alternative effects, and, more generally, epistemic qualification effects for public scientific testimony (Parker 2014; van der Bles et al. 2019, 2020). Likewise, I have noted problems with folk associations of scientific justification and crucial experiments as well as the problematic folk descendent of the *Nullius in verba* tradition. Finally, I have noted aspects of social cognition, such as stereotype effects and in-group/out-group dynamics, that may lead to over- and underestimation effects on laypersons' epistemic assessment of public scientific testimony.

Generally, a lot of trouble derives from the fact that we extend folk epistemological heuristics that may be apt for ordinary interactions to our assessments of sources of scientific knowledge (Gerken 2017a; O'Connor and Weatherall 2019). With this key insight about the folk epistemology of public scientific testimony in mind, we are better equipped to return to the issue of norms and guidelines for scientific expert testimony.

5.4 Norms and Guidelines for Expert Scientific Testifiers

In this section, I will articulate some of the norms that govern scientific expert testimony. To get started, I briefly return to the framework for normativity within which I will work.

5.4.a Norms and guidelines: As noted (in Chapter 2.4.c), norms are constitutively characterized in terms of a *standard*—the telos that the norm is responsive to. For example, epistemic norms are characterized in terms of truth. While scientific expert testimony has diverse aims and roles, truth remains a fundamental standard of epistemic norms of science. However, a more specific aim of public scientific testimony is that recipients' beliefs or credences in scientific hypotheses align with the strength of the scientific justification, and that this is due to an appreciation of science's epistemic authority. The latter conjunct expresses a *basing constraint*, which may impose presentational constraints on public scientific testimony. For example, it is widely accepted that scientific expert testimony should be simplified in a manner such that it is comprehensible to the relevant audience. This constraint is also related to the standard of truth. If the audience is incapable of comprehending the content of the testimony, they are typically incapable of forming the corresponding belief, and, a fortiori, of forming a true belief (Keohane et al. 2014). So, presentational constraints on scientific expert testimony may be motivated within a veritistic framework. With this point in mind, let us revisit the distinction between norms and guidelines (Chapters 1.5 and 2.4.c–d).

Whereas norms are objective benchmarks of assessment that the testifier may lack cognitive access to, guidelines are met only if they are *followed* by the testifier. Hence, guidelines may be thought of as rules of thumb that approximate the relevant norms. Consequently, I will first consider some epistemic and presentational norms and subsequently discuss whether they may perform double duty as a guideline or whether a guideline may be developed on the basis of them. The norms, and especially guidelines that I will articulate, have an ameliorative aspect insofar as they are contingent but principled recommendations for public scientific testimony that reflect its epistemic roles. But although I assume that there is a difference between the operative social norms of public scientific testimony and the epistemic norms that I will promote, the ameliorative project is not revolutionary. Rather, the epistemic norms may be taken as idealizations of the *best practices* in public scientific testimony. Central parts of the idealization consist in abstracting away from the aspects of the operative practice that reflect non-epistemic aims of science. Likewise, the idealization consists in abstracting away from epistemic compromises dictated by the media platform, the target recipients, the communication context, etc.

My endeavor to articulate norms and guidelines heeds a broad *pro tanto* methodological assumption that public scientific testimony should not be at odds with the nature of the relevant science but rather reflect it if feasible. This methodological assumption may be motivated, in part, by reflection on the roles of public scientific testimony (Burns et al. 2003; Keohane et al. 2014). Recall that one of public scientific testimony's central roles is to help recipients align their beliefs with the scientific justification, and that this is, according to the *basing constraint*, due to an appreciation of science's epistemic authority.

Moreover, scientific expert testimony often represents the scientific enterprise more generally, and it may serve a role of improving scientific literacy. Everything else being equal, these roles are better fulfilled if public scientific testimony reflects the underlying scientific practice than if it is at odds with it. For example, it is problematic if a scientific expert presents early inconclusive evidence for a novel hypothesis as if it provided evidence that conclusively established the hypothesis as scientific knowledge. Likewise, it is problematic if scientific expert testimony presents a hypothesis that is the subject of genuine scientific expert disagreement as if there is scientific consensus about it. These examples of public scientific testimony are problematic because they are at odds with the underlying science rather than reflective of it.

5.4.b An epistemic justification norm: Given the assumption that testimony is an assertive speech act, so is scientific expert testimony. So, given the further assumptions that assertive speech acts are governed by epistemic norms, so is scientific expert testimony.

Reflection on scientific expert testimony provides a rationale for assuming that its epistemic norm is structurally similar to the epistemic norm of intra-scientific testimony, NIST, according to which the context of the testimony determines how strong scientific justification is required (Chapter 4). Cases of unqualified scientific expert testimony that *p* are apt to illustrate this point. Consider a variety of the case *WIND SPEED* (from Chapter 4.2.b) in which the meteorologist, Anita, provides, on the basis of a hunch, the following scientific expert testimony to the public: "The storm will have wind gusts with a maximum speed of 100 km/h when it makes landfall." Clearly, Anita's hunch provides inadequate warrant for her scientific expert testimony. Since laypersons take such scientific expert testimony as authoritative, they might fail to take the appropriate precautionary measures.

Moreover, there is good reason to take the epistemic norm of scientific expert testimony to be a *sliding threshold* norm according to which the threshold degree of required warrant varies with the science communication context. The determiners of science communication context include stakes, availability of evidence, urgency, and the opportunity for qualified assertions (cf. Chapter 4.2.b). Another central parameter is the social role that the scientist plays in the science communication context (Khalifa and Millson 2020). In some contexts, action will be dictated by her scientific expert testimony, and in others it merely provides informal input to a public deliberation. Everything else being equal, the former science communication context requires better warrant than the latter.

Finally, it is reasonable to suppose that the kind of warrant required for scientific expert testimony is scientific justification. Recall, from Chapter 3.5, *Hallmark III*, according to which scientific justification is characteristically discursive justification, and the assumption that scientific testimony is properly based on scientific justification. Given these assumptions, it is natural to take scientific expert testimony to require scientific justification. Given the special authority of scientific expert testimony in public discourse, it is important that it be distinguished from non-scientific testimony by a feature that makes it epistemically authoritative. Scientific justification delivers on this account (cf. Chapter 3.3). Conjoining the assumption that scientific expert testimony must be properly based on scientific justification with the assumption that the epistemic threshold is contextually determined suggests an epistemic norm. I will call it the *Norm of Expert Scientific Testimony* (or "NEST" for short):

> **NEST**
>
> In a science communication context, SCC, in which S's scientific expert testimony that *p* conveys that *p*, S meets the epistemic conditions on appropriate scientific expert testimony that *p* only if S's scientific expert testimony is properly based on a degree of scientific justification for believing or accepting that *p* that is adequate relative to SCC.

Simplified, NEST is the claim that *the context determines the degree of scientific justification that scientific expert testimony must be properly based on*. NEST only concerns the epistemic conditions for appropriate scientific expert testimony. So, a scientist may violate prudential, conventional, or ethical norms although she meets NEST. Apart from the presentational norms that I will develop below, I will not say an awful lot about these norms. However, it is important to note that there may be conflicts, and even dilemmas, between epistemic and ethical desiderata (Gerken forthcoming a).

NEST explains why scientific experts are standardly expected to be able to articulate or point to some scientific justification for what they testify and why context determines how strong this scientific justification must be. So, I take NEST to be plausible *qua* epistemic *norm*. It less likely that it can also serve as a *guideline* for scientific experts in the public arena. As stated, it is a fairly complex philosophical thesis which invokes concepts and distinctions that scientists may not be sufficiently familiar with. For example, scientists may not have the conceptual distinction between epistemic conditions and overall conditions readily available in a manner required for following NEST as a guideline. Moreover, it may be a hard judgment for scientific expert testifiers to determine whether their testimony is properly based on an adequate degree of scientific justification. However, cruder approximations to NEST, such as "don't assert anything stronger than you have scientific justification for," may serve as guidelines for scientific expert testifiers.

A limitation of NEST is that it concerns individual scientific expert testifiers. So, it does not speak to group scientific expert testimony, and an extension to groups might require qualifications (Croce 2019a). But I conjecture that such a qualified epistemic norm of group scientific expert testimony may be modeled on NEST.

However, reflection on group scientific testimony raises the issue of what it takes for any public scientific testimony to be *properly based* on the required degree of scientific justification. Questions concerning the epistemic basing relation are notoriously thorny. However, an answer may be approached by noting that it lies between two extremes. Given that scientific collaboration often yields distributed justification, it would be too extreme to require that the testifier possesses every aspect of the required scientific justification. On the other hand, the mere presence of contextually adequate scientific justification in the scientific community will not by itself suffice if the testifier has no clue that this is so. The failure of these extreme views suggests that both the nature of the scientific justification and the science communication context partly specify the required basing relation. So, the testifier may fulfill the basing relation in importantly different ways. For example, she may fulfill it by possessing enough of the scientific justification herself. But she may also possess it by some familiarity with the nature of the existing scientific justification in the scientific community. However, it is debatable whether S may testify *qua* scientific expert *merely* on the basis of higher-order justification that there is adequate scientific justification. In such a case, the scientist straddles expert scientific testimony and science reporting. Below, I consider how a scientist may qualify her testimony when the relevant scientific justification lies outside her domain of epistemic expertise (Chapter 5.5).

More generally, in cases where an outright testimony that *p* would violate NEST, scientific experts can qualify their assertion in various ways. In cases where they only have higher-order justification, they may accompany their testimony with an indication of the nature and strength of the first-order scientific justification. Or so I will argue.

5.4.c A presentational justification norm: The norms governing public scientific testimony, and hence scientific expert testimony, are not exhausted by the epistemic norms. Since public scientific testimony is a public speech act, there is good reason to suppose that it is governed by presentational norms. Indeed, this assumption underlies much work in the science of science communication

(Keohane et al. 2014; Kahan 2015a; see also Keren 2018). For example, it is commonly accepted that expert scientific testimony must be simplified as to be *comprehensible* to the target audience.

Communication strategies for public scientific testimony have not always been explicitly conceptualized as norms or guidelines. So, this is an area in which philosophy may contribute by providing relevant distinctions between, for example, standards, norms, and guidelines (Whyte and Crease 2010; Nickel 2013; John 2018). Recall that public scientific testimony should facilitate that the public may align their beliefs with the scientific justification for scientific hypotheses due to appreciation of the epistemic force of science. Given this aim, normative requirements on presentation go beyond comprehensibility. This assumption is reinforced by the desideratum that public scientific testimony helps increase the public audience's understanding of the relevant scientific hypothesis and general scientific literacy (Huxster et al. 2018; Drummond and Fischhoff 2017b; Slater et al. 2019).

Consequently, I will articulate a norm that may further both the epistemic standard of truth and the idea that the recipients should appreciate the epistemic force of the science on which the testimony is based. I begin with a general norm for public scientific testimony. This general norm has species that apply to scientific expert testimony (this chapter) and science reporting (next chapter). I call the general norm for public scientific testimony the *Justification Explication Norm* (or "JEN" for short):

> ***Justification Explication Norm (JEN)***
>
> Public scientific testifiers should, whenever feasible, include appropriate aspects of the nature and strength of scientific justification, or lack thereof, for the scientific hypothesis in question.

Since this is a general norm of public scientific testimony and this chapter concerns scientific expert testimony, I will focus on a species of JEN that specifically concerns scientific experts. I call this norm *Justification Expert Testimony* (or, in acronym, "JET"):

> ***Justification Expert Testimony (JET)***
>
> Scientific expert testifiers should, whenever feasible, include appropriate aspects of the nature and strength of scientific justification, or lack thereof, for the scientific hypothesis in question.

In Chapter 6, I will argue that another species of JEN applies to science reporting, but for now I will consider the species that applies to scientific expert testimony, JET, and its ramifications. Specifically, I will, in Chapter 5.5, invoke JET to develop and motivate a guideline for scientific expert testifiers. Figure 5.1 gives an overview of what I will cover in this chapter:

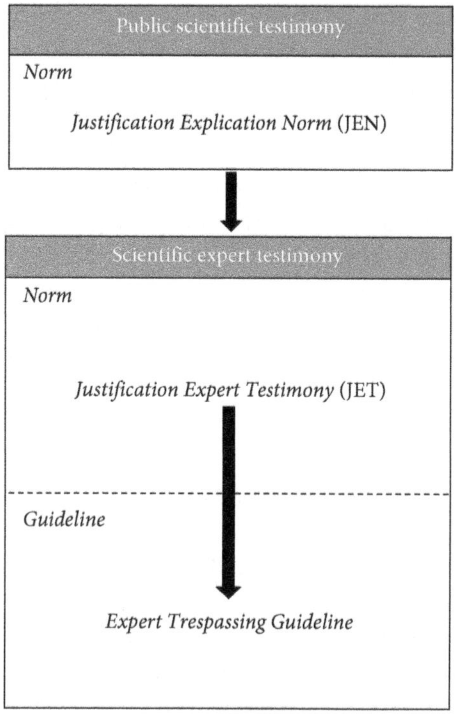

Figure 5.1 Norms for expert scientific testimony

So, the species of JEN that applies to scientific expert testifiers, JET, calls for explicating aspects of scientific justification. Naturally, the aspects of scientific justification that are appropriate to explicate varies with contextual features such as platform, intended recipients, etc. So, I will unpack JET by discussing some aspects of the nature and strength of scientific justification that are appropriate to include in various contexts of science communication.

One reason why explicating aspects of the scientific justification is valuable is that doing so generally furthers the aims and roles of scientific expert testimony. As argued in Chapter 3.5.c, it is truth-conducive for a group to possess discursive justification. But as the principle *Non-Inheritance of Scientific Justification* has it, scientific testimony does not by default transmit the kind or degree of scientific justification that the scientific expert testifier possesses (Chapter 2.2.b). So, merely testifying that *p* may not yield anything more than entitled belief among the recipients. In contrast, explicating the scientific justification for *p* may convey some degree of discursive justification. So, given that discursive justification is truth-conducive, explicating scientific justification may improve the epistemic position of the receiving public. Moreover, explicating scientific justification may facilitate that the public's beliefs align with the relevant scientific justification in accordance with the *basing constraint* requiring that uptake of scientific testimony should reflect some appreciation of its epistemic force.

However, JET is compatible with the basic task of facilitating entitled testimonial belief about the hypothesis in question. In many contexts, lay recipients do not acquire scientific justification from a simplified indication of the nature and strength of the relevant scientific justification. They may merely acquire a basic entitlement for the relevant hypothesis, and perhaps an appreciation that it is based on a scientific source that it is reasonable to defer to. Thus, JET does not compete with the ambition of providing laypersons with *basic* enlightenment. Yet, the best way to generate testimonial entitlement is sometimes to indicate the nature of one's justification even though the recipient will not acquire a similar justification. In some cases, a layperson recipient acquires testimonial warrant by appreciating that the testimony is based on scientific justification without appreciating any details of it. In such cases, I will say that the recipient requires *entitlement through appreciative deference*. I will elaborate on this idea in Chapter 7.4.c.

The further aims of generating public understanding of specific scientific hypotheses and general scientific methods are also well served by explicating the scientific justification for the relevant hypotheses—at least when this is done in accordance with a comprehensibility requirement. A potential long-term effect may be the sustaining, and even increasing, of public trust in scientific testimony.[26] Moreover, some empirical research suggests that explicating scientific justification is appreciable by lay recipients even on singular occasions (see Gerken 2020c, Chapter 6.3.c in this volume).

I am optimistic that JET—or an appropriately simplified version of it—may serve as a guideline for scientific expert testifiers. If so, the principle may help to curb problems of exaggerating, hyping, and overselling research. Given the incentive structure that scientists are subject to, they may be tempted to use public scientific testimony to garner visibility, and this may lead them to leave out important qualifications. JET counterweighs such sensationalist tendencies. Of course, it is important not to oversell JET's capacity to diminish overselling. Plenty of ways in which scientific experts may oversell their research remain. But if scientists are expected to indicate at least aspects of the nature and strength of the justification for a scientific hypothesis, it will be trickier to hype it disproportionally.

[26] See Hawley 2012 for a similar view and John 2018 for an opposing one.

JET is articulated with a feasibility clause, and the specification of it helps to indicate the contextually appropriate aspects of the nature and strength of scientific justification. There are two main reasons for the feasibility clause. First, the expert scientific testifier may not possess all the relevant scientific justification, and second, the science communication context might hamper the explication of scientific justification. The former obstacle may often be cleared since scientific testimony is discursive, and scientific expert testifiers in the relevant domain will therefore often be able to explicate its basic nature and strength. However, in cases of collaborative science, which involves distributed scientific justification, the scientific expert who is the spokesperson for the research group may only possess parts of the relevant scientific justification.[27] But even in such cases, the feasibility clause does not immediately kick in. A collaborating scientist is—with exceptions that may occur in radically collaborative research—often in a good position to acquire a basic grasp of the scientific justification.

Moreover, the requirement is deliberately articulated in terms of the strength and nature of the relevant scientific research rather than in terms of its minute details. For example, a biomedical scientist might explain, in an appropriately simplified manner, that the evidence for the finding is a randomized controlled study and indicate why this provides fairly strong scientific justification. Likewise, a scientific expert may indicate the strength of scientific justification in a justificatory hierarchy (Berlin and Golub 2014; Murad et al. 2016). For example, scientists may clarify that the justification derives from a single study and not a more reliable meta-analysis. In other contexts, it is appropriate to convey the strength of scientific justification in a less technical manner. For example, in a CNN report on findings further debunking the MMR vaccine-autism link, a medical expert is quoted: "At this point, you've had 17 previous studies done in seven countries, three different continents, involving hundreds of thousands of children" (CNN 2019).

Likewise, scientists may be clear when their evidence is weak or entirely inconclusive. For example, a story about the risks of (American) football published the week of the 2020 Super Bowl notes that studies indicate neurophysiological alternations in football players but that the evidence concerning the relationship of such alternations to cognitive and behavioral symptoms is much weaker (Wetsman 2020). In this context, a scientific expert in sports medicine is cited as indicating lack of scientific justification as follows: "We're quite a ways off from being able to look at imaging showing persistent changes without any symptoms and say why that matters" (Wetsman 2020). Similarly, in March 2020, President Trump touted hydroxychloroquine as a treatment for coronavirus, to which a lead member of the White House Coronavirus Task Force,

[27] Winsberg et al. 2014; de Ridder 2014a; Faulkner 2018; Huebner et al. 2018.

Dr. Fauci, responded: "The information that you're referring to specifically is antecdotal. It was not done in a controlled clinical trial. So you really can't make any definitive statement about it" (ABC News 2020). This statement clearly distinguished the epistemic strength of anecdotal evidence and clinical trials. Moreover, it was accompanied by an explanation of the nature of a clinical trial in laypersons' terms.

These are but a few examples of how scientific expert testifiers may indicate contextually appropriate aspects of the nature and strength of scientific justification. Thus, JET is true to the nature of science, or, at least, to the scientific justification underlying the reported hypothesis. The required grasp of the nature and strength of the relevant scientific justification is something that a scientific expert who agrees to testify is capable of acquiring in many cases—even cases of collaborative science. Thus, JET sets forth a non-trivial but typically feasible requirement on scientific expert testifiers.

The second obstacle to articulating the strength and nature of scientific justification concerns the science communication context. The media platforms for science communication operate in an attention economy that generally calls for simplicity and drama. Just as 1980s rock bands did not opt for titles such as *Seek and Revise*, *The Penultimate Countdown*, or *Welcome to the Subtropical Forest*, media platforms tend to opt for drama and simplicity in their science coverage (Valenti 2000; Figdor 2018). Insofar as scientific expert testifiers seek to contribute to widely visited general media platforms, they have to deal with their news criteria and formats. This puts constraints on the mode of presentation. But, as above, it would be mistaken to assume that the feasibility clause *easily* overrides the normative requirement of presenting the nature and strength of the relevant scientific justification. Even in *infotainment* formats such as TED talks, scientific justification is standardly explicated. For example, in a TED talk with more than three million views, primatologist de Waal supports the claim that being an alpha male is stressful by presenting evidence from glucorticoid levels (a stress indicator) in feces from baboons (de Waal 2017). Likewise, he supports the conclusion that successful alpha males are empathetic with evidence that chimpanzee alpha males score higher than any other group members in terms of consoling distressed group members (de Waal 2017). Part of what makes it appropriate to present this scientific justification is that it provides a good narrative for presenting scientific hypotheses. This is not to deny that it may be infeasible to present aspects of the scientific justification. Yet, scientific expert testifiers who fail to articulate any aspects of the nature and strength of the relevant scientific justification often fail (perhaps blamelessly) in meeting the normative ideal. Recall that I do not purport that norms such as JEN and JET align perfectly with the operative social norms in science communication. However, the fact that much of the best scientific expert testimony is in accordance with JET provides some reassurance that it is not an overly demanding normative ideal.

As mentioned, I will, in Chapter 6, argue that JET is but one species of the general norm, JEN, insofar as another species of JEN governs science reporting. So, I will revisit the question feasibility once the discussion has been extended from scientific expert testimony to journalistic science reporting. The same is true with regard to addressing the empirical evidence for and against JET. The norm is best discussed in comparison with alternatives, given that most of the relevant empirical research has concerned science reporting. So, I postpone a detailed empirical defense of JET to Chapter 6.3–4. However, it is worth noting—as a preview—that several studies indicate that laypersons' uptake of polarizing scientific hypotheses improves if they are presented alongside the scientific justification for them. Moreover, some large-scale studies found no evidence of backfire effects. Finally, empirical reasons suggest that JET is comparatively robust against psychological factors such as motivated reasoning, epistemic qualification effects, source bias, in-group/out-group cognition, etc. I return to these empirical considerations in Chapter 6.3–4.

5.4.d Concluding remarks on norms of scientific expert testimony: The norms of scientific expert testimony that I have developed are characteristic in that they require scientific justification. Being subject to such a normative requirement is part of what makes a given piece of testimony a piece of scientific testimony (cf. Chapter 3.1). As noted, a scientist who does not meet the requirement, set forth by NEST, for providing an outright testimony that *p* should, in general, either avoid testifying altogether or qualify their testimony. One type of qualification is naturally included by testifying in accordance with the general norm of public scientific testimony, JEN, which calls for explication of the nature and strength of the relevant scientific justification. Moreover, the species of JEN that concerns scientific expert testifiers, JET, has it that scientists are also subject to this requirement when testifying to the public *qua* scientific experts. In other cases, different types of qualification are called for. For example, the scientists may, in some cases, be required to indicate the limits of their expertise as a qualification to their testimony. I will now turn to such cases.

5.5 Scientific Expert Trespassing Testimony

There are qualifications that scientific expert testifiers should include when they testify about a domain of epistemic expertise that they are not epistemic experts in. This type of testimony may be called *scientific expert trespassing testimony* (Hardwig 1994; Gerken 2018a; Ballantyne 2019). In this section, I briefly revisit the issue in order to focus more on the role of scientific expert testimony in society.

5.5.a Scientific expert trespassing testimony: In the context of public interviews, etc., it is not uncommon that a scientific expert in one domain provides testimony about another domain that she is not an expert in (see Gerken 2018a; Ballantyne 2019 for cases). I provide the following characterization of the phenomenon—labeled "expert trespassing testimony" (Gerken 2018a):

Expert Trespassing Testimony

S's testimony that p is expert trespassing testimony iff
(i) S is an expert in D1 where D1 is a domain of expertise.
(ii) S is not an expert in D2 where D2 is a purported domain of expertise.
(iii) $p \notin$ D1.
(iv) $p \in$ D2.

According to (i) and (ii), expert trespassing testimony is something that only an epistemic expert in some domain can perform. This helps distinguish it from more general phenomena such as testimony on false authority. Clause (iii) specifies expert trespassing testimony as concerning a proposition outside one's domain of epistemic expertise, whereas clause (iv) helps avoid classifying *every* testimony outside S's domain of expertise as expert trespassing testimony. A paleontologist testifying that the library is open is not engaged in scientific expert trespassing testimony.

The characterization of expert trespassing testimony allows for a derivative characterization of the context in which it occurs:

Expert Trespassing Context

S's conversational context is an expert trespassing context iff a significant subset of the audience is likely or reasonable to regard S's expert trespassing testimony as expert testimony.

Thus, the present conceptualization clarifies that it is largely an audience-independent matter whether S is engaged in trespassing testimony, but largely an audience-dependent matter how problematic its consequences are. Assume, for illustration, that a sociologist, S, enters an astrophysics conference and begins to provide testimony about what happened a long time ago in a galaxy far, far away. In this case, the astrophysicists will quickly recognize that S is a layperson in astrophysics. An audience-dependent account does not characterize S's testimony as scientific trespassing testimony. But this seems wrong, and the present account provides a more reasonable diagnosis: S is providing scientific expert trespassing testimony but not in an expert trespassing context.

Before moving on, let me flag some restrictions: I will focus on cases in which the D1 expert testifies in another domain, D2, in which she has no expertise derivable from D1 or otherwise. This is in order to set aside subtle cases of general or transferable scientific competences. Moreover, I will focus on cases in which the expert is superior to laypersons as well as objectively reliable with regard to D1 but *on a par* with laypersons with regard to D2 (for elaboration, see Gerken 2018a).

5.5.b Troubles with scientific expert trespassing testimony: Scientific expert trespassing testimony can be epistemically problematic. In virtue of being epistemically problematic, it can be morally problematic. Thus, it makes sense to start with the epistemic troubles.

Scientific expert trespassing testimony may be epistemically problematic insofar as it may put the recipient in an *epistemically inhospitable circumstance*. These are, roughly, circumstances which somehow undermine S's ability to form truth-conducive beliefs. For example, a usually reliable person may assure me that he will vote for the dean's proposal although he plans to vote against it. This puts me in an epistemically inhospitable circumstance with regard to forming belief about his vote. To work with a generic notion of epistemically inhospitable circumstances, I will only assume that it is epistemically externalist in the following sense: S can be in an epistemically inhospitable circumstance even though the cognitive practice she engages in is, from her own perspective, epistemically reasonable (Gerken 2013a, 2013b).

Scientific expert trespassing testimony is epistemically problematic because it may put the recipient in epistemically inhospitable circumstances in virtue of being likely, and even reasonable, to give the testimony *disproportionate weight*. This concern may be put as a general argument (Gerken 2018a):

Argument from Disproportionate Weight

G1: If H is in a circumstance in which it is epistemically rational to give disproportionate weight to S's testimony that *p*, H is in a circumstance that is normally epistemically inhospitable with regard to *p*.

G2: In many expert trespassing contexts, H is in a circumstance in which it is epistemically rational to give disproportionate weight to S's testimony that *p*.

G3: In many expert trespassing contexts, H is in a circumstance that is normally epistemically inhospitable with regard to *p*.

G1 is a general epistemological thesis reflecting the idea that epistemically inhospitable circumstances may be such that a belief that is epistemically rational from a global perspective is not locally truth-conducive (Gerken 2013b, 2018a). G2 is motivated by cases in which an audience is reasonable to give trespassing testimony the weight of genuine scientific expert testimony (Gerken 2018a). Note that the epistemological sub-conclusion, G3, is the basis for the ethical assessment and the expert guideline that I will propose (see Gerken 2018a for further motivation).

However, the sub-conclusion that trespassing testimony is often epistemically problematic must be juxtaposed with the recognition that it is not invariably problematic. Since trespassing testimony may be properly based, its epistemic benefits may in some cases outweigh its epistemic costs. An important example involves intra-scientific trespassing testimony that is properly based on deference to a scientific expert. Consider S, who is an expert in D1 but not D2. On the basis of the testimony from S*, who S recognizes as an expert in D2, S believes that *p*, which is a member of D2 but not D1. If S goes on to provide intra-scientific testimony that *p* to his research group, he is providing intra-scientific trespassing testimony. However, since it is properly based, it may in some collaborative contexts be epistemically permissible to do so (Gerken Ms a). If intra-scientific trespassing testimony was never permitted, much scientific collaboration would be infeasible. If scientists could provide intra-scientific testimony that *p* only if they mastered the relevant scientific justification for *p*, the fine-grained division of cognitive labor that is central to the epistemic force of science would be severely hampered. Similarly, much public scientific testimony would be infeasible, if properly based trespassing testimony was never permissible (Gerken Ms a). This is not to say that properly based trespassing testimony is never epistemically problematic. For example, it may be reasonable to expect negative consequences if S cannot defend *p* against challenges by articulating the scientific justification for it. Assume, for example, that *p* concerns climate science and that lobbyists are expected to raise specific challenges to it in a setting where the lack of an immediate rebuttal will lead to widespread climate science skepticism. In such a context, S may not be in a good enough epistemic position to give trespassing testimony that *p* even though he bases it properly on genuine scientific expert testimony. Specifically, the testimony may require more discursive justification than S has in such a context. Thus, it is a highly contextual matter when intra-scientific testimony is epistemically problematic and the same goes for public scientific trespassing testimony. Given this qualification, I will consider the ethical problems of scientific expert trespassing testimony.

The idea that scientific expert trespassing testimony may be ethically problematic may be motivated in various ways (Gerken 2018a; DiPaolo forthcoming; Satta forthcoming). One way is the *Motivation from Harm*, which appeals to two assumptions. The first is that many of the cases in which expert trespassing testimony puts the hearer in an epistemically inhospitable circumstance could easily have been avoided. The second assumption is that it is often morally problematic to put someone in an epistemically inhospitable circumstance if this could easily have been avoided.

The second assumption requires qualifications to defend in full generality. So, here I will just defend it as it relates to the relevant type of case (Elgin 2001; Franco 2017 provide further perspectives). Assume that S is a scientific expert in astrophysics who testifies that the MMR vaccine causes autism in an expert trespassing context. In this case, S is clearly misleading the layperson audience to form a belief that may lead to harmful actions. I take it to be a desideratum to diagnose such a speech act as morally problematic.

Importantly, scientific expert trespassers may be well-intentioned, and I believe that many (although not all) are. Hence, it is worth distinguishing scientific expert trespassing testimony from posers, charlatans, Frankfurtian bullshitters, and impostors who speak on false authority without good intentions. But although good intentions may exculpate the trespassing scientists in some cases, there are other cases of culpable ignorance in which the scientist *should* have recognized that his layperson testimony would be mistakenly regarded as scientific expert testimony (see Gerken 2018b for a case and further explication of the *Motivation from Harm*).

A different motivation—the *Motivation from Trust*—trades on the idea that societies that pursue the ideals of deliberative democracies depend on scientific expert testimony and on public trust in scientific experts (Hawley 2012; Turner 2001; Wilholt 2013). But this requires that it is feasible for laypersons to identify scientific expert testifiers (Goldman 2001; Guerrero 2016). If the relevant social roles of expertise are obscured, the *societal* division of labor is compromised. But scientific expert trespassing testimony obscures the distinction between scientific expert testimony and testimony from activists, influencers, and public intellectuals who are not epistemic experts. Hence, it may compromise scientific experts' ability to play their proper role in the societal division of cognitive labor. For instance, it may proliferate *merely apparent* scientific expert disagreement, and this, in turn, may hamper the appeal to science in law-making or public deliberation (Hawley 2012). More generally, expert trespassing testimony may erode public trust in scientific expertise by contributing to the psychological causes of the *Challenge of Selective Uptake.*

5.5.c A guideline for scientific expert trespassing: In light of the characterization of scientific expert trespassing testimony and the diagnosis of the havoc it may wreak, I propose the following guideline for scientific expert testifiers:

> ***Expert Trespassing Guideline***
>
> When S provides expert trespassing testimony in a context where it may likely and/or reasonably be taken to be expert testimony, S should qualify her testimony to indicate that it does not amount to expert testimony.

This *Expert Trespassing Guideline* may be able to serve double duty as both a norm and a guideline for scientific expert testifiers. I label it a guideline in order to facilitate its use as such. Moreover, similar proposals have been set forth both in philosophy and in more official guidelines such as *The Singapore Statement on Research Integrity.*[28]

[28] Hardwig 1994: Resnik and Shamoo 2011. See Gerken 2018b on differences in formulation.

Note that *Expert Trespassing Guideline* reflects both NEST and JET. It reflects NEST in that the trespasser typically lacks the degree of scientific justification required in the contexts in question. Likewise, *Expert Trespassing Guideline* is congenial to JET's demand that the appropriate aspects of the strength and nature of the scientific justification be explicated. Not only do trespassers often fail to meet the relevant epistemic requirements, they also fail to indicate the epistemic status of the testimony.

Expert Trespassing Guideline may be defeated or overridden by a number of circumstances. Likewise, one may be excused from violating it in certain conditions. Clarifying the circumstances in which the guideline may be waived is part of rendering it applicable. So, I will briefly mention a non-exhaustive list of such circumstances (following Gerken 2018b; Ms A). In some science communication contexts, it may be infeasible to make the appropriate qualification. But it is important to not overestimate how difficult it is to meet requirements set forth by *Expert Trespassing Guideline*. A scientific expert may fulfill it by saying "Well—I am not a scientific expert in this area, but I personally believe that *p*." Harder contexts are those in which the qualification may be reasonably expected to have adverse effects (John 2018). Recall, from Chapter 5.3.d, the potential *epistemic qualification effect* according to which epistemic qualification to a testimony may in some contexts make it appear unreliable to laypersons. So, in some contexts, the qualification may leave the hearers in a worse epistemic circumstance than omitting it.

To address this type of conundrum, it is worth generalizing a bit. In some cases, following *Expert Trespassing Guideline* may be reasonably expected to have bad effects, and violating it may be reasonably expected to have good effects. In some such cases, *Expert Trespassing Guideline* may be waived. However, there is a troublesome Faustian element to trespassing in many cases (Goethe 1808/2014; Mann 1947/1999). In cases where the scientist intends to violate the *Expert Trespassing Guideline* for a greater good, she is, in effect, *manipulating* people's beliefs. So, even if the goodness of the end overrides the badness of the means, and the scientist's speech act is, therefore, overall a good one, the means are still bad considered in isolation. Cases in which meeting the *Expert Trespassing Guideline* can reasonably be expected to have bad overall consequences due to *qualification effects* may be diagnosed similarly: The good aspects of abiding by *Expert Trespassing Guideline* remain good although the speech act is not good all things considered.

Other science communicating contexts may provide dilemmas for scientific expert testifiers (Gerken forthcoming a). For example, a scientist may provide false expert trespassing testimony that leads to good consequences in virtue of the audience forming or consolidating a false belief. Likewise, a scientist's ability to provide genuine scientific expert testimony on an important matter may require that she first provides expert trespassing testimony on a related matter. Finally, it should be noted that there are expert trespassing contexts in which qualifying a scientific expert testimony in accordance with *Expert Trespassing Guideline* is not enough to render it appropriate and the expert ought to refrain from testifying on the subject matter altogether. Examples include cases in which formal scientific expert testimony is solicited in a domain distinct from the testifier's domain of scientific expertise. Such cases include many instances of expert witness testimony in courtroom trials (Satta forthcoming).

My aim here is not to diagnose every case but to make a general point for such diagnoses: The good features of meeting the *Expert Trespassing Guideline* and the bad features of violating it should figure in the overall assessment of the testimony. This point may be reinforced by the *Motivation from Trust* according to which scientists' expert trespassing testimony may have grave consequences for the epistemic authority of science. So, immediate good consequences of violating the *Expert Trespassing Guideline* must be considered in relation to the concern that scientific expert trespassing may erode public trust in scientific expert testimony. Bearing this point in mind, let us take stock.

5.5.d Concluding remarks on scientific expert trespassing testimony: The characterization of *Expert Trespassing Testimony* is compatible with assuming that many scientists who engage in it are well-intentioned. Thus, it may help distinguishing the phenomenon from related cases of ill-intentioned or indifferent lookalike experts.

Since expert trespassers are often well-intentioned, they may be receptive to changing their testimonial habits. This indicates the importance of training scientists in science communication. Furthermore, it may be beneficial if scientists acquire what Turner calls a "competence-competence" (Turner 2014: 280). Roughly, this is the meta-competence of being able to reliably determine whether a given task or question falls within one's competence or domain of epistemic expertise (Turner 2014; Gundersen 2018). Given that domains of expertise often overlap, and some expert competences are fairly generic, this is no trivial matter. But the reflections on scientific expert trespassing testimony suggest that it is an important one.

However, it would be mistaken to focus exclusively on the individualistic level and forget the social dimension. Generally speaking, it is a good thing that there are social incentives for scientists to engage in public scientific testimony. But news criteria are driven by the maximization of attention rather than truth, and this may increase scientific expert trespassing testimony. Indeed, scientists may face a dilemma between being unable to provide any scientific expert testimony and trespassing to some degree (Gerken forthcoming a). While there is no simple solution to this challenge, securing platforms in which scientists may provide scientific expert testimony on their own terms is an important task from a societal perspective (I return to this issue in Chapter 7.4).

Another important societal task is to enable a scientific culture with a division of *testimonial* labor in which scientists provide scientific expert testimony within their domain of epistemic expertise and otherwise defer to colleagues or qualify their testimony in accordance with the *Expert Trespassing Guideline.* Note that one may qualify by indicating that other scientists are the proper epistemic experts. For example, a scientist may answer a question as follows: "My view is that *p*, but since I am a sociologist, this is not my field. Consider asking a biologist."

In sum, the challenges associated with scientific expert trespassing testimony are many and complex. So, although *Expert Trespassing Guideline* offers a central component of the response to those challenges, a comprehensive response strategy must take place at both the individual level and at the societal level.

5.6 Concluding Remarks

I began this chapter by arguing that a central role of scientific expert testimony is to facilitate that the public may align their beliefs with the relevant scientific justification due to some appreciation of the epistemic force of science. However, scientific expert testimony faces the *Challenge of Selective Uptake* due to social and psychological mechanisms such as motivated cognition and other less recognized folk epistemological biases and misconceptions about science. These include misconceptions of science, salient alternative effects, epistemic qualification effects, and in-group/out-group dynamics, etc.

Reflection on the challenges for scientific expert testimony led to an epistemic norm thereof, NEST. This epistemic norm is complemented by a general presentational norm of public scientific testimony, *Justification Explication Norm* (JEN), and a species that concerns scientific expert testimony, *Justification Expert Testimony* (JET). On the basis of these norms, I turned to the more concrete problem of scientific expert trespassing testimony, which I distinguished from related forms of speaking on false authority. Inspired by JET, I set forth *Expert Trespassing Guideline*, which may help curb the problem of scientific expert trespassing testimony. However, this guideline is unlikely to address pseudo-scientific testimony. A broad strategy, then, would be to focus on training scientists to minimize expert trespassing testimony and training science reporters to seek to minimize the presence of fake scientists. However, this advice for science reporting must be juxtaposed with further challenges for science reporting. These challenges are the topic of the next chapter.

6

Public Scientific Testimony II

Science Reporting

6.0 Science Reporting and Science in Society

This chapter concerns a second main type of public scientific testimony which plays an important role in society—namely, *science reporting* by people, such as journalists, who are not typically experts themselves. Laypersons acquire much familiarity with scientific hypotheses from science reporting. In consequence, considerable effort has been devoted to articulate strategies for science reporting, and I will contribute to these efforts.

In Section 6.1, I briefly discuss the nature of science reporting, its roles in society, and some of the challenges for it. In Section 6.2, I consider the prospects and problems for some proposed principles for science reporting. In Section 6.3, I set forth an overarching principle of science reporting of my own—*Justification Reporting*. In Section 6.4, I defend it against empirical and philosophical concerns. In Section 6.5, I address the much-debated principle of *Balanced Reporting* and articulate a more restrictive version of it. In Section 6.6, I summarize the key conclusions.

6.1 Science Reporting and the Challenges for It

Scientists are often busier with producing science than with disseminating it to the public. Consequently, laypersons are not primarily informed by scientific expert testimony but by science reporting. Even when scientists themselves provide public scientific testimony, it is often curated, mediated, and shaped by journalistic practices and platforms. Thus, the enlightenment ideal of a scientifically informed public depends in large part on science reporting.

6.1.a Science reporting and third-hand scientific knowledge: The phrase 'science reporting' is often used broadly, for example, to include reporting about the business of science. However, I use it more restrictively to denote a type of *testimony* which is, according to the characterization, *Testimony*, an assertion that *p* which is offered as a ground for belief or acceptance that *p* on its basis (Chapter 2.1.a).

Scientific Testimony: Its roles in science and society. Mikkel Gerken, Oxford University Press. © Mikkel Gerken 2022.
DOI: 10.1093/oso/9780198857273.003.0007

Science reporting differs from scientific expert testimony in terms of the expertise of the testifiers—i.e., reporters such as journalists—who do not typically produce the research that is reported. Occasionally, science reporters are also scientific experts in the relevant domain, and in such cases, scientific expert testimony and science reporting overlap. But often, science reporters possess little or no epistemic expertise and contributory expertise (Chapter 1.2). Hence, a science reporter may be more of a layperson than an epistemic expert in the relevant domain. But insofar as science reporting takes the form of a testimony, that is, it is properly based on scientific justification, it qualifies as scientific testimony according to *Justification Characterization* (Chapter 3.1.d). However, science reporting may be based on scientific justification in a less direct manner than other types of scientific testimony. The fact that a more indirect basing relation may be proper for science reporting is part of what distinguishes it from other types of scientific testimony.

The knowledge that *p* obtained from testimony is often called "second-hand knowledge," and, when one has no other warrant for believing that *p*, "isolated second-hand knowledge" (Fricker 2006a; Lackey 2011; Benton 2016). Derivatively, the knowledge that the recipients of science reporting receive when things go well may be characterized as (isolated) third-hand knowledge. After all, laypersons may acquire it from science reporters who, in turn, acquire it from scientific experts. This type of testimonial knowledge is often nearly isolated since science reporters may only have a single direct source. I write 'nearly isolated' because science reporters typically have warranted background assumptions about the social structure of science and the scientists' place in it. But, in accordance with the principles *Non-Inheritance of Warrant* and *Non-Inheritance of Scientific Justification* (Chapter 2.2.b), the kind of warrant that the science reporter accrues from scientific expert testimony differs from the scientists' warrant. In circumstances where members of the general public lack understanding of science, they may not even be aware that the reported proposition, *p*, is scientifically justified. So, if the recipients acquire knowledge from the science reporting in a minimal background case, they may acquire *isolated* third-hand scientific knowledge. More commonly, perhaps, recipients of science reporting tend to be aware that the report is scientifically based, although they have little idea of the nature of the scientific basis. In many such cases, they acquire entitlement through appreciative deference (Chapter 7.4.c).

6.1.b The diverse roles of science reporting: Science reporting fulfills different roles in society. But it is worth highlighting the overarching enlightenment ideal of disseminating truth. More specifically, this may be the ideal of aligning the public's beliefs and credences with the state-of-the-art scientific justification on the basis of an appreciation of science's epistemic authority. Given this veritistic role, science reporting plays a fundamental role in promoting a scientifically informed public.

However, science reporting is bound by the parameters of the "attention economy" in which it operates: "Science journalism, again in ways typical of other types of journalism, seeks to hang stories on *traditional news pegs*... timeliness, conflict, and novelty" (Dunwoody 2014: 32; Nisbet and Fahy 2015). For example, a good deal of science reporting revolves around practical advice to the recipients. In fact, health and medicine dominate science reporting (Bauer 1998). One study found that 70 percent of science stories in elite US newspapers were health-related (Pellechia 1997). Some brands of science reporting aim to entertain as much as to inform. Such *infotainment* stories tend to involve dramatic topics (mummies, extraterrestrials, pandas) or remarkable discoveries (water on Mars) or health (obesity, sex). So, despite science reporting's central role in disseminating scientific research, the selection of content and the stylistic format of the reporting are determined by a range of factors (Angler 2017). My focus will be on the format.

As a general rule, it is often important to scientists to maximize accuracy, whereas science reporters may be less incentivized to maximize accuracy if it compromises newsworthiness and accessibility. Although visibility has become increasingly important for scientists, the different incentive structures that scientific experts and science reporters are subject to may give rise to conflicts (Figdor 2010; Valenti 2000). In caricature, scientists value narrow and quantitative information conveyed with technical terminology and careful qualifications, whereas journalists value broad and qualitative information conveyed in simple language without qualifications (Valenti 2000). While this picture is a caricature, it contains grains of truth. Moreover, the caricature may influence scientists' conception of journalists and vice versa. Finally, it reflects the institutional structures and incentives that scientists and journalists operate under. For a journalist, it may be less than a few hours between initiating a story and publishing it. But for a scientist, it may be more than a few years between initiating an article and publishing it. So, although differences between scientific experts and science reporters are sometimes exaggerated, their differing incentives and overall roles in society sometimes complicate public scientific testimony.

These complications are closely related to the social and psychological obstacles to public scientific testimony considered in Chapter 5.2–3. So, I will revisit this issue as it relates to the distinctive constraints on science reporting.

6.1.c Laypersons' uptake of science reporting: The psychological and social factors that impact science reporting are similar to those that impact public scientific expert testimony. Consequently, I will rely on the survey from Chapter 5.2–3 and focus on how these socio-psychological factors bear on the distinctive roles of science reporting.

Science reporting also faces the *Challenge of Selective Uptake*, that laypersons who generally accept public scientific testimony reject it when it concerns select scientific hypotheses that are equally well-warranted (Chapter 5.2.b). Moreover, varieties of motivated cognition, such as identity-protective cognition, are central, but not unique, psychological causes of this challenge (Kahan et al. 2011; Kahan et al. 2012; Lewandowsky et al. 2018).

The configuration of platforms for science reporting may amplify some of these effects. For example, many news platforms are aligned with certain political views and reach specific demographics. In the US, political alignment may be the primary mode of categorizing a media outlet (Nisbet and Fahy 2015). Consider the characterization *right-wing radio* or the fact that a political podcast might advertise itself as *progressive*. The fact that media platforms are more or less explicitly tied to social groupings may lead to echo chambers and, perhaps, associated polarization.[1] Relatedly, it may socially reinforce identity-protective reasoning at the individual level and contribute to in-group/out-group dynamics. Specifically, the strength of warrant for a reported hypothesis may be overestimated if the source is regarded as sharing the recipients' social values and underestimated in the opposite case—cf. the folk epistemological principles *Epistemic Overestimation* and *Epistemic Underestimation* (Chapter 5.3.e). Moreover, group membership may be signaled by publicly trusting a source associated with the values of a group and by publicly doubting a source associated with opposing or orthogonal values. So, the fact that science reporting takes place on platforms that align with social demographics may exacerbate the *Challenge of Selective Uptake* in virtue of amplifying cognitive biases and social dynamics (Lazer et al. 2018; Iyengar and Massey 2019).

More generally, a polarized media landscape that is constrained by an attention economy may reinforce some of the folk psychological obstacles to scientific expert testimony. For example, the misconception that it is irrational to rely on science reporting without understanding the underlying issues oneself may be especially damaging in a polarized media landscape with insufficient indicators of reliability of the competing channels. Likewise, the news value of reporting remarkable discoveries may reinforce a tacit misconception of science as proceeding via crucial experiments rather than via gradually accumulating evidence (Kovaka 2019).

In sum: There is a complex interplay between, on the one hand, the media reality that science reporters face and, on the other hand, the psychological biases and social pressures that impact laypersons' uptake of science reporting.

6.1.d Concluding remarks on roles and challenges for science reporting: These introductory remarks should—if nothing else—make it clear that the roles of science reporting and the challenges for fulfilling these roles are extremely

[1] See, for example, Sunstein 2009; Pariser 2011; Frost-Arnold 2014; Rini 2017; Boyd 2018, forthcoming; Alfano and Klein 2019; Iyengar and Massey 2019; Sullivan et al. 2020; Alfano and Sullivan 2021.

complex. In particular, there are serious challenges to science reporting's fundamental role of contributing a reasonable alignment between the best available scientific justification and the public's beliefs. Many of these challenges arise from how the noted psychological biases and social pressures interact with the media landscape. Consequently, it is a desideratum for principles of science communication to minimize the impact of psychological biases and social pressures.

6.2 Some Models of Science Reporting

Science reporting is critical to the well-functioning of societies that pursue the ideals of democracy, but it faces serious challenges. Consequently, recent years have witnessed a wealth of empirical and conceptual research on science communication, much of which has resulted in diagnoses of the challenges and general strategies for science reporting (see, e.g., Jamieson et al. 2017). So, as background for my own proposals, I will start out with a brief tour of this corner of the marketplace of ideas.

6.2.a Deficit models and their limits: A natural thought is that public misunderstandings about science simply result from a lack of information about the relevant scientific hypotheses. In short, when public belief is at odds with science, the reason is a simple information deficit among the public.

This diagnosis of the challenges for science reporting motivates a fairly straightforward communication strategy: Given that the lay public simply lacks information about the relevant scientific hypotheses, all that needs to be communicated is the scientific hypotheses in question. For example, science reporters may simply report that there is anthropogenic global warming (AGW), that the measles, mumps, and rubella (MMR) vaccine does not cause autism, or that social distancing slows the spreading of coronavirus.

A small terminological caveat: In principle, every science communication strategy that seeks to provide information that the audience is expected to lack may, in a broad sense, be labeled a deficit model. So, to avoid classifying every science communication strategy as a deficit model, I use the phrase in a narrower sense to denote science communication strategies of reporting the bare scientific hypothesis. Hence, I articulate the narrow sense of the deficit models as follows:

> *Deficit Reporting*
>
> Science reporters should, whenever feasible, merely report the scientific hypotheses that meet a reasonable epistemic threshold.

Thus, the *Deficit Reporting* principle reflects an optimistic enlightenment picture according to which recipients will rationally update their beliefs once they receive relevant information from an epistemically superior source. *Deficit Reporting* principles differ in how the idea of a *reasonable epistemic threshold* is specified. But these differences will matter little for the most serious criticism of *Deficit Reporting*, which concerns its capacity to address the *Challenge of Selective Uptake*.[2] A central theme in the criticism is the charge that *Deficit Reporting* is not responsive to basic psychological mechanisms involved in laypersons' uptake of public scientific testimony. As noted in Chapter 5.3, motivated cognition and identity-protective cognition are central among these psychological mechanisms. In particular, laypersons' uptake of science communication concerning divisive topics that bear on social values such as political or religious values is likely to be affected by such psychological mechanisms (Kahan 2016, 2017). So, in these cases, it is highly questionable that science reporters simply need to relieve laypersons from an information deficit. Thus, *Deficit Reporting* appears to be deficient as a general science communication strategy. In particular, it appears to be insensitive to the types of cases which manifest the *Challenge of Selective Uptake*.

In consequence, *Deficit Reporting* is widely regarded as reflecting a simplistic—perhaps even naïve—enlightenment ideal according to which the public at large simply needs to be told that a given scientific hypothesis is true in order to update their beliefs accordingly. This makes for a simple science communication strategy but one that is not sensitive to the psychological mechanisms of laypersons' uptake. In particular, it is insensitive to the evidence which suggests that it may be more important to hold on to a false belief that is central to a recipient's social identity than to exchange it for a true one. Such a change in view may be socially costly in the extreme compared to the limited personal gain of improving one's ratio of true to false beliefs. For example, the immediate personal advantage of believing that there is anthropocentric climate change may for some people be miniscule compared to the social costs of adopting this belief. Consequently, human cognizers are not epistemically ideal cognizers in the sense that the pursuit of true belief may be trumped by a range of non-cognitive factors. *Deficit Reporting* is largely insensitive to the epistemically non-ideal aspects of human psychology. So, considered as a general science communication strategy, it is inadequate. In particular, it is not inapt with regard to the types of cases in which social values are at odds with the aim of maximizing true belief and minimizing false ones.

I agree with this standard critique according to which *Deficit Reporting* is not apt to address the troublesome cases and that it is, therefore, not a viable general science communication strategy. But it is worth juxtaposing this reasonable criticism with the point that *Deficit Reporting* may be quite appropriate for many scientific issues. Laypersons tend to bring their umbrella when the meteorologist says that it will rain. More generally, laypersons are not socially or emotionally invested in many

[2] Miller 2001; Sturgis and Allum 2004; Weber and Stern 2011; Keren 2018.

scientific issues, and science reporting about such issues is less likely to face problems of selective uptake. This is not to deny that science reporting in accordance with *Deficit Reporting* will be ineffective in many cases. However, a general assessment should consider both these problem cases and the cases in which the general public is likely to trust a sparse science reporting of the relevant hypotheses. So, while *Deficit Reporting* errs on the side of simplicity as a general science communication strategy, it is also important to avoid erring on the side of overkill in criticizing it. In many cases, most laypersons do not need science reporting that goes beyond what *Deficit Reporting* recommends.

That said, the task remains of articulating a general science communication strategy that may both serve to address basic information deficits and to address the cases that constitute the *Challenge of Selective Uptake*. So, I will now turn to a prominent communication strategy.

6.2.b Consensus reporting: Given that merely reporting the hypothesis, *p*, appears to be inadequate as a science communication strategy, it is natural to ask what science reporters may add. A prominent answer is that they may add information about scientific consensus. One reason for this idea is that scientific consensus that *p* may be taken as a reasonably accurate proxy for a strong scientific justification for *p*. Moreover, it may seem fairly easy for science reporters to communicate consensus information and easy for the layperson recipients to appreciate it. This suggests a strategy that I will call *Consensus Reporting* and articulate as follows:

> ***Consensus Reporting***
>
> Science reporters should, whenever feasible, report the scientific consensus, or lack thereof, for a reported scientific hypothesis.

A good deal of optimism has surrounded *Consensus Reporting*, and some science communication efforts—such as www.consensusproject.com have been centered around it (Kahan 2015b; Merkley 2020).

Consensus Reporting may be thought to address the *Challenge of Selective Uptake* insofar as beliefs that there is scientific consensus that *p* have been argued to be *gateway beliefs* to beliefs that *p* (van der Linden et al. 2015, 2016, 2017; Lewandowsky et al. 2018). The gateway belief hypothesis is partly motivated by empirical studies. The most intensely studied example of *Consensus Reporting* concerns science communication about AGW in the US. This case is important because there is a robust scientific consensus concerning AGW, whereas the US public is divided along political lines on the issue. One line of motivation for *Consensus Reporting* comes from finding an inverse correlation between accepting that AGW is real and denying or underestimating the scientific consensus that this is so (McCright et al. 2013; Hamilton 2016).

On the basis of such correlational studies, proponents of *Consensus Reporting* argue that it is apt to address challenges of science communication such as the *Challenge of Selective Uptake*. The key thought is that if the belief that there is scientific consensus that *p* leads to the belief that *p*, the selective rejection of scientific testimony may be diminished by consensus information. That is, reporting that there is scientific consensus that *p* is thought to ensure that laypersons who generally accept public scientific testimony also accept scientific testimony that *p*.

Proponents of *Consensus Reporting* are, of course, aware of the problems in using correlational studies as evidence for assuming that consensus information *causes* improved reception of science. So, they have sought to show experimentally that exposure to consensus reporting of scientific claims results in a revision of participants' beliefs or credences. Several studies found significantly changed attitudes for a group exposed to consensus reporting compared to a control group (Lewandowsky et al. 2013; van der Linden et al. 2015). Another study found that the public underestimated the degree of scientific consensus and that "receiving the information about scientists' beliefs raises respondents' beliefs that climate change is already underway and that it has been caused by human activity by 6 and 5 percentage points, respectively" (Deryugina and Shurchkov 2016). Similarly, a study by Bolsen and Druckman found that consensus reporting had a positive significant effect on perceived consensus about climate change (Bolsen and Druckman 2018). Although they found no direct effect of consensus reporting on belief about climate change itself, they note that belief about consensus is a strong predictor of belief in climate change. So, they take the results to support the gateway model (Bolsen and Druckman 2018; but see Section 6.2.c for criticism).

Thus, *Consensus Reporting* has emerged as an intuitively appealing and straightforwardly implementable science communication strategy, which has found wide application in science reporting about divisive issues. In particular, *Consensus Reporting* has governed much reporting about climate science (Kahan 2015b; Kovaka 2019).

6.2.c Some limitations of *Consensus Reporting*: While *Consensus Reporting* may be the most deployed strategy for communicating science about controversial issues, it has not gone without criticism. I will add to this criticism. However, my aim is not to refute the idea of consensus reporting, but to argue that it cannot stand alone.

Consider, for a start, the assumption motivating *Consensus Reporting* that scientific consensus tends to be explained by strong scientific justification and that it, therefore, indicates reliability. While there is some truth to this, bare consensus reports are not in general constitutively associated with science or reliability. This point was put starkly in a speech by Crichton: "In science consensus is irrelevant. What is relevant is reproducible results. The greatest scientists in history are great precisely because they broke with the consensus" (Crichton 2003). Crichton's diagnosis may be an overstatement since it overlooks the social aspects of scientific consensus and may even encourage a *great white man fetish* (cf. Chapter 5.3.c). It may also be questioned whether reproducibility is generally necessary for reasonable acceptance (Fidler and Wilcox 2018). However, the history of science contains a number of cases where the consensus rejected an epistemically superior hypothesis. Famous examples include the germ theory of disease (Gaynes 2011) and continental drift theory (Oreskes 1999). So, a less categorical idea in the vein of Crichton's quip is that scientific consensus indicates reliability only if it is properly based on scientific justification.

Another line of criticism of *Consensus Reporting* involves the idea that lack of consensus need not reflect badly on scientific hypotheses and that critical minorities are central to the scientific process.[3] So, focusing science reporting on consensus may delegitimize it in areas with a lack of consensus. As an example of why this is problematic, consider pre-consensus findings concerning AGW. These were legitimately scientific and very much worth reporting. Likewise, viewpoints of certain minorities are worth taking seriously because their challenge to a consensus may be owed to their specific perspective. Generally, requiring scientific consensus for reporting a scientific hypothesis would be an overly radical restriction.

Another motivation for *Consensus Reporting* which should be critically scrutinized is the idea that consensus information is easy for science reporters to provide and easy for lay recipients to appreciate. One concern with these assumptions is an important distinction between consensus that is based on a shared epistemic basis and mere agreement, which may be based on irrational factors (Miller 2013). Determining whether a consensus belongs to the former or latter type can be very demanding, as Miller exemplifies in a case study (Miller 2016). So, sorting out whether a scientific consensus is based on shared or mutually supporting scientific justification requires some familiarity with the relevant scientific justification. Likewise, local disagreement may indicate truth-conduciveness of a wider consensus on related issues (Dellsén 2018b). But untangling local expert disagreement from a wider consensus may require expertise that science reporters lack. Hence, science reporters will sometimes have a hard time determining whether there is wide scientific consensus for a given hypothesis and whether it is held for the right epistemic reasons (Intemann 2017).

Turning to the recipient side, it is unclear that the transition from the belief that there is scientific consensus that *p* to the belief that *p* is as smooth as the gateway belief model suggests. Slater et al. put the point as follows: "It is one thing to be able to recognize cases of scientific consensus. It is quite another to recognize the

[3] Longino 1990, 2002; Beatty and Moore 2010; de Melo-Martín and Intemann 2018.

epistemic significance of such consensus" (Slater et al. 2019: 255).[4] As noted, laypersons may negatively assess scientific consensus due to a *great white man fetish* according to which a lone genius singlehandedly overthrows the dogmatic consensus of the scientific community (Chapter 5.3.c). Some science education may misleadingly focus on mythic versions of such cases (Galileo, Darwin, etc.). For example, it is inaccurate to characterize Galileo's Copernicanism as suppressed by a consensus among astronomers (Machamer 2005). But this has not stopped climate science skeptics such as Texas's governor Perry and Congressman Cruz from invoking the co-called *Galileo gambit* of comparing climate science skeptics to Galileo (Salzberg 2015). But even without organized science denial, the misconception that scientific progress owes to individual geniuses may lead motivated cognizers to resist the step from scientific consensus that *p* to belief that *p* (Kovaka 2019; Slater et al. 2019).

Moreover, some laypersons may take scientific consensus to indicate little more than shared social values of the scientific community (Kahan 2015b). At worst, some laypersons may even take it as evidence of conspiracy among an elite class of scientists (Castanho et al. 2017; see also Van Prooijen and Jostmann 2013). Indeed, one science denialist strategy seeks to exploit this by explicitly recognizing consensus among scientists and appealing to it as evidence of collusion, conspiracy, or contamination of the relevant science by the scientists' values. A notable example is President Trump's claim that climate scientists "have a very big political agenda" (BBC 2018b).

Some of the empirical evidence supports the idea that consensus reporting may contribute to such us-and-them social cognition. Some recipients may see the scientific community as an out-group, and this may trigger folk epistemological heuristics such as *Epistemic Underestimation*, which leads recipients to underestimate the source's epistemic position (Chapter 5.3.c). Relatedly, some of the evidence of backfire effects indicates that they may be generated by consensus reporting. For example, Cook and Lewandowsky found, in a cross-cultural study, that among US participants, consensus reporting generated "belief polarization with a small number of conservatives exhibiting contrary updating" (Cook and Lewandowsky 2016: 177). Specifically, strong believers in free markets *lowered* their acceptance of AGW upon receiving consensus information (Cook and Lewandowsky 2016). Similarly, Bolsen and Druckman compared the effects of consensus reporting to a politicized consensus message and found that doing so nullified the positive and significant effect of consensus messaging (Bolsen and Druckman 2018). Moreover, they found no positive effect of consensus messaging among "high knowledge Republicans" but rather a backfire effect for this group (Bolsen and Druckman 2018). In consequence, they concluded that "*politicizing*

[4] See also Keren 2018; Intemann 2017; de Melo-Martín and Intemann 2018.

science eliminates the positive impact of a consensus message" (Bolsen and Druckman 2018: 394; original italics). The worry is that consensus reporting may inadvertently politicize the science and thereby compromise its own impact.

Further empirical research on backfire effects is required (Wood and Porter 2019). However, the noted findings provide some concerning evidence that reporting scientific consensus is ineffective, and perhaps even counterproductive, in the very cases that it is meant to address—namely, cases of science reporting on divisive issues. If *Consensus Reporting* is primarily effective in conveying information about divisive scientific issues to an agreeing audience, it may be a suboptimal strategy for addressing the *Challenge of Selective Uptake.*

Furthermore, the initial empirical evidence for *Consensus Reporting* has been challenged on empirical grounds (e.g., Kahan 2017; see response by van der Linden et al. 2017). Kahan notes that the control groups in some studies alleged to support *Consensus Reporting* received no relevant information (van der Linden et al. 2015; Deryugina and Shurchkov 2016). Such studies may provide evidence that public scientific testimony that *p* alongside consensus information is more effective than no public scientific testimony that *p*. But such studies do not provide evidence that it is more effective to report consensus about *p* than to report that *p* without consensus information. Evidence for an effect of *Consensus Reporting* requires a control group that receives the same substantive report without the additional consensus information. Moreover, Deryugina and Shurchkov found no effect of *solely* reporting consensus (Deryugina and Shurchkov 2016). What they found was an effect when they added detail to the consensus reporting. But this effect was non-lasting, and they found no effect on beliefs about the necessity of climate policy and no effect on willingness to donate to climate initiatives (Deryugina and Shurchkov 2016). So, the studies that are sometimes taken to support *Consensus Reporting* do not do so in a clear-cut manner.

Moreover, several other studies fail to provide evidence that consensus reporting produces any effect. For example, Dixon and colleagues compared consensus reporting to appeal to social values (cf. Chapter 6.2.e) and found that only the latter had an effect which was not boosted by consensus reporting (Dixon et al. 2017: 526). Although we should be careful to avoid concluding too much from null results, they reinforce the concern that the empirical evidence for the effectiveness of *Consensus Reporting* is far from clear-cut.

6.2.d Concluding remarks on *Consensus Reporting*: The problems with *Consensus Reporting* are not negligible. In particular, it is a concern that *Consensus Reporting* may, to some laypersons, contribute to problematic in-group/out-group social cognition rather than indicating reliability. It is probably an overreaction to conclude that *Consensus Reporting* should not play any role in science communication. But *Consensus Reporting* is inadequate as a self-standing science communication principle. Perhaps it may be improved by clarifying *why* scientific consensus is a strong indicator of reliability. Laypersons may not appreciate that scientific consensus generally occurs only when the scientific justification for a hypothesis is strong. So, on divisive topics, *Consensus Reporting* alone is unlikely to resolve the *Challenge of Selective Uptake* and related challenges.

This raises a question as to how *Consensus Reporting* may be integrated with other forms of science reporting, and I will return to this once I have set forth my own proposal.

6.2.e Value-based reporting: *Value-Based Reporting* is the increasingly prominent idea that science reporting should appeal to the recipients' social values. This idea is motivated by the empirical research indicating that social values affect laypersons' uptake of science reporting. Here is a broad articulation of the idea:

> ***Value-Based Reporting***
>
> Science reporters should, whenever feasible, report a scientific hypothesis in a manner that appeals to the social values of the intended recipients.

The idea of value-based reporting is intertwined with the debates about the value-free ideal of science that I have opted not to thematize in this book (Douglas 2009; Brown 2020). Consequently, I will restrict myself to a few tentative points about *Value-Based Reporting*, recognizing that there is a larger discussion to be had.

The empirical motivation for *Value-Based Reporting* derives from the evidence for identity-protective reasoning surveyed in Chapter 5.3.a–b. For example, backfire effects may be driven by social distance. Social distance may be characterized as the perception or experience of difference from another person or group in terms of group identity, affect, intimacy, familiarity, etc. (Eveland et al. 1999; Boguná et al. 2004; Stephan et al. 2011). One study compared responses to science reporting in high and low social distance conditions (Hart and Nisbet 2012). The "social distance" manipulation concerned whether the social identity of the groups described as facing consequences of climate change was similar to or different from that of the participants. Specifically, the consequences were described as affecting "residents of Upstate New York" (low) and "residents of South of France" (high). In the high social distance condition, there was a backfire effect in terms of a "decreased support among Republicans for climate mitigation policy" (Hart and Nisbet 2012: 716).

Similarly, Zhou found backfire effects among Republicans when climate change reporting was framed in a way that encouraged support for government action or personal engagement against climate change (Zhou 2016). Identity-protective cognition is postulated as an explanation of these findings precisely because they involve cues of social identity (Sherman and Cohen 2006; Kahan 2016, 2017).

In a study comparing principles resembling *Deficit Reporting*, *Consensus Reporting*, and *Value-Based Reporting*, Dixon and colleagues found that the last of these was significantly more effective than the former two (Dixon et al. 2017 following Campbell and Kay 2014). They tested various types of value-based reporting targeting both religious and political views. For example, one condition highlighted a free-market solution to climate change. This had a positive effect on participants who self-identified as "conservative" and "very conservative" compared with a similar testimony with no such messaging—i.e., in accordance with *Deficit Reporting*.

According to *Value-Based Reporting*, science reporting should exploit these effects. For example, climate science reporting should include social identity clues that Republicans can appreciate when they are the intended recipients (Hart and Nisbet 2012). More generally, Hart and Nisbet suggest that science reporters "may be effective by focusing on messages that target specific segments of the public" (Hart and Nisbet 2012: 717). Hart and Nisbet suggest that this may involve utilizing tools such as *audience segmentation analysis* (Maibach et al. 2008). Likewise, Dixon et al. conclude that "emphasizing free market solutions to climate change might be a particularly effective messaging strategy for generating more favorable beliefs among conservative Americans" (Dixon et al. 2017: 529). One strand of such a strategy is labeled *identity affirmation* and consists in showing the target recipient group "that the information in fact supports or is consistent with a conclusion that affirms their cultural values" (Kahan et al. 2011: 169). A different strand is labeled *narrative framing* and consists in "crafting messages to evoke narrative templates that are culturally congenial to target audiences" (Kahan et al. 2017: 170). Finally, *Value-Based Reporting* may be implemented by *pluralistic advocacy*, which consists in ensuring that experts representing diverse social values are among the scientists who are reported as accepting the hypothesis (Kahan et al. 2010, 2011: 169).

Thus, *Value-Based Reporting* may be implemented in different ways. These must be assessed empirically and philosophically in a piecemeal manner. Given that these strands of *Value-Based Reporting* are still being worked out, I restrict myself to some general concerns.

Value-Based Reporting does not seem viable as a general science communication strategy. The different social groups of recipients are different in virtue of diverging, and sometimes downright conflicting, social values. Moreover, there is evidence that both liberals and conservatives are about equally prone to motivated cognition (Kahan 2013; Frimer et al. 2017). Hence, implementing *Value-Based Reporting* in terms of identity affirmation and targeted messaging does not permit for a uniform science reporting to the general public. I take this to be a significant cost of some implementations of *Value-Based Reporting*. Furthermore, *Value-Based Reporting* is not fully general in that it is tailor-made to address "hot topics" that are divisive along social lines. So, unless we want to apply *Value-Based*

Reporting across the board, science reporters also need principles for science reporting on topics that are not socially divisive. But it may be problematic to report divisive and non-divisive issues in radically different ways. One reason for this is that the public might find value-based science reporting less credible if it differs too starkly from ordinary science reporting. Another reason is that *Value-Based Reporting* on select topics may contribute to politicizing science. By framing science reporting in terms of non-cognitive values, one runs the risk of turning a factual question into a value question for recipients who might not (yet) think of it as such.

A related concern is that science reporting should reflect the high epistemic standards of science rather than deteriorate into cheap marketing stunts similar to those invoked to peddle cars and credit cards. Even if it is truth that is marketed, it would be problematic if science communicators in effect tricked people into believing the truth. This would be in conflict with the *basing constraint* according to which the laypersons' alignment of their beliefs to the strength of the scientific justification should be due to some appreciation of the epistemic authority of science (Chapter 5.1.c). Moreover, while *Value-Based Reporting* may be an effective strategy in isolated cases, it is not clear that it is viable as a general strategy. At least, there is a risk that recipients will soon realize that they are a targeted segment and that the reporting is framed to convince them. As a consequence, trust in scientific institutions might dwindle. Radical implementations of *Value-Based Reporting* may, in Whyte and Crease's metaphor, result in poisoning the well of public trust in science (Whyte and Crease 2010). Finally, there is a risk that science reporting caters to people's opinions to a degree where it becomes incapable of critically informing them in a manner that permits for rational change of opinion.

These worries concern more extreme implementations of *Value-Based Reporting* than many of its proponents have in mind. But noting the problems of the most radical implementation of the strategy indicates fundamental concerns that may constrain less radical implementations. The general lesson is that there is a fine line between minimizing biases and exploiting them in an illegitimate manner (Persson, Sahlin, and Wallin 2015). Fortunately, proponents of the strategy are aware of these concerns: "long-term reliance on targeted climate change messaging could negatively impact trust in scientific institutions and authorities involved in strategic messaging. Therefore, it is important to consider that targeting cannot and should not be the sole method for addressing conservative climate change skepticism" (Dixon et al. 2017: 531).

I largely agree with this assessment. The concerns do not show that *Value-Based Reporting* has no place in science reporting. A carefully developed *Value-Based Reporting* strategy may be an important component in science communication—especially in cases that are already politicized. However, the noted concerns provide reasons for regarding it as a secondary communicative strategy to be discerningly invoked or as complementing science reporting rather than being a part of it. Due to the risk of deteriorating into mere marketing, *Value-Based Reporting* should rarely, if ever, be a self-standing mode of science reporting. These sub-conclusions raise the questions: What principle of science reporting should *Value-Based Reporting* be secondary to? What principle of science reporting may stand on its own? I will answer these questions in terms of a positive proposal and then indicate the role that *Value-Based Reporting* may play in it (Section 6.3.d).

6.3 Justification Reporting

In this section, I will set forth a norm of science reporting which I will argue may also serve as a *guideline*, albeit a highly abstract one that requires supplementation for its implementation. The basic idea is that science reporters should not merely communicate the scientific *findings*, but also the strength and nature of the scientific justification for them. Consequently, I call the principle *Justification Reporting.*

6.3.a Introducing justification reporting: Recall that I, in Chapter 5.4.c, advocated a general presentational norm for public scientific testimony, *Justification Explication Norm* (acronymized "JEN"):

> ***Justification Explication Norm (JEN)***
>
> Public scientific testifiers should, whenever feasible, include appropriate aspects of the nature and strength of scientific justification, or lack thereof, for the scientific hypothesis in question.

Whereas I have argued that this general norm has a species applying to scientific expert testifiers, *Justification Explication Testimony* (JET), I will now argue that JEN also has a species that applies to science reporting. I call this species *Justification Reporting* and articulate it as follows:

> ***Justification Reporting***
>
> Science reporters should, whenever feasible, report appropriate aspects of the nature and strength of scientific justification, or lack thereof, for a reported scientific hypothesis.

As in the case of JET, I will argue that *Justification Reporting* may serve as the basis for more concrete guidelines for science reporters and, in particular, guidelines that pertain to the journalistic ideal of balanced reporting. But before initiating the discussion of *Justification Reporting* and its ramifications, it may be helpful to provide a wee map of the relationship between the norms and guidelines (Figure 6.1).

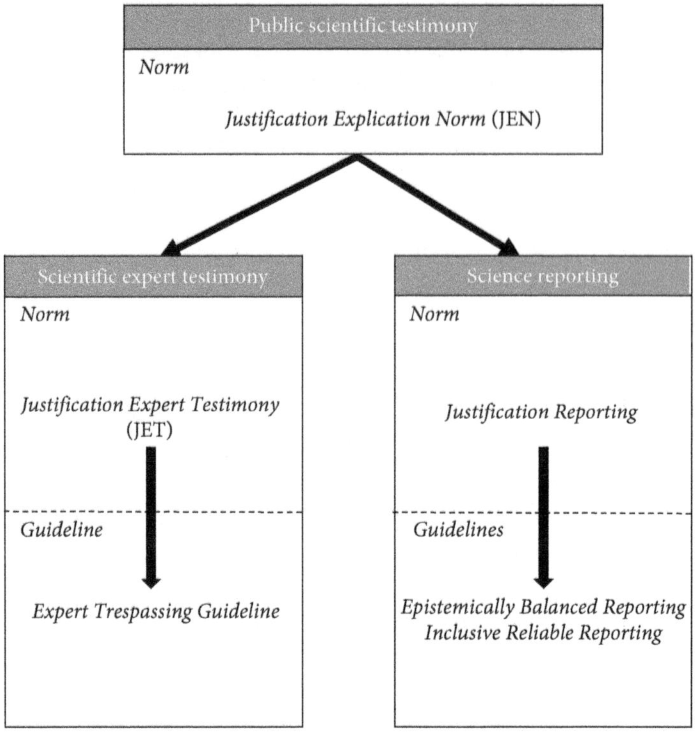

Figure 6.1 Norms for public scientific testimony

With this overview in place, I will unpack *Justification Reporting*. It follows the genus, JEN, in calling for an explication of appropriate aspects of the relevant scientific justification. So, broadly speaking, the principle reflects an enlightenment tradition that values a scientifically informed public. This involves a broad methodological assumption that public scientific testimony should not be at odds with the nature of the relevant science but reflect the nature of what is reported on—namely, scientifically justified hypotheses and theories (cf. Section 5.4.a). I sloganize this idea as *public scientific testimony in the scientific image* (Gerken 2020c).

The qualification 'appropriate' was tacit in my original articulation (Gerken 2020c). This qualification indicates that there is a communicative task in explicating key features of the scientific justification in a manner that may be appreciated by the expected or intended recipients. Consequently, it is highly contextual which aspects of the nature and strength of justification are appropriate to report. But they may include the type of evidence that favors *p* and an indication of the strength of this evidence compared to other types of evidence. I will give some examples as I unfold *Justification Reporting*.

I will start by contrasting *Justification Reporting* with *Deficit Reporting*, according to which all that needs to be communicated is the relevant hypothesis (Chapter 6.2.a). Although *Deficit Reporting* is compatible with requiring some threshold of scientific justification for the hypothesis, it does not recommend that this justification be reported. This contrasts starkly with *Justification Reporting*, which recommends reporting the hypothesis as well as aspects of the scientific justification for it. Of course, doing so addresses a broader informational deficit but, contra *Deficit Reporting*, *Justification Reporting* does not recommend communicating that *p* simply by reporting that *p*. So, while both principles reflect a broad enlightenment ambition, they do so in categorically different ways. Whereas *Deficit Reporting* focuses on reporting the scientific hypotheses, *Justification Reporting* focuses on the scientific justification for these hypotheses. Distinguishing between these approaches is important because they are easily conflated, and the critique that compromises the former does not typically compromise the latter. In particular, I will argue that *Justification Reporting* is more promising than *Deficit Reporting* in handling the fact that layperson recipients are not epistemically ideal cognizers. Rather, they are prone to motivated cognition as well as a wide variety of other psychological and social influences on their uptake of science reporting.

Another principle in the literature that superficially resembles *Justification Reporting* is called *Weight-of-Evidence Reporting*.[5] The principles share a focus on reporting the strength of evidence. But on closer examination, *Weight-of-Evidence Reporting* approaches recommend doing so via consensus reporting. In fact, two proponents, Dunwoody and Kohl, recommend a relabeling: "Although this concept has been labeled "weight of evidence" in past studies, we relabel it here "weight of experts" to more accurately capture its emphasis on communicating the distribution of expertise rather than evidence per se" (Dunwoody and Kohl 2017: 339). In contrast, *Justification Reporting* recommends reporting the strength of scientific justification per se by explicitly reflecting its nature and strength.

While these differences may seem subtle, they matter greatly for implementation. For example, *Justification Reporting* recommends that if a hypothesis is based on randomized controlled trials, this should be part of the science reporting. How this should be done is in part dependent on the expected audience and context of communication. So, as the qualification 'appropriate aspects' indicates, *Justification Reporting* should be supplemented with pedagogical requirements according to which the nature and strength should be explained in laypersons' terms. In many cases, it may be overly ambitious to expect that the recipients can appreciate the scientific justification. But they may often appreciate that the hypothesis is backed by scientific justification that is superior to alternatives

[5] Dunwoody 2005; Dixon and Clarke 2013; Clarke et al. 2015; Dunwoody and Kohl 2017.

such as doing one's own research on the matter, and this may be sufficient for entitled testimonial belief through appreciative deference (Chapter 7.4.c).

Moreover, *Justification Reporting* calls for an indication of the strength of scientific justification. For example, it could be explained, again in a simplified manner, why randomized controlled trials provide fairly strong scientific justification. Moreover, it is not too hard to indicate its approximate place in the evidential hierarchy (Berlin and Golub 2014; Murad et al. 2016). For example, in March 2020 when President Trump tooted chloroquine as a "game changer" in treating coronavirus, several public media reported that the scientific basis for this claim was weak and distinguished it clearly from scientific justification-based clinical trials (see, e.g., CNet 2020). Likewise, science reporting may distinguish the scientific basis provided by a single study from a more reliable meta-analysis. This is not uncommon. For example, CNN reported an article publishing "a huge umbrella study that looked at over 200 meta-analyses of the health benefits of coffee and that found drinking three to four cups of black coffee a day provides the most health benefits overall" (CNN 2020). So, I do not see *Justification Reporting* as revisionary but as distilled from the best practice in actual science reporting.

In this manner, *Justification Reporting* is true to the nature of science or, at least, to the relevant scientific justification. In contrast, none of the other science communication strategies recommend reporting that the scientific justification consists in a randomized controlled trial, a computer assisted mathematical proof, structured focus group interviews, or whatever the case may be. So, *Justification Reporting* is distinctive in doing so. I return to the relationship among the various principles in Chapter 6.3.d. But first I will outline some motivations for *Justification Reporting*.

6.3.b Philosophical motivations for *Justification Reporting*: *Justification Reporting* features a 'whenever feasible' clause and unpacking this aspect of the principle may serve to motivate it. This discussion doubles as a defense of the principled feasibility of the sister norm for scientific experts, JET. Below in Section 6.4.e, I will turn to practical obstacles and argue that although they are serious, they are far from devastating.

A key reason to think that it is in principle feasible to implement *Justification Reporting* has to do with the nature of scientific justification discussed in Chapter 3.3. In particular, it is important to recall *Hallmark 2*, according to which scientific justification is gradable, and *Hallmark 3*, according to which scientific justification is discursive. Since scientific testimony, and hence science reporting, is based on scientific justification, and since scientific justification is discursive, it is in principle possible to articulate it. Likewise, since scientific justification is gradable, it is often possible to indicate its strength. For example, a science reporter may report, in laypersons' terms, that a study's sample size is low or that a finding of a mixed-methods study is supported by the convergence of two independent strands of evidence. So, given the nature of scientific justification, it is in principle possible to articulate appropriate aspects of its strength and nature in an approximate manner. So, *Justification Reporting* is a normative ideal that may be approximated. Yet it provides some specific guidance in how this should be done—namely, by explicating appropriate aspects of the nature and strength of the relevant scientific justification. In consequence, it may serve as the basis for more specific guidelines for science reporters who consider context, audience, and platform.

Justification Reporting does not appeal to a paternalistic authority that *Consensus Reporting* may signal. As noted, focus on scientific consensus might encourage motivated cognition or even doubts about the motives of the scientists. In contrast, *Justification Reporting* provides some first-order epistemic reasons to accept the hypothesis. In effect, the recipients who are exposed to scientific justification are encouraged to form a decision on this basis. While this is a hypothesis for empirical investigation, it is not unreasonable to hypothesize that such a non-paternalistic communication that does not merely appeal to authority may over time contribute to trust in science (Hawley 2012: 74 makes a similar point). Thus, *Justification Reporting* reflects that the recipients are not epistemically ideal agents and that they may be largely ignorant about the nature of science.

Importantly, *Justification Reporting* provides a natural way to make epistemic comparisons between scientific justification and unscientific alternatives. Assume for example, that the strength and nature of the scientific justification for a hypothesis regarding the *infection fatality rate* of a virus is explicated in a piece of science reporting. Given such an explication, one may clarify that although the scientific justification is associated with uncertainty, it is comparatively superior to a line of unscientific justification for a competing hypothesis that might be salient to recipients (cf. Chapter 3.3). In this manner, the recipients may gain an appreciation for science even when it cannot deliver highly reliable results. In fact, a mode of science reporting that consistently involves a focus on the nature and strength of scientific justification may reasonably be thought to contribute to an improved science literacy in the public. Of course, this is a hypothesis that must be empirically investigated. But if it is correct, it speaks in favor of *Justification Reporting* as a long-term strategy.

Relatedly, *Justification Reporting* meets the *basing constraint*, according to which the public should not align their beliefs with the scientific justification out of fear, manipulation, etc., but out of some degree of appreciation of the epistemic strength of science. As the principle *Non-Inheritance of Scientific Justification* (Chapter 2.2.b) has it, the recipients will not thereby acquire a similar justification. But they may acquire testimonial entitlement through *appreciative deference* to the source that possesses this justification. I return to this idea in Chapter 7.4. So, *Justification Reporting* is comparatively non-manipulative and forthright in that it simply communicates the same epistemic reasons that are convincing to scientists. Again, I hypothesize that this aspect of the principle may promote a sustained trust in scientific institutions and science reporting.

6.3.c Empirical motivations for *Justification Reporting*: While *Justification Reporting* reflects some of the best science reporting out there, it has only recently been articulated as a distinctive strategy (Gerken 2020c). Consequently, available empirical research does not directly investigate the principle in its present formulation. Nevertheless, indirect empirical support for *Justification Reporting* may be extrapolated from existing empirical research, given some auxiliary assumptions. For example, studies that show an impact of reporting scientific explanations support *Justification Reporting.* After all, to provide a candidate scientific explanation of why the hypothesis is true is akin to articulating the scientific justification for it (Hyman 2015: 135–6; McCain and Poston 2014; McCain 2015).

Further evidence comes from a study which examined the effects of brief explanations of the greenhouse effect on laypersons' beliefs about AGW (Ranney and Clark 2016). To get a baseline, Ranney and Clark conducted a series of experiments which provided evidence that the US participants had little idea about the mechanics of global warming. On this basis, they investigated the effects of providing scientific justification in the form of explanations. For example, participants were, in one experiment (Experiment 3), exposed to a 400-word explanation of the mechanics of the greenhouse effect. The participants' acceptance of AGW was measured on a scale from 1 to 9. The result was a statistically significant increase in acceptance immediately after the intervention: "Global warming acceptance ratings increased significantly from a 6.3 pretest mean to a 6.6 posttest mean ($z = 3.45$; $p = 0.001$)" (Ranney and Clark 2016: 59). To explore whether the effect was lasting, participants were tested again after four days, and the effect was found to persist. Importantly, the study provided no indication of polarization across the political spectrum (Ranney and Clark 2016: 59). Ranney and Clark replicated the experiment in a variety of ways and found increases in acceptance of AGW following brief interventions "with mechanistic information" to be "+14 percent, +12 percent, +11 percent, and (after a 3-day delay) +15 percent" (Ranney and Clark 2016: 67). Furthermore, they found an increase of +15 percent with an intervention where participants were exposed to representative statistics, which also included information about consensus (Ranney and Clark 2016: 67). Again, none of the experiments provided any evidence of polarization (Ranney and Clark 2016: 68).

Ranney and Clark draw two conclusions that support *Justification Reporting.* First, explicating scientific justification in terms of a mechanism for a scientific hypothesis can increase laypersons' credence in it. Second, they found no evidence of polarization across the political spectrum. Conservatives increased their level of acceptance of AGW as much as liberals did. This supports the hypothesis that *Justification Reporting* is not particularly prone to trigger motivated cognition. The former finding gives a clue as to how scientific justification may be presented in a manner appreciable by laypersons—namely, in the guise of explanatory mechanisms.

I have focused on Ranney and Clark's findings because they are instructive and because space does not permit for a comprehensive survey of the empirical literature. But several studies provide similar evidence for *Justification Reporting*. For example, Johnson prompted Republicans and Democrats to focus on the power of arguments' mechanistic explanations and found that this manipulation reduced biased evaluation as well as differences between Republicans and Democrats (Johnson 2017). In consequence, Johnson recommends that science communicators should include brief mechanistic explanations (Johnson 2017).

Other studies motivate *Justification Reporting* more indirectly by finding correlations among appreciation of relevant scientific justification and acceptance of divisive scientific hypotheses such as AGW. For example, a study found evidence that, among Swiss laypersons, "knowledge increases public concern about climate change independent of cultural worldviews" (Shi et al. 2015: 2097). A cross-cultural study of lay populations from six countries found a similar effect (Shi et al. 2016). Shi et al. suggest that *causal* information is particularly appreciable for a lay audience, and this is broadly congenial to the mentioned findings that scientific justification in mechanistic terms is something that affects laypersons' uptake of science reporting. Likewise, a study of Australian laypersons found "an additive effect of specific climate change knowledge on beliefs about the causes and consequences of climate change" (Guy et al. 2014). Similar effects were found in a study of US undergraduate students' "climate literacy," which includes awareness of scientific justification pertaining to AGW: "basic climate literacy appears to reduce polarization between Republicans and Democrats by increasing the chances that Republicans will become more concerned about AGW (Democrats are already concerned, more or less regardless of their level of climate literacy)" (Bedford 2015: 195). Another large-scale study using a demographically representative survey ($N = 1100$) investigated acceptance of evolutionary theory and found that knowledge of evolution theory significantly predicted acceptance of it (Weisberg et al. 2018a). This persisted even when controlling for religious and political views: "contrary to some similar work on climate change (Kahan et al. 2012), we found a relationship between increasing knowledge and increasing acceptance at all levels of religiosity and political ideology" (Weisberg et al. 2018a: 217). Lombrozo et al. also found that "accepting evolution is significantly correlated with understanding the nature of science, even when controlling for the effects of general interest in science and past science education" (Lombrozo et al. 2008: 290).

Not all of these studies directly support *Justification Reporting*. For example, there is a difference between how understanding the relevant science affects the uptake of science reporting and how including scientific justification in science reporting affects the uptake of it. But although *Justification Reporting* has not been tested directly, available experiments provide some indirect support of it (and yet further examples are noted in Ranney and Clark 2016 "General Discussion"). The mentioned studies augment the two points supporting *Justification Reporting*: First, the findings support the idea that familiarity with explanations that reflect relevant scientific justification is conducive to proper uptake of public scientific testimony. Second, explicating scientific justification does not appear to encourage polarization or backfire effects (see also Wood and Porter 2019).

On a more constructive note, the body of empirical research provides important hints about how to explicate the nature of scientific justification in a manner that may be appreciated by lay audiences. In particular, some studies indicate that scientific justification that is explicated in causal or mechanistic terms is particularly compelling to lay audiences (Ranney and Clark 2016; Johnson 2017; Shi et al. 2015, 2016). Relatedly, laypersons may be attracted to reductive explanations (Hopkins et al. 2016, 2019; Weisberg et al. 2018b). Furthermore, other textual or pictorial cues of "scientificness" may increase trustworthiness (Thomm and Bromme 2012; Hopkins et al. 2016). Of course, these psychological effects offer opportunities for manipulation. That said, if indicators of scientificness appropriately represent the relevant scientific justification, *Justification Reporting* is hardly manipulative. However, it is important to explore differences in audiences appreciating the scientific justification or merely appreciating that it is there.

I re-emphasize that, although the noted findings lend some support to *Justification Reporting*, the empirical work is far from conclusive—in part because its support is indirect. So, awareness of the limitations of the empirical research is key.

6.3.d The relationship between *Justification Reporting* and the other approaches: Although *Justification Reporting* should be central to science reporting, it may often combine fruitfully with *Deficit Reporting*, *Consensus Reporting*, and *Value-Based Reporting*.

While I agreed with the criticism of *Deficit Reporting* as a general principle, I also noted that there are science communication contexts in which it is unproblematic since many scientific hypotheses are not divisive. So, in some contexts, *Deficit Reporting* may govern science reporting that the pollen count will increase next month or that marine biologists have identified a new kind of deep sea fish. Thus, there is a fairly natural division between *Deficit Reporting*, which may govern basic reporting of non-divisive issues, and *Justification Reporting*, which is apt for divisive issues or topics where the epistemic basis for a hypothesis matters for public deliberation. However, some degree of consistency in science reporting may be desirable. Relatedly, it should not be ruled out that including scientific justification in science reporting about non-divisive cases may open up avenues for science reporting about divisive cases. For example, clarifying that the weather forecast rests on model-based justification may make it easier to appeal to model-based justification in science reporting of climate change. This diachronic perspective is particularly relevant because the recipients are epistemically imperfect consumers of scientific testimony. So, this is another sense in which *Justification Reporting* is sensitive to the cognitive nature of recipients in a manner that *Deficit Reporting* is not.

While *Consensus Reporting* has its place, the criticism of it suggests that it is generally better to integrate it with *Justification Reporting* than to provide *bare* consensus information. As highlighted, a worry about bare consensus reporting is that it merely appeals to the authority of scientists and may encourage the us-and-them social cognition governed by *Epistemic Underestimation*, according to which both social stereotypes may lead evaluators to underestimate the scientists' epistemic position (Chapter 5.3.e). It may be a folk epistemological misconception that scientists are a social group that marches in synch. To counter this, consensus information may include an explanation of *why* scientific consensus indicates reliability. This would typically include a discussion of the strength and nature of the justificatory basis for the consensus. Thus, *Consensus Reporting* may be improved by being integrated with *Justification Reporting*. However, the latter is more fundamental in the sense that the relevance of scientific consensus derives from its indication of strong scientific justification.

Value-Based Reporting may also have a place in science reporting about divisive issues. Recall that a central worry with *Value-Based Reporting* is that it is consistent with exploiting the sort of biases that it is supposed neutralize. However, the manipulative character of this type of implementation may be curbed by integrating it with *Justification Reporting*. If value-based reporting is required to revolve around an explication of the nature and strength of scientific justification, it is less likely to deteriorate into mere marketing. For example, *Justification Reporting* is compatible with *pluralistic advocacy*, which highlights to the recipients that experts who represent diverse social values are providing the relevant scientific justification (Kahan et al. 2010, 2011). For example, anecdotal evidence of *why* a member of a skeptical group came to change her mind may naturally include a presentation of the relevant scientific justification that convinced her (NPR 2019). However, *Value-Based Reporting* should serve *Justification Reporting* insofar as its central function should be that of minimizing biases that hamper recipients' ability to assess the relevant scientific justification.

All in all, *Justification Reporting* may be regarded as central in a package of approaches to science reporting. A general more strategic rationale augments this assumption. The rationale concerns the ongoing battle against science denialism (Dunlap et al. 2011; Intemann and Inmaculada 2018; O'Connor and Weatherall 2019). Specifically, I hypothesize that *Justification Reporting* is harder to undermine than implementations of *Deficit Reporting*, *Consensus Reporting*, and *Value-Based Reporting*. For example, *Deficit Reporting* is extremely vulnerable to misleading defeaters since the recipients do not acquire resources to reject them from the testimony. In contrast, *Justification Reporting* may provide the recipients with reasons to reject some—but of course not all—misleading defeaters. Of course, *Justification Reporting* may offer science deniers the option of providing undercutting defeaters which target the scientific justification. But although this may be done in various ways, it is a fairly demanding task for the science skeptic because it requires her to engage the relevant scientific justification.

The presence of counter-consensus messaging suggests that *Consensus Reporting* is fragile with regard to organized science denial (Dunlap et al. 2011). One reason for this is that it is fairly cost-free to disseminate anti-consensus messaging insofar as the media platforms are in place. Laypersons who are exposed to conflicting consensus messaging may be prone to *Epistemic Overestimation*—the overestimation of the epistemic position of those of their own in-group—and may therefore maintain their antecedent attitudes (Chapter 5.3.e). In contrast, *Justification Reporting* may be harder to undermine insofar as this requires counter-justification. This is by no means denying the possibility of raising unreasonable doubts about scientific justification and fabricating pseudo-scientific justification.[6] There are plenty of examples of well-funded pseudo-scientific reporting that should be recognized as serious challenges. Nevertheless, providing convincing pseudo-scientific justification is typically labor-intensive. This is especially so if the audience is scientifically literate. So, I hypothesize that it is easier to fabricate anti-consensus reporting than to fabricate pseudo-scientific justification. *Value-Based Reporting* may be even more problematic because the adversaries of science are often highly familiar with the social values of the groups that are skeptical about science. For example, science skepticism may be driven by political or religious motives. Everything else being equal, political and religious leaders are better equipped than scientists and science reporters to debate in political and religious terms. Thus, *Value-Based Reporting* appears to recommend conducting the debate where the science skeptics excel. In contrast, *Justification Reporting* recommends conducting the debate where scientists excel—namely, in the domain of scientific justification. Hence, *Justification Reporting* gives reporting of genuine science an important home turf advantage, whereas *Consensus Reporting* and *Value-Based Reporting* appears to give the home turf advantage to science skeptics. This provides an additional reason to regard *Justification Reporting* (as well as JET for scientific experts) as central to public scientific testimony.

None of these comparisons suggests that *Justification Reporting* provides the one final strategy against socio-psychological biases or organized science

[6] Oreskes and Conway 2010; Dunlap et al. 2011; Biddle and Leuschner 2015; Intemann 2017; Intemann and Inmaculada 2018; O'Connor and Weatherall 2019.

skepticism, but only that it should be a central approach. So, I will turn to some of the limitations of *Justification Reporting*.

6.4 Justification Reporting—Objections, Obstacles, and Limitations

Despite the virtues of *Justification Reporting*, it is not a silver bullet that can resolve all the challenges of science reporting. In order to assess the pros and cons of *Justification Reporting*, I will briefly consider some objections to it and some of its limitations (most of the discussion will also apply to the norm of scientific expert testimony, JET).

6.4.a Motivated cognition and identity-protective reasoning: The power of motivated cognition—perhaps particularly in the guise of identity-protective reasoning—should not be underestimated. Consequently, it is a concern that reporting the strength and nature of scientific justification may have an adverse effect on laypersons' uptake. Fortunately, since *Justification Reporting* highlights epistemic considerations over non-epistemic ones, it is unlikely to contribute to political divisiveness, although it is unlikely to overcome existing divisiveness. However, the surveyed empirical work provides some grounds for cautious optimism—even with regard to divisive cases. Moreover, some studies indicate that cognitive effort and ability are not invariably deployed for motivated cognition. For example, Pennycook and Rand's participants with high scores on a cognitive reflection test (CRT) rated partisan fake news as less accurate than participants with low CRT scores—*even* for headlines aligning with their political ideology (Pennycook and Rand 2019). On this basis, they suggest that susceptibility to partisan fake news is better explained by lack of reasoning than by motivated reasoning (Pennycook and Rand 2019). This may speak in favor of a mode of science reporting, such as *Justification Reporting*, that is likely to trigger more reflective processing.

So, while the challenges from motivated cognition should not be underestimated, they are better understood as obstacles for *Justification Reporting* than as a refutation of it. Given the tenacity of motivated cognition, there is probably no science communication strategy that is capable of entirely overcoming its effects. So, given that *Justification Reporting* appears to do comparatively well, it should not be taken as a strike against it that it cannot neutralize motivated cognition. However, motivated cognition is likely to interact with other social and psychological effects. So, in the remainder of this section, I will consider motivated cognition interacting with such obstacles to science reporting in accordance with *Justification Reporting*.

6.4.b Backfire effects: Some evidence indicates backfire effects in which some recipients become less inclined to believe that *p* upon receiving public scientific testimony that *p* (Chapter 5.3.b). As noted, this body of evidence is far from univocal, and studies supporting *Justification Reporting* indicate that it is not particularly susceptible to backfire effects. Of course, we should be ever cautious about concluding too much from null results. But some of these studies are fairly impressive in scope. For example, Wood and Porter conducted a large (10,100-participant) study of fifty-two divisive issues (Wood and Porter 2019). However, they found a positive effect of corrections across the political spectrum: "Among liberals, 85 percent of issues saw a significant factual response to correction, among moderates, 96 percent of issues, and among conservatives, 83 percent of issues" (Wood and Porter 2019: 143). Moreover, they found no evidence of backfire effects: "No backfire was observed for any issue, among any ideological cohort. The backfire effect is far less prevalent than existing research would indicate" (Wood and Porter 2019: 143).

In many cases, this study tested science communication consistent with *Justification Reporting* given that the conditions involving "factual corrections" often indicated the relevant scientific source or justification. For example, in the condition concerning abortion, the correction began "Statistics from the Center for Disease Control tell a different story" and thus hinted at scientific justification based on statistical analysis of empirical data. So, overall, the study provides some evidence that *Justification Reporting* is fairly robust against identity-protective reasoning and backfire effects given that it aligns with other studies (e.g., Ranney and Clark 2016: 68; Bedford 2016; Weisberg et al. 2018a).

6.4.c Science literacy and misconceptions about science: Another *prima facie* challenge for *Justification Reporting* are the findings that science literacy may increase polarization (Chapter 5.3.b). I advertised *Justification Reporting* by suggesting it could, over time, help increase the lay public's science literacy. But if increased science literacy merely enables motivated cognition and ultimately increases polarization, it might be objected that this apparent benefit is in fact problematic (Taber and Lodge 2006; Kahan et al. 2017).

With regard to this objection, I am not inclined to be concessive. As already noted, several studies found that increased trust in science correlated with varieties of scientific literacy.[7] Although these studies do not establish that motivated cognition is never augmented by scientific literacy, they indicate that the problematic effects of science literacy are so irregular that we should continue to promote science literacy. If the problematic effects of science literacy provided a good reason to abandon *Justification Reporting*, they would provide an even better

[7] Lombrozo et al. 2008; Guy et al. 2014; Shi et al. 2015, 2016; Bedford 2015; Weisberg et al. 2018; Wood and Porter 2019.

reason to abandon science education. I take this to be a *reductio* of the idea that the problematic effects of science literacy provide a good reason to abandon *Justification Reporting*.

The main lesson is that further research is needed to indicate the particular conditions under which specific kinds of scientific literacy may raise problems for *Justification Reporting*. That way, we can devise empirically informed strategies for how to implement it or supplement it as to minimize these problems. So, on this issue, I uphold an enlightenment tradition according to which a contribution to public understanding of science is generally a good thing (Slater et al. 2019). That said, the aim of *Justification Reporting* may simply be that of enabling the recipients to form or sustain entitled testimonial belief though *appreciative deference* to science (Chapter 7.4.c). But in any case, *Justification Reporting*'s ability to contribute to public understanding of science is a point in its favor. This is particularly so if it is correct that prominent folk misconceptions about science are damaging for a reasonable uptake of public scientific testimony among laypersons.

For example, the misconception that crucial experiments are central to science may be mitigated by justification reporting. At least, focus on the kind and strength of, for example, the data and models involved in climate science makes it natural to clarify that crucial experiments are not available in this type of scientific research. This point is naturally articulated in explaining that model-based reasoning provides the kind of justification that amounts to the comparatively strongest justification that may reasonably be hoped for. Something similar is true with regard to the misconception that "doing one's own research" is more rational than trusting public scientific testimony. This more or less articulate presupposition is the folk version of the *Nullius in verba* sentiment and *Justification Reporting* offers some resources for minimizing its damaging effects. This misconception may reflect the phenomenon of *source bias*, which is the inclination to favor some sources over other ones that are epistemically equal or superior (Chapter 5.3.c). This inclination is a concern insofar as we tend to prefer direct (perceptual) observation over inferential or statistical sources. After all, in many domains of scientific inquiry, complex models, computer simulations, or statistical inferences provide the best, and sometimes the only available, scientific justification (Winsberg 2001, 2012, 2018; Bailer-Jones 2009). So, explicating the best scientific justification may not be most convincing to the recipients. This is also a serious concern but one that may be mitigated by highlighting that the reported scientific justification is the best or sole justification in the domain in question. In fact, science reporting that articulates the kind and strength of the relevant scientific justification in laypersons' terms provides a natural platform for indicating that the issue is not one that laypersons can research on their own. In addition to the potential efficacy on particular occasions, this mode of science reporting may build science literacy over time. In particular, it might help diminish the problematic folk presupposition that reflects that uptake of scientific testimony is contrary to the nature of science.

6.4.d Folk epistemological biases and epistemic qualification effects: Some of the folk misconceptions about science go hand in hand with folk epistemological biases such as source bias or epistemic focal bias. For example, the misconception that science always provides knowledge or proof is problematic given *epistemic qualification effects*, according to which qualifications that flag epistemic fallibility or limitations of a scientific hypothesis may be detrimental to laypersons' uptake of it (Chapter 5.3.d). Examples of epistemic qualifications include reporting the limitations of statistical power, stipulations of a model, or specific confounders of a study. This is a concern for *Justification Reporting*, which recommends indicating epistemic qualifications in reporting a scientific hypothesis, *p*. The worry is particularly serious in divisive cases in which epistemic qualifications may trigger or further motivated cognition.

However, being upfront about limitations of scientific justification may be beneficial despite epistemic qualification effects. One study found that ascriptions of expertise were lowered whether a flaw was disclosed by own admission or by external criticism (Hendriks et al. 2016). But, importantly, "ascriptions of integrity and benevolence were higher when admitted vs. when introduced via critique" (Hendriks et al. 2016: 124). Importantly, epistemic qualifications about, for example, confounders to a study indicate limitations, not flaws. So, the study does not show that including such epistemic qualifications negatively impacts assessments of expertise. Moreover, some findings suggest that expert sources who include epistemic qualifications are regarded as more persuasive (Jensen 2008; Karmarkar and Tormala 2010). However, the evidence on the issue is mixed insofar as other studies did not find any mode of communication of uncertainty to *increase* recipient trust (van der Bles et al. 2020).

Yet, it is not unreasonable to think that negative consequences of epistemic qualification may be mitigated by explaining, in laypersons' terms, that the inconclusive scientific justification is the best or only justification available. This may be done in a way congenial to *Justification Reporting*—i.e., via low-key explanations about the nature of science. For example, it may be explained that *p* cannot be proven by a crucial experiment but must be justified in a cumulative manner. Likewise, it may be useful to minimize the use of binary epistemic vocabulary such as 'scientific proof' or 'scientific knowledge.' In particular, it may be best to eschew the term 'knowledge' in much science reporting given that it is extremely basic in our folk epistemology and, hence, likely to be processed associatively (Gerken 2017a, forthcoming c; Phillips et al. forthcoming). Specifically, heavy reliance on the term 'knowledge' in science reporting may trigger unfortunate folk epistemological heuristics that exhibit overly skeptical responses to salient alternatives (Chapter 5.3.d; Gerken 2017a).

In contrast, numerical qualifications may trigger a more reflective mode of reception that is less susceptible to overly skeptical epistemic qualification effects. In one survey it was found that "technical uncertainty," which occurs when uncertainty is conveyed as quantified error ranges and probabilities, had positive effects or null effects but no negative effects on uptake (Gustafson and Rice 2020). Similarly, van der Bles and colleagues conclude that "the provision of numerical uncertainty—in particular as a numeric range—does not substantially decrease trust in either the numbers or the source of the message" (van der Bles et al. 2020: 7680). This contrasts with verbal qualifications: "Verbal quantifiers of uncertainty, however, do seem to decrease both perceived reliability of the numbers as well as the perceived trustworthiness of the source" (van der Bles et al 2020: 7680).

However, not all verbal quantifiers are equal. In particular, there may be important differences between epistemic qualifications that recipients process associatively and those that they process reflectively. As mentioned, some studies suggest that verbally highlighting that scientific justification comes in degrees may be beneficial to recipient uptake.[8] Recall also that Lombrozo and colleagues found evidence that understanding of the provisional nature of science correlated with acceptance of evolution theory across the political spectrum (Lombrozo et al. 2008). On this basis, they provide a list of aspects of scientific justification that are reasonable to include in teaching and, I presume, in science reporting as well (Lombrozo et al. 2008: 292). In this regard, it is also worth recalling the studies indicating that familiarity with relevant scientific justification was conducive to uptake.[9]

Generally, there is reason to think that at least some verbal presentations of the strength and nature of the scientific justification may affect recipients more like a numerical presentation—even if it is done verbally. Yet, it is important to acknowledge that *some* types of epistemic qualifications in science reporting may, in *some* contexts, negatively impact the recipients' uptake (van der Bles et al. 2019, 2020; Gustafson and Rice 2020). On the other hand, when scientific uncertainty is very low, reporting the strength and nature of scientific justification offers the opportunity to convey that this is the case. Moreover, science reporting that focuses on scientific justification offers a very natural way to convey that unscientific alternatives are associated with far higher uncertainty. So, although further empirical study is called for, epistemic qualifications may be naturally embedded in justification reporting in ways that might counter their negative effects on uptake.

However, given the limited and mixed evidence regarding epistemic qualifications, it is important to further investigate how different epistemic qualifications bear on uptake of science reporting and how they interact with motivated

8 Jensen 2008; Jung 2012; Ranney and Clark 2016; Johnson 2017.

9 Shi et al. 2015, 2016; Bedford 2015; Weisberg et al. 2018a; Wood and Porter 2019.

cognition. But it is hardly too soon to worry that science communicators may face dilemmas whether to include an epistemic qualification that they can reasonably expect to negatively impact recipient uptake (Gerken forthcoming a).

6.4.e Feasibility troubles?: The strategy of providing low-key philosophy of science within science reporting may reignite skepticism about the feasibility of *Justification Reporting*. As noted, what I regard as best practice of existing science reporting often includes explications of the nature and strength of scientific justification. But while this illustrates the feasibility of *Justification Reporting* in some contexts, the *practical* aspects of feasibility remain to be addressed.

In general, feasibility is more of a concern for *Justification Reporting*, which applies to science reporters, than for JET, which applies to scientific experts. As mentioned in Chapter 5.4.c, a scientific expert can—with important exceptions—often acquire a basic grasp of the scientific justification that is sufficient for the purpose of expert scientific testimony. But science reporters are typically further removed from the relevant scientific justification than the scientific experts. This is a practical challenge because scientific justification is often *esoteric* in the sense that it is difficult for a science reporter to acquire it well enough to report it with reasonable accuracy (Goldman 2001). Or so goes the objection.

Although it is not feasible for science reporters to heed *Justification Reporting* in some cases, it is important to avoid overestimating the difficulty of doing so. Science journalists are often scientifically literate and are skilled in communicating complex scientific information by invoking simple indicators of the nature and strength of scientific justification. A prominent example is the *Levels of Evidence Pyramid* popular in the medical sciences (see Berlin and Golub 2014; Murad 2016 et al. for recent discussions). While epistemologists and philosophers of science may find much to criticize in such devices, they are often apt for science communication. Just as science reporters may need to simplify the hypothesis, they may also need to simplify the scientific justification for it. In fact, science reporters often excel in providing such simplifications by invoking clever analogies, metaphors, comparisons, etc.

A related obstacle for implementing *Justification Reporting* is the fact that on many media platforms, it is only feasible to report the scientific finding and not the relevant scientific justification for it. As mentioned, science journalists are, like most other journalists, operating in an attention economy. Often, reporting the mere finding, as opposed to the scientific justification for it, will be more conducive to generate clicks, views, shares, etc. (Dunwoody 2014; Figdor 2017).

Although these obstacles for implementing *Justification Reporting* are serious, they are not devastating since the principle only requires that *appropriate aspects* of the scientific justification be reported. It is a highly contextual matter how to do so as it depends on the media platform and the expected audience. Hence, science journalists will often be best positioned to adjudicate what the appropriate aspects are. Generally speaking, there is a division of labor between those who help articulate norms of science communication and those who implement them (Angler 2017). It would amount to philosophical hubris to dictate very specific rules of implementation.

But although I have articulated *Justification Reporting* as a norm, I think that it may serve double duty as a highly general guideline. After all, it provides some general but fairly concrete requirement on implementation—explicate the nature and strength of the scientific justification—while allowing for flexibility in how to do so. In many contexts, a minimal sketch of the scientific justification will be appropriate. For example, merely indicating the relevant justification ("climate scientists rely on complex models representing a large number of observations in justifying that *p*") may be better than simply reporting that *p*. Moreover, it is often feasible to include indications of the comparative strength of the scientific justification ("Scientists rely on models because they are the most reliable sources of evidence for or against *p* that are available to us").

The two parenthetical explications of scientific justification in the paragraph above amount to a total of thirty-eight words that will be comprehensible to many laypersons. So, they exemplify that it is often feasible to indicate aspects of the nature and strength of scientific justification for the reported hypothesis. In many contexts, a simplified explanation of how models work and why they are more reliable than alternative sources of justification may be added. Science reporters are skilled in reporting aspects of the scientific justification in a simplified manner that is appreciable by their audience (Thomm and Bromme 2012). So, although I do not purport that *Justification Reporting* is reflecting a general practice in contemporary science reporting, it is inspired by the best currently practiced science reporting. This fact augments it against concerns that it is invariably infeasible.

6.4.f Concluding thoughts on *Justification Reporting*: Allow me to conclude the case for *Justification Reporting* with a bit of *anecdata*: Over the course of writing this book, I have presented *Justification Reporting* to both philosophers and science journalists. Strikingly, concerns about feasibility have mainly been raised by the philosophers, whereas science journalists have often responded by thinking constructively about the implementation of *Justification Reporting*. For example, scientific justification may be presented in terms of the process of the investigation. This is a form of journalistic storytelling that may fit well-established types of narrative structures. Likewise, one may combine an anecdotal report of someone from a skeptical group who ended up trusting the science by explicating the scientific justification that convinced her (NPR 2019).

In general, science journalists need new narratives to replace the traditional focus on individual scientists that may encourage an unfortunate *great white man fetish* (Section 5.3.b). *Justification Reporting* may be a principled source of alternative narratives that revolve around the relevant scientific justification. However, there is a division of labor where philosophers contribute to abstract norms and guidelines of science reporting, whereas journalists contribute to implementing them.

My aim here is merely to counter the worry that *Justification Reporting* is invariably too demanding. Skilled science reporters possess a brand of *interactional expertise* that allows them to comprehend scientific justification well enough to report on it in a manner appreciable to laypersons (Collins and Evans 2007). In some instances, science reporters need so-called T-shaped expertise, which consists of a mix of broad scientific literacy and more specific epistemic or contributory expertise (Chapter 1.2.c). Although many researchers-gone-journalists possess T-shaped expertise, it is generally hard to acquire. So, societies that pursue ideals of deliberative democracy should invest in appropriate training for science reporters. Just as it is a societal challenge to secure platforms for such in-depth science reporting, it is a societal challenge to ensure that there are science reporters capable of providing it. I return to these challenges in Chapter 7.

In general, it is important to assess norms and guidelines of science reporting both in terms of their immediate effects on recipient uptake and in terms of the wider role of science reporting in society. I have argued that *Justification Reporting* can stand its ground with regard to laypersons' immediate uptake, and it also strikes me as reasonable with regard to the role of science in society. In my initial introduction of *Justification Reporting*, I sloganized it as *public scientific testimony in the scientific image* (Gerken 2020c). The idea behind the slogan is that *Justification Reporting* helps to ensure that the core aspects of the scientific grounds for decision-making are included in science reporting. So, while *Justification Reporting* has important limitations, it is nevertheless a candidate centerpiece of the norms of science reporting in a society that strives for enlightenment values. In particular, it sits well with the ideal of scientifically informed public deliberation.

In Chapter 7, I will revisit this broader picture. But at this point, I will consider whether *Justification Reporting* may earn its keep by informing a specific and very difficult debate about science reporting—the debate about balanced reporting of science.

6.5 Balanced Science Reporting

In this section, I briefly consider the prevalent journalistic principle of *Balanced Reporting* and propose a revision to its application to science reporting. Much of my presentation is a condensation of (Gerken 2020d), but I will highlight how my proposal is based on *Justification Reporting* and, ultimately, the general norm of public scientific testimony, JEN. Thus, I aim to integrate the specific proposal into the broader framework that I have advanced.

6.5.a The question of balance: The idea that journalistic reporting should be balanced in the sense of representing both sides of a dispute in a neutral manner is a longstanding journalistic principle. Versions of the idea of balanced reporting have considerable practical ramifications. For example, it is reflected in editorial guidelines of news organizations, such as the BBC: "We must do all we can to ensure that 'controversial subjects' are treated with due impartiality in all our output" (BBC 2018a). Ironically, it even used to figure in Fox News's slogan "Fair and Balanced." The principle has been articulated in many ways by proponents and opponents.[10] However, the following articulation of the idea is fairly representative:

> ***Balanced Reporting***
>
> Science reporters should, whenever feasible, report opposing hypotheses in a manner that does not favor any one of them.

The notion of *favoring* amounts to presenting the hypothesis in a way that may reasonably be expected to lead to audience acceptance of the hypothesis. While *Balanced Reporting* reflects the scientific value of neutrality, it has been criticized for being a problematic principle for *science* reporting. In particular, it has been argued that *Balanced Reporting* about science gives rise to biased reporting because treating opposing views equally is misleading when they are not equally well warranted.[11] This misleadingness of *Balanced Reporting* may be articulated as a conflict with the idea that since objective journalistic reporting is committed to truth, only the most reliable sources should be consulted. Applied to science reporting, this journalistic principle may be articulated as follows:

> ***Reliable Reporting***
>
> Science reporters should, whenever feasible, report the most reliably based hypotheses and avoid reporting hypotheses that are not reliably based.

Of course, *Reliable Reporting* is not always possible to follow in practice but, as with *Balanced Reporting*, it may be regarded as an ideal to pursue. However, the two principles are incompatible. In particular, *Balanced Reporting* is incompatible with the second conjunct of *Reliable Reporting*. So, we are faced with the following question:

[10] Nelkin 1987: 19; Entman 1989: 30; Dixon and Clarke 2013: 360; Dunwoody 2014: 33; Dunwoody and Kohl 2017: 341.

[11] Gelbspan 1998; Boykoff and Boykoff 2004; Mooney and Nisbet 2005; Figdor 2013, 2018; Simion 2017; Weatherall et al. 2020.

> ***The Question of Balance***
>
> How should *Balanced Reporting* and *Reliable Reporting* be balanced in science reporting?

A prominent proposal is to completely ban *Balanced Reporting* in science reporting whenever it conflicts with *Reliable Reporting*. For example, in a criticism of *Balanced Reporting*, Boykoff approvingly suggests "that we may now be flogging a dead norm" (Boykoff 2007: 479). A similar trend is found in public debates such as Read's op-ed in *The Guardian* "I won't go on the BBC if it supplies climate change deniers as 'balance'" (Read 2018). This op-ed was subsequently supported by fifty-seven politicians, scientists, and writers (Porritt et al. 2018).

While banning *Balanced Reporting* would resolve the conflict, it would amount to subjecting every epistemically subpar scientific hypothesis to a general no-platforming policy, according to which it should not figure in science reporting. But it may be overly restrictive to bar minority scientific criticisms of dominant scientific views from the public debate. This is not to deny that no-platforming is appropriate in some contexts. Holocaust deniers do not deserve a balanced hearing in public scientific testimony about World War 2 history. But, in other cases, hypotheses defended by a legitimate scientific minority on the basis of limited scientific justification may deserve a hearing in public deliberation. Such hypotheses may lack in scientific justification because they are novel or because they are pursued by a scientific minority that lacks adequate resources for providing scientific justification. So, although they may not (from an epistemic perspective) deserve to be reported in accordance with *Reliable Reporting*, completely excluding them may be an overreaction to the problems with *Balanced Reporting*. For the same reason, the second conjunct of *Reliable Reporting*, according to which science reporters should altogether avoid reporting unreliably based scientific hypotheses, may be overly categorical. Indeed, the *Question of Balance* is an instance of general tension between scientific authority and respect for diversity.

My overall diagnosis is that both *Balanced Reporting* and *Reliable Reporting* are overly general and too categorical. So, rather than completely abandoning one of them, I will propose that both principles may be revised in a unified manner that (i) resolves the conflict between them, (ii) preserves the features that originally motivate them, and (iii) minimizes their undesirable features.

6.5.b The balancing act: To ensure a unified revision of *Balanced Reporting* and *Reliable Reporting*, I take *Justification Reporting* as an important principled basis. *Justification Reporting* requires that science reporters indicate the degree and nature of the scientific justification and doing so is a straightforward way to favor an epistemically superior hypothesis. Hence, I propose the following revision of *Reliable Reporting*:

Inclusive Reliable Reporting

> Science reporters should, whenever feasible, report hypotheses in a manner that favors the most reliably based ones by indicating the nature and strength of their respective scientific justifications.

Inclusive Reliable Reporting retains the positive requirement to favor reliably based hypotheses in science reporting but, as opposed to the original principle, it permits reporting differing perspectives and critical opposition. However, *Inclusive Reliable Reporting* does not require or even encourage science reporters to report epistemically inferior hypotheses. Indeed, it is consistent with no-platforming such hypotheses in some science communication contexts. Thus, *Inclusive Reliable Reporting* is, in one sense, weaker than *Reliable Reporting* because it permits the reporting of less reliable sources and views. But, in another sense, it is stronger in virtue of requiring that the most reliable hypotheses be favored in a specific manner—namely, by reporting the nature and strength of the relevant scientific justification.

The restriction of *Balanced Reporting* may be similarly guided by *Justification Reporting*'s requirement of explicating the nature and strength of the relevant scientific justification. Here is the result:

Epistemically Balanced Reporting

> Science reporters should, whenever feasible, report opposing hypotheses in a manner that reflects the nature and strength of their respective scientific justifications or lack thereof.

Epistemically Balanced Reporting retains the attractive property of *Balanced Reporting* that a wide range of scientific perspectives on an issue *may* be reported. But it does so without the misleading appearance of epistemic equality between epistemic non-equals. In particular, it does so in accordance with *Justification Reporting*'s requirement of explicating the nature and strength of the reported hypotheses.

Given that *Epistemically Balanced Reporting* and *Inclusive Reliable Reporting* are not merely compatible but congenial in reflecting the common underlying principle, *Justification Reporting*, there is no conflict between them. Moreover, I have argued that the revisions preserve the features that originally motivated *Balanced Reporting* and *Reliable Reporting*. Finally, the restrictions help to minimize the problematic aspects of the original principles—i.e., that *Balanced Reporting* is epistemically misleading and that *Reliable Reporting* no-platforms too many hypotheses.

6.5.c Implementations and remaining challenges: Before concluding, I will say a few things about the challenges associated with implementing the revised principles, focusing on *Epistemically Balanced Reporting*. Clearly, the proper implementation of *Epistemically Balanced Reporting* is a highly contextual affair that depends on the specific news story, media platform, target audience, etc. Hence, the principle falls short of an *editorial* guideline for science reporters. However, *Epistemically Balanced Reporting* may arguably be doing double duty as a norm and as a highly general guideline that may constrain more specific editorial guidelines. Hence, I classified it as a guideline in Figure 6.1. That said, it is important to recognize that developing yet more context-specific implementable editorial guidelines is something that journalists are better equipped to do than philosophers are.

However, some of the challenges for an implementation may be considered at a fairly general level, and here philosophers may contribute. I have already spoken to the challenge of feasibility by arguing that science journalists are fairly capable of articulating the basic nature and strength of scientific justification for the reported hypothesis. However, *Epistemically Balanced Reporting* involves the further requirement that they can do so in comparative terms. In some cases, such as those involving scientific expert disagreement, this may be hard to do. But the principle is generally implementable, and even in cases of expert disagreement, science journalists have some resources at hand (Goldman 2001; Rolin 2020). So, at least, *Epistemically Balanced Reporting* provides an important ideal that science reporters may pursue.

A different challenge arises from motivated cognition. Since *Epistemically Balanced Reporting* requires that if a politically divisive hypothesis is reported, the nature and strength of the scientific justification for it should be explicated. A concern, then, is that motivated reasoners will overestimate the explicated scientific justification. Likewise, the *reporters* may be influenced by motivated cognition. Although this concern highlights a real limitation of *Epistemically Balanced Reporting*, it is hardly a sufficient reason to abandon it. First of all, *Epistemically Balanced Reporting* does not entail that all such politically divisive hypotheses make the minimal threshold of scientific justification that a no-platforming principle might require. Secondly, *if* cases such as anti-vaxxing, climate science denial, and such are reported, *Epistemically Balanced Reporting* is, in effect, prescribing a rebuttal. That is, the scientific justification for the relevant hypotheses will be described as debunked by the (also presented) strong scientific justification for the opposing hypotheses. I take rebuttals to be an important form of science reporting. Hence, it is a point in favor of *Epistemically Balanced Reporting* that it is compatible with rebuttals in science reporting. Finally, the empirical evidence supporting *Justification Reporting* also supports *Epistemically Balanced Reporting*, and, although this remains to be empirically investigated, I speculate that the comparative presentation of scientific justification may be beneficial for laypersons' uptake. However, I re-emphasize that I do not purport that the tenacious problems arising from motivated cognition will be eliminated by *Epistemically Balanced Reporting*. Rather, the discussed empirical work suggests that it makes for a comparatively promising way of addressing them.

6.5.d Epistemic balance and no-platforming: As noted, *Epistemically Balanced Reporting* is compatible with no-platforming policies for hypotheses that do not meet a minimal threshold of scientific justification (Simpson and Srinivasan 2018; Levy 2019; Peters and Nottelmann forthcoming). For example, an epistemically balanced journalistic treatment of a given hypothesis is compatible with denying the proponents of the hypothesis a platform to promote it. Furthermore, *Epistemically Balanced Reporting* only requires that *if* an epistemically inferior hypothesis is reported, this is done by indicating that the strength and nature of its scientific justification is inferior. Thus, *Epistemically Balanced Reporting* does not require that flat Earth theories be reported in a story about tectonic plate movement.

It is a limitation of *Epistemically Balanced Reporting* that it does not entail a principled account of when to platform and when to no-platform testimony about various hypotheses. Since this is a grand question that intersects with foundational debates in political philosophy, I will not pursue such principles here. However, I assume that failing to meet a minimal threshold of scientific justification may be a reason to no-platform a hypothesis. Thus, I accept that it may be permissible, and perhaps even required, to no-platform some hypotheses in some science communication contexts. On the other hand, I also accept that even hypotheses that are backed by a significantly lesser degree of scientific justification than the dominant scientific hypotheses may be reported if this happens in an epistemically balanced manner. Finally, I take it that laypersons' expected uptake may be a factor that figures in an editorial decision about whether a hypothesis with limited scientific justification should be reported in an epistemically balanced manner or not at all.

Interestingly, recent evidence suggests that in the case of climate change, the science reporting has evolved to a point where climate science skepticism is more rarely presented in a balanced manner (McAllister et al. 2021). This development in climate science reporting may be an instance of how operative social norms can evolve to approximate objective ones more closely. Yet, the ethos of balanced reporting continues to influence science reporting more generally. But the change in operative social norms regarding climate change gives hope that conceptualizing and criticizing a journalistic practice may contribute to positive changes to that practice.

6.5.e Concluding remarks on *Epistemically Balanced Reporting*: In sum, *Epistemically Balanced Reporting* and *Inclusive Reliable Reporting* provide a response to the *Question of Balance* that is unified in that both principles are shaped by *Justification Reporting*. Although these principles may serve as general guidelines, there is a further step in developing contextually implementable guidelines such as editorial guidelines. However, both principles are more concrete and more plausible than the influential principles that they replace—i.e., *Balanced Reporting* and *Reliable Reporting*.

6.6 Concluding Remarks on Science Reporting

As in the case of scientific expert testimony, I have considered science reporting in relation to the socio-psychological obstacles for laypersons' reasonable uptake that empirical research has uncovered. This perspective gives rise to a rather critical, although not thoroughly dismissive, assessment of *Deficit Reporting*, *Consensus Reporting*, and *Value-Based Reporting*. However, the criticism provides the basis for a positive proposal, *Justification Reporting*, which should be central to science reporting, although it might on many occasions be fruitfully combined with the other principles. Finally, I applied *Justification Reporting* as a unifying principle in addressing the *Question of Balance*. A central upshot of doing so is the principle *Epistemically Balanced Reporting*. But, as so often in philosophy, a new proposal raises new problems, such as the problem of articulating a principled no-platforming policy in science reporting. This problem leads to the intersection between philosophy of science and political philosophy, and this is the area that the final chapter of the book is concerned with.

PART IV

SCIENTIFIC TESTIMONY IN SCIENCE AND SOCIETY

The fourth and final part of the book consists in Chapter 7 and a brief Coda. In Chapter 7, I unite the specific sub-conclusions that I have argued for throughout the book in a more general picture. The purpose of the Coda is to briefly indicate how this picture relates to important debates about cognitive diversity and epistemic injustice. Thus, Part IV seeks to synthesize the book's overarching themes and to point ahead to some novel areas of research on scientific testimony.

Chapter 7 begins with arguments for two theses about intra-scientific testimony, *Methodology* and *Parthood*, which jointly help to outline a positive *testimony-within-science* picture. I then argue for two theses about public scientific testimony, *Enterprise* and *Democracy*, which jointly indicate a principled alliance between science and societies that pursue the ideals of deliberative democracy. On this basis, I discuss the role of public scientific testimony in the *societal* division of cognitive labor. Specifically, I argue for the importance of *appreciative deference* to public scientific testimony and develop a novel norm for laypersons' uptake of public scientific testimony.

The Coda highlights in a brief and highly selective manner how scientific testimony relates to cognitive diversity and epistemic injustice. After characterizing these notions, I consider their relationship to intra-scientific testimony and public scientific testimony in turn. While my main aim is to briefly indicate areas for future research, I suggest that that battling epistemic injustice for cognitive diverse groups is central to the broader goal of aligning scientific expertise and democratic values.

7
The Significance of Scientific Testimony

7.0 The Place of Scientific Testimony in Science and Society

In this concluding chapter, I synthesize some of the sub-conclusions that I have argued for throughout the book and draw some general conclusions about the significance of scientific testimony. In particular, I unify the various arguments that scientific testimony is a genuine part of the scientific method. Moreover, I consider the roles of public scientific testimony in society.

In Section 7.1, I consider the roles of scientific testimony in science and society by way of articulating four theses that I will argue for. In Section 7.2, I provide arguments for two theses concerning the roles of intra-scientific testimony within the scientific enterprise. In Section 7.3, I provide arguments for two theses concerning the roles of public scientific testimony in society. In Section 7.4, I revisit the roles of public scientific testimony in the societal division of cognitive labor. In Section 7.5, I conclude by drawing some broader substantive and methodological lessons.

7.1 Scientific Testimony as the Mortar of the Scientific Edifice

The purpose of this section is to outline some general theses about the significance of intra-scientific testimony in science. I begin doing so with some slogans and context. On this basis, I articulate two specific theses about intra-scientific testimony that I will argue for.

7.1.a Slogans and metaphors: Effective scientific collaboration calls for norms that govern intra-scientific testimony—or so I argued in Chapter 4. Similarly, norms that govern public scientific testimony are important for public trust in science—or so I argued in Chapters 5 and 6. These arguments and their conclusions are important in their own right. But they also indicate how important intra-scientific testimony is to science and how important public scientific testimony is to society at large. Back in the Introduction, I contrasted the slogan of the Royal Society—*Nullius in verba*—with a slogan of my own: *Scientific testimony is the mortar of the scientific edifice.*

The slogan is, of course, metaphorical and, hence, merely apt to convey the general idea of substantive arguments. But the metaphor is apt to suggest that

Scientific Testimony: Its roles in science and society. Mikkel Gerken, Oxford University Press. © Mikkel Gerken 2022.
DOI: 10.1093/oso/9780198857273.003.0008

while testimony is a part of science, it is a different type of part than, for example, observations. Rather, scientific testimony is the part of science that makes such observational building blocks hang together (compare Miller and Freiman 2020). The slogan also indicates something that I will argue for in this chapter: Scientific testimony is key to the scientific enterprise's ability to sustain itself (Section 7.3).

However, the slogan and the edifice metaphor are misleading in other regards. For example, they are both static, whereas testimony is a part of a dynamic scientific process. So, I will move from the introductory metaphor to arguments for some more substantive theses.

7.1.b Two theses about intra-scientific testimony: The first two theses that I will argue for concern the relationship between intra-scientific testimony and the scientific enterprise. Here they are:

> *Methodology*
>
> The distinctive norms governing intra-scientific testimony are vital to the scientific methods of collaborative science.
>
> *Parthood*
>
> Intra-scientific testimony is a vital part of collaborative science.

Since *Methodology* and *Parthood* are articulated with the term 'vital,' they may be spelled out by specifying this term as a quasi-technical one. As a first approximation, *vitality* contrasts with, on the one hand, *essence* and, on the other hand, *enabling conditions*.

Being a vital part of something else is weaker than being an essential or necessary part of it. As the difficulties with providing a demarcation criterion of science have shown, it is doubtful that any practices, norms, or methods are essential to science (Hansson 2017). Science comes in many disciplinary varieties, and even within disciplines science has shown itself to be a movable beast that transforms itself radically over time. However, the assumption that very few, if any, parts of the scientific practice are essential to it does not entail that all of its parts are equally central to it.

For example, the contingent fact that a given scientific practice takes place at universities is less central to it than the contingent fact that it involves systematic observation and data analysis. Their being a part of science is explained by their being directly conducive to its epistemic functions. So, while a vital part of X need not be essential to X, it amounts to much more than being a mere enabling condition for X. Likewise, a vital part of something is not a peripheral part that can be easily replaced. The presence of a vital part of X is explained in a principled manner by its central contribution to the core function of X. For example, I will argue that intra-scientific testimony is a vital part of collaborative science insofar as it is a contingent but principled contributor to core functions of such science—namely, its epistemic functions. So, to say that something is a vital part of science is not an attempt to define science but rather to elicit the principled reasons why it is practiced the way it in fact is.

For illustration, recall Neurath's metaphor of a ship being continuously rebuilt at sea (Neurath 1921/1973). A structural feature of the ship, such as the keel, is a vital part of it. This is so even if the keel was not a part of the original structure and even if it may be replaced with another type of keel or abandoned in future keelless incarnations of the ship. But as the ship is in fact built, the keel is central to its purpose of sailing forward in a stable manner. While the keel may not be irreplaceable, it would wreck the ship if it were removed without substantive reworkings of the hull. So, the keel is a *vital part* of the ship, not a mere enabling condition for its existence (such as a dry dock) or its successful operation (such as wind). And while it may be a contingent part of the ship, it is a vital part because its presence is explained by its central contribution to a constitutive function of the ship (sailing). In contrast, another part of the ship, such as a galleon figure, is non-vital to it because its presence is not explained in terms of the ship's core function.[1]

Furthermore, some X can be vital for some Y although X is not a part of Y. For example, kidneys are vital for detoxicating blood, although kidneys are not part of detoxicated blood. This is so even if blood toxins could be eliminated by an implanted dialysis machine. This possibility does not compromise the idea that in a body where kidneys are actually detoxicating blood, they are vital to doing so. However, I will argue that the practice of intra-scientific testimony is a vital *part* of scientific practice, not merely vital to scientific practice. So, the closest, albeit still imperfect, analogy with kidneys is this: Just as a principled specification of how a well-functioning body actually works involves a specification of the kidneys' contribution, a principled specification of how science actually works involves a specification of intra-scientific testimony's contribution. So, ascribing vital part-hood does not amount to ascribing essence. Yet, saying that something is a vital part of X contributes to an account of X that is principled even though it may be metaphysically contingent. Given this specification of what vitality amounts to, I will return to the two theses that I will argue for: *Methodology* and *Parthood.*

Methodology concerns the types of norms that are distinctive of intra-scientific testimony, such as the norms of intra-scientific testimony and uptake articulated in Chapter 4. Importantly, *Methodology* does not entail any commitment to a single, unified methodology that demarcates science. It merely claims that some distinctive norms governing intra-scientific testimony are methodological norms that are vital to scientific practice. For example, violators of testimonial norms may engage in science in a sub-optimal manner, and egregious violators may fail

[1] Modulo cases in which the ship's central purpose is decorative, ceremonial, or some such.

to engage in science altogether. This point is compatible with assuming that science without intra-scientific testimony is possible. The point is merely that if intra-scientific testimony contributes to the epistemic force of a scientific practice, then failing to draw on it without a suitable replacement or blatantly violating its norms may amount to flirting with pseudo-science. A comparison with established norms of observation, data analysis, etc. may convey the idea. Not all scientific investigations require data analysis, but when a data analysis increases the epistemic force of a scientific investigation, it should be deployed in accordance with the norms governing it or be replaced with a suitable alternative. These norms may include more or less explicit criteria for selecting an appropriate statistical test (see UCLA Statistical Consulting Group 2021). Sometimes such norms are so explicit that they coincide with concrete guidelines. But even for something as concrete as statistical analysis, norms governing more subtle choices may only be manifested as a disciplinary practice. Given that the pursuit of truth is central to science, the operative social norms distinctive of science have an epistemic dimension. A scientist may be criticized on epistemic grounds if she violates the norms for selecting a suitable statistical test. This indicates that the norms governing the choice of statistical test contribute to a truth-conducive data analysis.

I will argue that something similar is true of intra-scientific testimony and the norms governing it. Moreover, I will suggest that the norms governing intra-scientific testimony are about as vital to scientific methodology as those governing scientific experiment design, instrumentation, observation, data analysis, etc. I use the phrase 'about as vital' because I will only attempt broad comparisons on the basis of the contribution to the epistemic aims of science. Specifically, I will argue that the contribution of the norms of intra-scientific testimony to the *epistemic force* of collaborative science is roughly on a par with the contribution of the norms governing, for example, data analysis.

The thesis *Parthood* claims that intra-scientific testimony is a vital part of collaborative science, and this claim may also be understood comparatively: The role of intra-scientific testimony in the scientific process is about as fundamental as the roles of observation, instrumentation, experimentation, data analysis, etc. So, according to *Parthood*, intra-scientific testimony is not merely an add-on or an enabling condition to the scientific practice but also a genuine part of it. Moreover, it is not any old part of it but a *vital* part that is approximately on a par with observation, data analysis, etc.

As noted, the claims about vitality fall short of claiming essential or constitutive features of science. But both theses mark contingent but principled truths about collaborative science. The contingency arises given that it is possible to engage in scientific practice without intra-scientific testimony. Such solitary science may have occurred in the early history of some scientific discipline. However, historical studies suggest that even early science differed from the myth of the solitary (white, male) genius. Interestingly, these studies include accounts of the Royal

Society (Shapin 1994; Moxham 2019; Bennett and Higgitt 2019). Yet some episodes in the history of science may have been largely solitary. Looking forward rather than backward, we can imagine how artificial intelligence may take over parts of science, leaving human scientists and their need to communicate behind. So, parts of science have been, and may become, fairly independent of intra-scientific testimony. However, *Methodology* and *Parthood* only claim that intra-scientific testimony and its norms help characterize collaborative science in a principled manner.

In sum, the theses *Methodology* and *Parthood* jointly provide an explicit contrast to the *Nullius in verba* tradition. They explicate the suggestion made from various angles in Chapters 1–4: Namely, that science is, with few exceptions, collaborative in a way that is characterized by a division of cognitive labor which, in turn, involves intra-scientific testimony and norms thereof.

7.2 Intra-Scientific Testimony and the Scientific Enterprise

In this section, I will outline arguments for the theses *Methodology* and *Parthood*. These arguments draw heavily on sub-conclusions of the previous chapters. Once the arguments are on the table, I will consider some objections that help to further illuminate the theses.

7.2.a An argument for *Methodology*: The argument for methodology is structurally fairly simple. Here it is:

Argument for Methodology

Collaboration's Contribution

M1: Scientific collaboration contributes immensely to the epistemic force of science.

Distinctive Norms

M2: The epistemic contribution of scientific collaboration depends on distinctive norms of intra-scientific testimony.

Methodological Norms

M3: If scientific collaboration contributes immensely to the epistemic force of science, and the epistemic contribution of scientific collaboration depends on distinctive norms of intra-scientific testimony, these norms are vital to the scientific methods of collaborative science.

M4: Scientific collaboration contributes immensely to the epistemic force of science, and the epistemic contribution of scientific collaboration depends on distinctive norms of intra-scientific testimony. (M1, M2)

M5: The distinctive norms of intra-scientific testimony are vital to the scientific methods of collaborative science. (M3, M4)

The conclusion, M5, is the thesis *Methodology*. For simplicity of presentation, the consequent of M3 uses the anaphoric phrase 'these norms' instead of the lengthy phrase 'distinctive norms of intra-scientific testimony' which occurs in the antecedent. But absent unforeseen presentational problems, equivocations, etc., the argument is valid, and the premises may be considered. The premises M1 and M2 are substantive theses in their own right which have been motivated as such in the previous chapters. So, I will briefly motivate each premise and indicate where the more extended motivation may be found.

The first premise (M1—*Collaboration's Contribution*) is the assumption that scientific collaboration based on the division of cognitive labor contributes immensely to the epistemic force of science. I use the term 'immensely' because I do not mean *fully* or even *primarily* but rather that scientific collaboration is one of the main reasons why science is epistemically forceful. This assumption is motivated by the considerations in Chapter 1.3.b and Chapter 3.5.c. So, I will defer to these sections and the extensive literature motivating the core idea: Scientific collaboration based on a fine-grained division of labor not only allows for considering more data than any single scientist ever could. It also allows for a degree of hyper-specialization that is epistemically truth-conducive when properly structured. So, when scientists collaborate in this structured manner, the epistemic benefits of such collaboration are often immense. Note that *Collaboration's Contribution* is contingent because it is a contingent, albeit highly principled, fact that scientists collaborate. So, the contingency of M1 explains the noted contingency of *Methodology*.

The second premise (M2—*Distinctive Norms*) is the assumption that scientific collaboration based on the division of cognitive labor depends on norms governing intra-scientific testimony for its epistemic force. This assumption is motivated in Chapter 4.5 (drawing on Chapters 1.5 and 2.4.c–d and 4.4.a). Scientific collaboration without intra-scientific testimony (or a proper replacement for it) is likely to be so scattered and ineffective that its epistemic advantages are severely compromised. For example, intra-scientific testimony contributes to merging distinct specialized sub-investigations into a sustained overarching investigation. However, this part of the scientific practice must also be governed by truth-conducive norms such as those governing providers of intra-scientific testimony. Similarly, the appropriate consumption of intra-scientific testimony requires that it is subject to norms of uptake that help to secure a truth-conducive collaborative scientific process. Such epistemic norms are distinctive norms of intra-scientific testimony, in contrast to non-distinctive norms pertaining to, for example, politeness. My proposed epistemic norm for providing intra-scientific testimony, NIST, is, moreover, distinctive in that it requires scientific justification. As emphasized, my proposals of norms of intra-scientific testimony are idealizations that seek to capture the most reasonable aspects of operative practices. But one need not accept the specifics of my proposals to accept that scientific collaboration depends on epistemic norms of providing and receiving intra-scientific testimony. The claim of M2 is merely the more general one that scientific collaboration depends epistemically on distinctive norms of intra-scientific testimony due to their role in securing its epistemic benefits.

The third and final premise (M3—*Methodological Norms*) is a conditional. Its antecedent has it that the epistemic force of science is partly but centrally explained by scientific collaboration which requires norms governing intra-scientific testimony. According to M3, this entails *Methodology*—i.e., that the distinctive norms governing intra-scientific testimony are vital to the scientific methods of collaborative science. Although M3 is not directly motivated previously in the book, a motivation from it may be drawn from the idea that norms can be vital for a practice (Chapters 1.5.b, 2.3.c–d, 4.2.a, and 4.5). For example, it is reasonable to suppose that norms distinctive of scientific data gathering are vital to empirical science insofar as the epistemic contribution of data gathering is structured by such norms (Chapters 2.4.c–d and 4.5). For example, the norm of double blinding in randomized clinical trials serves epistemic functions insofar as it helps to curb placebo effects among the participants and confirmation bias among the researchers. The norm of double blinding is a clear example in part because it is institutionalized as a concrete guideline (Aschengrau and Seage 2020). But even more tacit norms that lack corresponding concrete guidelines may serve epistemic functions. According to M3, the same goes for norms of intra-scientific testimony. Recall that the relevant notion of vitality only requires that the vital methodological norms help characterize collaborative science in a contingent but principled manner. Given this idea of vital norms, it seems plausible that the norms governing a scientific practice that help explain the epistemic force of collaborative science are among the distinctive norms vital to its methods. To make the abstract point more concrete, consider the epistemic norm of intra-scientific testimony, NIST, which helps to explain the epistemic force of collaborative science (Chapter 4.2). Moreover, NIST is sensitive to the nature and aims of science by requiring scientific justification. One might quarrel with the specific articulation of NIST. After all, it is an idealization of operative social norms in part because it seeks to capture the epistemic dimensions of social practices that do not only pursue truth but also novelty, visibility, citations, etc. However, it is harder to deny that a distinctive norm of intra-scientific testimony that plays a significant role in furthering the epistemic aims of science is among the norms that are vital to the methods of scientific collaboration. M3 reflects this idea.

This concludes my (recap of) motivation of the premises of *Argument for Methodology*. As noted, the argument brings the sub-conclusions that I have argued for previously in the book to bear on the overarching thesis,

Methodology. Before considering some objections, let us consider the structurally similar argument for *Parthood*.

7.2.b An argument for *Parthood*: The argument for *Parthood* shares its structure with the argument for *Methodology*.

Argument for Parthood

Collaboration's Contribution

P1: Scientific collaboration contributes immensely to the epistemic force of science.

Testimony's Contribution

P2: Intra-scientific testimony is an epistemically vital part of scientific collaboration.

Collaboration Parthood

P3: If scientific collaboration contributes immensely to the epistemic force of science, and intra-scientific testimony is an epistemically vital part of scientific collaboration, then intra-scientific testimony is a vital part of collaborative science.

P4: Scientific collaboration contributes immensely to the epistemic force of science, and intra-scientific testimony is an epistemically vital part of scientific collaboration. P1, P2

P5: Intra-scientific testimony is a vital part of collaborative science. P3, P4

This argument shares the structure of *Argument for Methodology* as well as the broad idea that the role of intra-scientific testimony in contributing to the epistemic goals of science makes it a vital part of science. Moreover, the first premise, P1, *Collaboration's Contribution*, is identical to the first premise of the preceding "sister argument," so I will move on to P2 and P3.

The premise P2, *Testimony's Contribution*, has it that intra-scientific testimony is an epistemically vital part of scientific collaboration (Chapters 1.3.c and Chapter 3.5.c). Recall that the scientific collaboration in question is based on a fine-grained division of cognitive labor. But, generally speaking, these distinct parts of the investigations must be brought to bear on each other in order for the division of labor to be epistemically beneficial. This is standardly done by way of intra-scientific testimony. Communicating the findings, obstacles, progress, etc. to one's collaborators is part and parcel of scientific collaboration. As highlighted in Chapter 4, scientists are held accountable for this part of the scientific collaboration in a relevantly similar manner to the way they are held responsible for making observations and analyzing them. This strongly suggests that intra-scientific testimony is a central part of the run-of-the-mill scientific collaboration.

Furthermore, specialized scientists who conduct separate parts of a unified investigation would be hard-pressed to collaborate in a manner with distinctive epistemic benefits without intra-scientific testimony. Just as distinct observations must be analyzed to contribute to a scientific investigation, its specialized aspects must be brought together to contribute to it. Intra-scientific testimony is the part of a science collaboration that does so. In this regard, the metaphor that intra-scientific testimony is the mortar of collaborative science is not too far off the mark. Although the mortar is a different part of a house than the bricks, it is still a vital part. Similarly, although intra-scientific testimony is a different part of the collaborative practice than gathering data, it is still a vital part of it. In addition to the noted unifying role, intra-scientific testimony is also a part of the early stages of scientific collaboration. It is a part of developing ideas, narrowing down a workable focus, developing study designs, and making procedural plans. All of these practices are core parts of scientific collaboration and standardly involve intra-scientific testimony. So, scientific collaboration involves intra-scientific testimony and depends on it for its epistemic force. This suggests that it is not merely a part of scientific collaboration but an epistemically vital part of it. This is P2.

The premise P3, *Collaboration*, is a conditional. Its antecedent claims that intra-scientific testimony is an epistemically vital part of scientific collaborations which contributes immensely to the epistemic force of collaborative science. According to P3, this claim entails *Parthood*—viz., the claim that intra-scientific testimony is a vital part of collaborative science. Importantly, the consequent of P3, *Parthood*, is only a claim about the vitality of intra-scientific testimony rather than a claim about essence or constitution. Hence, it is compatible with some strains of science in which intra-scientific testimony plays little or no role. It only requires that intra-scientific testimony is ordinarily an epistemically vital part of science. So, in a nutshell, P3 has it that since intra-scientific testimony is a part of scientific collaboration which contributes significantly to its epistemic force, it is reasonably regarded as a vital part of collaborative science itself. This rationale does not require that *vital parthood* is transitive. Rather, it is an abductive rationale that appeals to the centrality of intra-scientific testimony's contribution to the epistemic force of science through its role in collaboration. Hence, an interesting line of objection to P3 would be to reject that a central aim of science is epistemic. After all, the sketched rationale for P3 appeals to intra-scientific testimony's contribution to the epistemic aims of science. The idea that epistemic aims are central to science is one of the broadly realist background assumptions of the book that I will not defend here (Psillos 1999; Chakravartty 2011). However, it may even be that P3 can be motivated on pragmatic grounds insofar as intra-scientific testimony arguably plays a similar role in contributing to the practical aims of science as the epistemic role that I have argued that it fulfills. However, given the assumption that science has important epistemic aims which intra-scientific testimony is central to pursuing, P3 seems plausible.

The arguments *Methodology* and *Parthood* are structurally similar. Moreover, they both appeal to the idea that intra-scientific testimony is partly but very significantly contributing to scientific collaboration's immense contribution to the epistemic force of science. So, in order to clarify the theses *Methodology* and *Parthood* and the arguments for them, I will consider some objections.

7.2.c The objection from lack of generality: I have touched on the first objection already. It is the concern that science may in principle take place in solitude, and hence, without intra-scientific testimony. For example, a lone genius on a deserted island may engage in scientific investigation. Likewise, types of scientific collaboration that need not involve testimony may perhaps be imagined. Does this show that intra-scientific testimony is not vital to collaborative science?

In response, the first thing to recall is that vitality is weaker than essence in that it only requires a contingent but principled characterization of science and scientific methodology. Thus, neither *Methodology* nor *Parthood* entails that intra-scientific testimony is necessary for collaborative science or that the norms governing it are necessary for scientific methodology.

Moreover, the possibility of science without testimony does not establish that intra-scientific testimony is not vital to the collaborative scientific practice. A principled characterization of collaborative science as it is in fact practiced must include intra-scientific testimony. To see this, consider an analogy with the central role that randomized controlled trials play in the medical sciences (Worrall 2007; Roush 2009; Cartwright 2010). One need not be a critic of randomized controlled trials to acknowledge that legitimate medical science need not always involve them. On the other hand, even critics of randomized controlled trials may acknowledge that they, and the norms governing them, are vital to the medical sciences. In many cases, a research group, G, would not live up to the state-of-the-art contemporary medical science if G did not perform randomized controlled trials in a particular way—e.g., by double blinding, performing statistical analysis, etc. Thus, randomized controlled trials contribute to a contingent but principled characterization of contemporary medical science as it in fact practiced for principled reasons.

At this juncture, some opponents may object that randomized controlled trials are nevertheless not a vital part of the medical sciences and that the distinctive norms governing them are not among the norms vital to the methodology of the medical sciences. But such an objection would be more plausible if we were trying to define medical science rather than trying to characterize the principled reasons why it is practiced the way it is. Other opponents may object that randomized controlled trials and the norms governing them are decidedly more central to the medical sciences than intra-scientific testimony and the norms governing it are to collaborative science. Such an objection strategy must establish a principled asymmetry with intra-scientific testimony. So, I will briefly consider whether this may be done.

7.2.d The objection from enabling conditions: It might be objected that while intra-scientific testimony is important to the scientific process, it is not vital but merely enabling. This objection may be fueled by postulating asymmetries between intra-scientific testimony and other parts of science, such as data collection. For example, it may be argued that since intra-scientific testimony is posterior to data collection, it is not a vital part of the scientific practice but merely an enabling part. This is not a good argument. Data analysis is also temporally posterior to data gathering, but this does not show that data analysis is not a vital part of science or that the norms governing it are not among those vital to the scientific method.

A more challenging version of the argument does not focus on temporal order but on explanatory priority. This argument begins with the assumption that data collection is explanatorily primary to intra-scientific testimony since the former is required for the latter to take place. The objector may highlight that the content of intra-scientific testimony is routinely about collected data, whereas it is exceptional that scientists collect data about intra-scientific testimony. So, it is suggested that intra-scientific testimony is *asymmetrically dependent* on parts of the scientific practice that are more commonly recognized as part of the scientific method—e.g., data collection.

But on closer reflection, the asymmetry is not particularly robust. There are many cases in which intra-scientific testimony is both prior to, and a prerequisite for, other aspects of the scientific process such as data collection. For example, scientists may have to explain the relevance and feasibility of a novel research project to the collaborators as an initial stage of their collaboration. Typically, this involves at least some claims that are based on scientific justification. So, the early stages of scientific collaboration often involve intra-scientific testimony. For example, the process of designing the study whereby data is to be collected is often a collaborative affair that requires intra-scientific testimony. Furthermore, larger research projects that seek to synthesize many studies and use mixed methods must also be coordinated by way of intra-scientific testimony prior to data collection. Finally, even the context of discovery often requires intra-scientific testimony. Generating new ideas is part of science, and often ideas arise from interlocution about findings, theories, and applications, etc. While part of the context of discovery may be speculative, other parts of it are based on scientific justification (Schickore 2018). So, intra-scientific testimony figures already at this stage of the scientific process. At the later stage when a new idea is generated, experimental studies must be designed, and this part of scientific collaboration also standardly requires intra-scientific testimony. Likewise, more specific procedural aspects of the investigation—such as coding and categorization of data—must be coordinated by way of intra-scientific testimony. So, in many cases, proper data collection is dependent on intra-scientific testimony in preceding stages of the scientific process.

In the discussed cases, the intra-scientific testimony does not appear to be *merely* enabling scientific practice. Rather, it seems to be as much a part of the ordinary practice of science as the subsequent data collection, data analysis, etc. Perhaps it might be objected that, according to my characterization of scientific testimony, the testimonies in early stages of a collaboration are intra-scientific testimony only if they are properly based on scientific justification. If so, goes the objection, such intra-scientific testimony is ultimately resting on previous observation, etc. In response to this objection, it may be noted that intra-scientific testimony is also involved in previous scientific work. So, the broader picture is not one in which proper science must take place before intra-scientific testimony may occur. Rather, the picture that emerges is one in which intra-scientific justification and other aspects of the scientific practice are extremely intertwined. So, the arguments from temporal or explanatory primacy fail to establish that intra-scientific testimony is a mere enabling condition.

The dismantling of these negative arguments for treating intra-scientific testimony as a mere enabling condition may be supplemented with a positive line of reasoning for treating intra-scientific testimony as a vital part of collaborative science. This line of reasoning draws on Longino's arguments that assertions in favor of a scientific hypothesis are required for scientific objectivity (Longino 1990). Roughly, this is because scientific objectivity requires critical examination by the wider scientific community. This, in turn, requires that the hypothesis, and perhaps its epistemic basis, is communicated broadly within it, defended against objections, etc. According to Longino, this communal scrutiny of hypotheses, in large part, is what distinguishes scientific justification in terms of objectivity (Longino 1990). Given this picture, intra-scientific testimony is a vital part of the communal examination that is, in turn, an epistemically vital part of the scientific practice broadly conceived. Thus, a reasonably broad understanding of collaborative scientific practice and its methodology will include intra-scientific testimony as one of its vital parts.

7.2.e The theses about intra-scientific testimony in conclusion: Jointly, *Methodology* and *Parthood* contribute to a *testimony-within-science* image which starkly contrasts with the *science-before-testimony* tradition which remains a part of the folk conception of science. Such a folk conception, according to which a scientist operates in isolation without central reliance on intra-scientific testimony, may encourage misconceptions such as the *great white man fetish* (Chapter 5.3.c). But although the theses may primarily be at odds with folk conceptions of science, even philosophers of science often consider the nature of scientific collaboration in abstraction from intra-scientific testimony. In contrast, *Methodology* and *Parthood* locate intra-scientific testimony at the center stage, rather than in the periphery, of collaborative science. The explicit arguments articulated here help to pinpoint this centrality as vital rather than essential on the one hand. But they also help to pinpoint it as vital rather than as a mere enabling condition on the other hand. In this manner, the arguments help characterize the significance of intra-scientific testimony within science. Moreover, the explicit arguments elicit the *reasons why* intra-scientific testimony is vital. In a nutshell, the arguments help reveal that intra-scientific testimony fills an important functional role of contributing to the epistemic force of collaborative science. Thus, a comprehensive characterization of collaborative science and its vital methodological norms must include a characterization of intra-scientific testimony and the norms governing it. By considering the epistemic norms of producing and consuming intra-scientific testimony, I have only begun the work of developing a *testimony-within-science* account. For example, it must be explored whether there are distinctive norms of intra-scientific testimony that pertain to various aspects of presentation—e.g., how much simplification is permitted or required, whether technical terms must be glossed, etc.

Finally, it is important to recognize that intra-scientific testimony is vital because this indicates inadequately recognized sources of fallibility. For example, if the structural incentives to provide and consume intra-scientific testimony in accordance with reasonable norms are corrupted, the truth-conduciveness of scientific collaboration may be compromised. Thus, articulating the two theses about intra-scientific testimony contributes to the understanding of collaborative science and its methodology.

7.3 Public Scientific Testimony in Society

In this section, I will outline arguments for two theses concerning public scientific testimony—*Enterprise* and *Democracy*. Once the arguments for these theses are on the table, I will respond to some objections to them and begin to consider some of their wider ramifications.

7.3.a Two theses about public scientific testimony: The two theses concerning the relationship between public scientific testimony and the ideals of deliberative democracy that I will argue for are stated as follows:

Enterprise

Public scientific testimony is critical for the scientific enterprise in societies pursuing the ideals of deliberative democracy.

Democracy

Public scientific testimony is a critical part of societies pursuing the ideals of deliberative democracy.

Both *Enterprise* and *Democracy* are articulated in terms of the idea of being *critical*. I say that factors are *critical* when they are contingently and partly, but nevertheless importantly and non-arbitrarily, enabling the continued existence of something. Thus, being critical for something is less demanding than being vital for it insofar as critical factors may only amount to enabling conditions. Since criticality is gradable, I use outright ascriptions of it when the degree of criticality of Y for X exceeds a comparatively high threshold. This may be the case if removing Y without replacing it would eliminate X or X's ability to function properly. For example, adequate food is critical to perform some physical task even though it might be replaced with an intravenous drip. But if adequate food is eliminated without being replaced, the ability to complete the task is severely compromised or even eliminated. However, adequate food is *merely* an enabling condition for it. Hence, it is critical for the physical task rather than a vital part of it or vital for it. Note that the critical entity may both be a part of the phenomenon that it is critical for or merely a resource for it.

By arguing that Y is critical to X, one does not contribute as significantly to a principled characterization of the nature of X as when one argues that Z is a vital part of X. My hip is *vital* to my ability to traverse an icy wilderness, whereas my boots are only *critical* for it. While neither are irreplaceable, my hip is integral to my walking across terrain in virtue of being part of me, whereas my boots are not. Nevertheless, an account of my pedestrian achievement should include an account of the roles of my hips and my boots. Moreover, deficiencies in either hip or boots may centrally explain why my progress was slowed down, etc. Analogously, the theses *Enterprise* and *Democracy* exemplify that we may learn something important about both Y, X, and their relationship by arguing that Y is critical for X.

The thesis *Enterprise* indicates how important public scientific testimony is for the scientific enterprise. But since it is cast in terms of criticality, it does not postulate that public scientific testimony is a necessary or even a vital part of the scientific enterprise. Rather, *Enterprise* is the claim that public scientific testimony is central in *enabling* the scientific enterprise in societies that pursue deliberative democracy's ideals. This marks one important difference between intra-scientific testimony and public scientific testimony. Nevertheless, public scientific testimony may be said to be a part of science in the more restricted sense of being critical for it.

The thesis *Democracy* concerns the relationship between public scientific testimony and societies that pursue the ideals of deliberative democracy. The thesis is not that public scientific testimony is an essential, irreplaceable, or even vital part of such societies. Rather, *Democracy* has it that the pursuit of deliberative democracy involves informing the public about policy-relevant matters and that public scientific testimony, as a contingent but non-arbitrary matter of fact, is a central way of doing so.

I restrict both theses to societies pursuing ideals of deliberative democracy primarily to retain a reasonable focus and because my arguments appeal to aspects of these ideals. A society pursues ideals of deliberative democracy insofar as it is to some extent governed by the ideals of deliberative democracy, such as the ideal of an informed electorate. For example, many democracies fund education, support research, and promote free media by appealing to epistemic ideals of deliberative democracy. This is so even in democracies where the distinctively deliberative ideals and the pursuit thereof are under pressure (Achen and Bartels 2017). Yet, what distinguishes a deliberative democracy from other types of democracy is the idea of an electorate that is reasonably well informed about policy-relevant matters, and these include scientific hypotheses. So, in a nutshell, *Democracy* is the claim that public scientific testimony is central in the pursuit of the ideals of a distinctively *deliberative* democracy.

Enterprise and *Democracy* are weaker theses than *Methodology* and *Parthood* insofar as they only postulate critical, rather than vital or constitutive, relations between public scientific testimony and the ideals of deliberative democracy. But taken together, the two theses nevertheless indicate an important bidirectional relationship between public scientific testimony and deliberative democracy. On the one hand, the ideals of an informed electorate indicate that public scientific testimony is critical to the scientific enterprise as *Enterprise* has it. On the other hand, public scientific testimony is central in the pursuit of the ideals of deliberative democracy as *Democracy* has it. Thus, an important aspect of the broader alliance between science and deliberative democracy may be elucidated by clearly articulating *Enterprise* and *Democracy* and providing explicit arguments for them.

7.3.b An argument for *Enterprise*: Recall that the thesis *Enterprise* is the claim that public scientific testimony is critical for the scientific enterprise in societies that pursue the ideals of deliberative democracy. Recall that the term 'critical' expresses the idea that the continued existence of the scientific enterprise partly but centrally requires public scientific testimony as an enabling condition. Here is an argument for *Enterprise*:

The Argument from Self-Sustainment

S1: The scientific enterprise is, in large part and for principled reasons, sustaining itself in societies pursuing ideals of deliberative democracy by providing epistemically authoritative public scientific testimony.

S2: If the scientific enterprise is, in large part and for principled reasons, sustaining itself in societies pursuing ideals of deliberative democracy by providing epistemically authoritative public scientific testimony, then public scientific testimony is critical for the scientific enterprise in societies pursuing the ideals of deliberative democracy.

S3: Public scientific testimony is critical for the scientific enterprise in societies pursuing the ideals of deliberative democracy. (S1, S2)

The Argument from Self-Sustainment is straightforwardly valid as it only consists of a single modus ponens inference. So, I will proceed to motivate its premises.

The premise, S1, is the claim that in societies pursuing ideals of deliberative democracy, the scientific enterprise is, for principled reasons, sustaining itself by providing epistemically authoritative public scientific testimony (Anderson 2011; Keren 2018). This is a contingent claim and S3—i.e., *Enterprise*—inherits this contingency. However, the mentioned principled reasons concern the principled connection between science and deliberative democracy's ideal of an informed electorate. In consequence, many societies pursuing ideals of deliberative democracy fund scientific research (Cohen 1989; Kitcher 2011). So, the contingency of the claim does not mean that it is arbitrarily true. Rather, the pursuit of the ideals of deliberative democracy would be epistemically compromised, even as an ideal, if public scientific testimony was absent and not replaced with a comparable source of warrant for policy-relevant hypotheses. Even theorists who take public reason to trump individual expertise tend to accept that science provides an indispensable input to public deliberation (Landemore 2012). Moreover, if the epistemic experts cannot transfer some of their epistemic authority to the deliberating public, deliberative democracy is unlikely to be an epistemically reasonable alternative to an epistocracy, according to which the epistemic authorities should rule (Estlund 2008; Kitcher 2011). Public scientific testimony is the premier way of securing such a transfer (Wilholt 2013; Irzik and Kurtulmus 2019). Finally, there is the question of legitimacy (Peter 2008; Grasswick 2010). To the extent that science is comparatively value-neutral, public scientific testimony may provide comparatively objective verdicts on matters of public interest. (I write 'comparatively objective' in recognition of the arguments against the value-free ideal of science (Douglas 2009, 2015; Winsberg 2018; Brown 2020)). So, public deliberation based on public scientific testimony may stand a reasonable chance of striking the balance between epistemically reasonable and politically legitimate verdicts. Hence, societies pursuing ideals of deliberative democracy have a principled reason to fund science, and many societies that self-identify as democracies do so. So, while this discussion of the relationship between deliberative democracy and science only scratches the surface of this major issue, it helps motivate S1. Specifically, it identifies some of the principled reasons why the scientific enterprise in large part sustains itself through public scientific testimony in societies that pursue deliberative democracy's central ideals.

The premise S2 has it, in a nutshell, that if the scientific enterprise sustains itself through public scientific testimony in societies that pursue ideals of deliberative democracy, public scientific testimony is *critical* for the scientific enterprise in such societies. This premise is motivated by reflection on what it takes for a practice to be a critical part of a larger enterprise. Recall that being critical only requires that public scientific testimony is contingently required to partly sustain science in societies that pursue ideals of deliberative democracies as an enabling condition. Hence, playing a large—contingent but principled—part in self-sustainment is a plausible sufficient condition for being critical. This general idea is no less plausible in the case of public scientific testimony. What makes this a plausible sufficient condition is the 'for principled reasons' component of the antecedent of S2 (i.e., S1). To see this, assume that public scientific testimony enabled the continued existence of science because it was aesthetically pleasing to philanthropes. In this case, little could be concluded about whether it is critical for societies pursuing deliberative democracy's ideals since this is not a reason that reflects a principled connection between public scientific testimony and the ideals of deliberative democracy. In contrast, S2 has it that public scientific testimony is critical because the pursuit of ideals of deliberative democracy requires epistemically authoritative testimony, and it helps to address this need.

Thus, articulating and motivating *The Argument from Self-Sustainment* does not merely serve to motivate *Enterprise*. It moreover helps to clarify the sense in which public scientific testimony is critical to science in societies that pursue the ideals of deliberative democracy and to explain why this is so.

7.3.c An argument for *Democracy*: Recall that the thesis *Democracy* is the claim that public scientific testimony is critical for societies pursuing the ideals of deliberative democracy. Here is an argument for this conclusion:

The Argument from an Informed Public

D1: Pursuing the ideals of deliberative democracy involves fostering an informed public.

D2: If pursuing the ideals of deliberative democracy involves fostering an informed public, all democratically legitimate practices which make a major contribution to an informed public are critical for societies pursuing the ideals of deliberative democracy.

D3: Public scientific testimony is a democratically legitimate practice which makes a major contribution to an informed public.

D4: All democratically legitimate practices which make a major contribution to an informed public are critical for societies pursuing the ideals of deliberative democracy. (D1, D2)

D5: Public scientific testimony is critical for societies pursuing the ideals of deliberative democracy. (D3, D4)

Since the *Argument from an Informed Public* is also structurally straightforward, I will turn to motivating its premises.

The first premise, D1, has it that pursuing the ideals of deliberative democracy involves fostering an informed public. Roughly, part of what is distinctive of *deliberative* democracy is that the electorate is sufficiently informed about politically relevant issues to engage in reasonably deliberative practices as a basis for their choices. Consequently, I take the notion of an *informed public* to involve the formation and sustainment of beliefs by a source that normally produces a degree of warrant over some threshold. There is room for reasonable debate about how high this threshold must be and whether it varies with practical factors (see Gerken 2017a, 2018d on the latter question). However, these debates need not be settled to argue that if public beliefs about scientific hypotheses are entirely ill-informed, then public deliberation will be epistemically compromised. Even if deliberation among laypersons is truth-conducive, I conjecture that it is not generally capable of turning unreliable inputs about scientific hypotheses into reliable outputs about them. Here the "garbage in, garbage out" principle likely rules. Note that D1 only claims that the pursuit of the *ideals* of deliberative democracy requires fostering an informed public. So, D1 does not involve the contingent and debatable claim that actual deliberative democracies are well-functioning in virtue of fostering an informed public. Relatedly, D1 is compatible with assuming that a deliberative democracy that fails to foster an informed public is the only preferable or justifiable political ideal. So, generally, D1 is a fairly weak thesis that speaks to a distinctive epistemic aspect of deliberative democracy.

The second premise, D2, has it that if pursuing the ideals of deliberative democracy involves fostering an informed public, all democratically legitimate practices which make a major contribution to an informed public are *critical* for societies pursuing ideals of deliberative democracy.

Let me say a bit about the sense in which I use the qualification 'major contribution' that figures in the argument. I assume that contribution to an informed public is major in the relevant sense only if it provides germane input for public deliberation regarding a societally important topic. It is not easy to give a principled characterization of what societally important topics amount to. But the broad idea is that it is the type of topic that the electorate is concerned with *qua* electorate. For example, it may be a topic that is relevant for policymaking or judicial decisions. These topics contrast with issues that are not relevant for policymaking. For instance, although information about sports leads to ample public deliberation, it is not generally germane to the relevant kind of public deliberation. Hence, sports TV networks do not provide a major contribution to an informed public even though they provide ample input to deliberation about sports in the public sphere. In contrast, information about vaccine efficacy or climate change is germane to the relevant types of public deliberation. In addition to requiring that the input must be germane to public deliberation of societal issues, I assume that a contribution to an informed public is a major contribution only if it constitutes epistemically weighty input for public deliberation. The idea of an epistemically weighty input involves the comparative claim that the input needs to be well warranted compared to input by alternative sources. Furthermore, the input is epistemically weighty only if it consists of a great magnitude of relevant input, uniquely relevant input, or input of especially great relevance. So, a practice that provides only a miniscule amount of germane but non-exceptional input to public deliberation does not provide a major contribution to an informed public.

The plausibility of D2 is partly due to the fact that being critical does not entail being irreplaceable. Recall that a practice can be critical without being necessary. To be a critical resource to a society pursuing deliberative democracy's ideals, a practice merely has to make a major contribution to fulfilling a role that is required for the well functioning of deliberative democracy. So, since being critical does *not* require being irreplaceable, it is hard to see how a practice that makes a major contribution to fulfilling the role of informing the public would not also be critical for societies pursuing deliberative democracy.

The qualification to *democratically legitimate* practices is included in D2 because there may be democratically illegitimate ways to foster an informed public. Since these should not be regarded as critical for deliberative democracy, D2 is restricted accordingly. But given the restriction to democratically legitimate practices, it is hard to think of a practice that makes a major contribution to an informed public that is not critical for societies pursuing deliberative democracy. Note, finally, that the argument could be developed without the universally quantified consequent of D2. But since I have not come across a counterexample, and no one has pressed one on me, I will opt for the universally quantified consequent of D2 in the present formulation of the argument. If someone comes up with a practice that makes a major contribution to an informed public, but which is not critical for societies pursuing deliberative democracy, it may teach a valuable lesson which may serve as the basis for a substantive restriction.

The third premise, D3, is the claim that public scientific testimony is a democratically legitimate practice which makes a major contribution to an informed public. I take it to be uncontroversial that the public scientific testimony makes a germane contribution insofar as it often concerns issues that are directly relevant for policymaking or issues that the electorate is concerned with *qua* electorate. Moreover, public scientific testimony makes a comparatively epistemically weighty contribution to an informed public. First of all, public scientific testimony is the comparatively superior source of input to public deliberation in many domains (Chapter 3.3). Furthermore, public scientific testimony provides much of the relevant input, some of which no other sources provide.

A key point for the motivation of D3 is that a practice can make a major contribution without being irreplaceable. So, I do not assume that other sources of information could not fill the social role of public scientific testimony. However, if public scientific testimony were dispensed with and not replaced, the public would be significantly less informed to the point where the electorate would be epistemically compromised. In societies that pursue the ideals of deliberative democracy, the various kinds of public scientific testimony clearly play a partial but major role in fostering an informed public. Motivation for this assumption derives partly from the noted epistemic superiority of scientific justification and thereby public scientific testimony which is properly based on it. Given that public scientific testimony is, in many policy-relevant areas, epistemically superior to other available sources, it will generally contribute to a reasonably well-informed public about those areas. Interestingly, this point highlights the importance of avoiding trespassing testimony of the sort discussed in Chapter 5.5. But here, the key point is that D3 is motivated by the arguments that public scientific testimony is a superior epistemic source about many policy-relevant areas than alternative sources that are available to the public.

One objection to D3 has it that public scientific testimony is not a democratically legitimate practice. This might be seen as an instance of more general worries about aligning scientific authority and democratic values (Kitcher 2003, 2011). However, such worries often concern the allocation of scientific resources rather than public scientific testimony. A more specific worry relates the arguments from inductive risk according to which scientists must invoke non-epistemic values in accepting and communicating hypotheses.[2] However, arguments against science's value-free ideal are rarely taken to compromise the democratic legitimacy of public scientific testimony. So, here I will just flag the issue and note that although public scientific testimony may of course be abused in non-democratic ways, it does not appear to be intrinsically at odds with democratic values. If so, D3 appears to be a plausible assumption.

This concludes the motivation for the premises of the *Argument from Informed Deliberation* and thereby the case for the thesis *Democracy*.

7.3.d How public scientific testimony and scientific investigation are interwoven: The theses *Enterprise* and *Democracy* jointly indicate how important public scientific testimony is to the scientific enterprise and to societies that pursue the ideals of deliberative democracy (see also Jasanoff 2004a). A description of science as a self-sustaining human enterprise would be inadequate without a characterization of public scientific testimony and the norms governing it. Public scientific testimony and scientific practice are closely interwoven.

For example, public scientific testimony is not merely relevant once scientific investigations are completed. This *science-before-testimony* picture is misleading because public scientific testimony may be required by core scientific practices such as data collection. The biomedical sciences provide an example. In this field,

[2] Douglas 2009, 2015; Steele 2012; John 2015a, 2015b; Brown 2020.

a lot of the data collection standardly requires *informed consent* by participant laypersons (Manson 2007; Beauchamp 2011). Informed consent, in turn, requires a species of public scientific testimony. Even the initial process of recruiting participants from the right demographic may require public scientific testimony. The putative participants must be informed about the aims, nature, and risks of the research project. While not all of these testimonies qualify as scientific testimony, some of them do. This testimonial interaction takes place temporally prior to data collection and analysis. As a matter of contingent but deep-seated social fact, scientists in societies pursuing ideals of deliberative democracy can, in some areas, only acquire data by recruiting participants and acquiring consent. But recruiting participants and acquiring consent often require public scientific testimony. So, as a contingent but highly principled fact in societies pursuing ideals of deliberative democracy, data collection and analysis are asymmetrically dependent on public scientific testimony. In slogan: *No public scientific testimony, no data.*

Another example comes from ecology and environmental sciences, which have seen an increase in *citizen science* in which the public partakes in the scientific process such as data collection.[3] Again, public scientific testimony is standardly required to ensure that the citizen scientists are able to contribute in an appropriate and structured manner. A final example that sits between radically collaborative research and citizen science is a paper in genomics that recruited 900 undergrads to evaluate and weigh evidence (Leung et al. 2015). Clearly, scientific testimony plays a critical role in the conduction of such a study, but it involves aspects of both intra-scientific and public scientific testimony. Given how diverse citizen science is, it arguably involves both intra-scientific testimony, public scientific testimony, and combinations thereof. In contrast, the case of informed consent typically only involves public scientific testimony—albeit its own distinctive kind.

The general lesson from these cases is that even public scientific testimony is far more closely interwoven with scientific practice than the *science-before-testimony* picture would have it. This observation reinforces *Enterprise* by indicating different concrete ways in which public scientific testimony is critical for the scientific enterprise. In some cases, public scientific testimony may be seen as part and parcel of the scientific practice similar to, for example, data collection. Cases of informed consent may exemplify this. In this case, the norms governing public scientific testimony for the purpose of obtaining informed consent are reasonably seen as *scientific* norms. In other cases, public scientific testimony is a mere enabling condition. In yet other cases, public scientific testimony is in a gray zone between being an integral part of the scientific practice and merely enabling it. Thus, the distinction between science and science communication is much more gradual, messy, and complex than is commonly thought.

[3] Dickinson, Zuckerberg, and Bonter 2010; Irwin 2015; Kullenberg and Kasperowski 2016; Allen 2018.

7.3.e The theses about public scientific testimony in conclusion: Taken together, *Enterprise* and *Democracy* clarify an important alliance between public scientific testimony and societies that pursue ideals of deliberative democracy. In particular, the theses jointly indicate subtle interdependencies among public scientific testimony, scientific research, and such societies. The media is often thought to be the fourth estate or pillar of democracy given the importance of open critical debate and public deliberation. But it should not be forgotten that the media is granted this status because of its role in enlightening the public. However, the media is often merely mediating science. Hence, the *source* of enlightenment is often varieties of public scientific testimony. So, if you will allow me to return to the architectural metaphors, public scientific testimony is at the base of several pillars of deliberative democracy. The media is one but, as noted, the legislative and the judicial pillars also partly rest on public scientific testimony (compare Jasanoff 1990, who regards science advisers as a "fifth branch").

Thus, *Enterprise* and *Democracy* contribute to a principled account of the role of science in society. Although the theses are important in their own right, they also raise questions about the societal roles of public scientific testimony. For example, they raise questions about the conventions concerning the public uptake of public scientific testimony, the public responsibility for securing platforms for it, and the (editorial) guidelines for such platforms. In what remains of the book, I will cursorily explore some of the ramifications of the theses.

7.4 Scientific Testimony in the Societal Division of Labor

In the final sections, I will move beyond the specific arguments and speed things up with some more brisk reflections about the wider ramifications of the four theses and the other conclusions of the book. I will take a social perspective on the consumer side of public scientific testimony. Specifically, I will consider a recipient norm of uptake of public scientific testimony. More generally, I will consider a societal division of labor that enables laypersons' acquisition of entitled testimonial belief on the basis of a stance that I will call *appreciative deference*.

7.4.a Scientific testimony and deliberative democracy: Since the theses *Enterprise* and *Democracy* indicate mutual dependencies between science and the wider society, they raise novel issues about the societal division of cognitive labor (Chapter 1.4.b). A simple *science-before-testimony* picture would suggest the following societal division of labor: First, science investigates some purely factual matters and, after passively receiving public scientific testimony, the public makes the relevant normative deliberations on this basis.

However, this picture is inaccurate. As noted, the extensive criticism of the tenability of science's value-free ideal suggests that non-cognitive values figure in the scientific practice (Douglas 2009, 2015; Steele 2012; Winsberg 2018; Brown 2020). Moreover, *Enterprise* and *Democracy* suggest that the testimonial engagement between scientists and laypersons should be a two-way street. One approach to this idea is Kitcher's *well-ordered science*, according to which scientific resources should be determined in a manner that reflects an idealized hypothetical democratic deliberation (Kitcher 2001, 2011). This idea has also received its fair share of criticism (Brown 2013; Douglas 2013; Keren 2013). But one need not adopt the idea of well-ordered science to accept that the obligations of securing a reasonable role for science in society also fall upon the recipients of public scientific testimony. Indeed, *Enterprise* and *Democracy* indicate this much insofar as they jointly indicate an alliance between public scientific testimony and the ideals of deliberative democracy. Together the two principles indicate that the scientific enterprise has a special role of contributing to an informed public through public scientific testimony in societies which pursue ideals of deliberative democracy. However, this role requires that there is a reasonable public uptake of public scientific testimony. So, the principles help illuminate an underappreciated aspect of the societal division of cognitive labor between science and the public: the consumer side of public scientific testimony.

7.4.b Public uptake of public scientific testimony: My focus on the consumer side of public scientific testimony reacts to two misguided presuppositions. One consists in the presupposition that it is primarily up to the testifiers to ensure that recipients have a reasonable uptake of public scientific testimony. The other consists in the presupposition that it is primarily up to the recipients to ensure that their uptake of public scientific testimony is epistemically reasonable.

How can these presuppositions both be misguided, you ask? Because it is to a large extent up the society as whole, rather than to the *individual* testifiers and recipients, to sustain a social environment in which laypersons may acquire warranted testimonial belief through public scientific testimony. This is the case both on a global scale as well as in more local areas. For example, Levy argues that it can be reasonable to regulate an epistemically inhospitable environment for health choices (Levy 2018).

When considering the recipient side of the matter, it is natural to consider obligations of the individual recipients of scientific testimony. For example, prominent accounts consider the epistemic vigilance that individual laypersons may exercise.[4] Likewise, Irzik and Kurtulmus set forth strong individual conditions on warranted trust and enhanced warranted trust in public scientific testimony (Irzik and Kurtulmus 2019, 2021). However, they recognize that meeting them is a challenge for many individuals and consider broader social conditions as well.

[4] Hardwig 1994; Goldman 2001; Sperber et al. 2010; Anderson 2011; Grundmann forthcoming.

I will pursue this latter social route since many recipients of public scientific testimony are ill-equipped to fulfill demanding individual vigilance requirements.[5] Indeed, there is an important structural point to be made: *Those who are least epistemically equipped to make warranted epistemic judgments about the contents of public scientific testimony also tend to be those who are least epistemically equipped to assess the source of it.*

Throughout the book, I have highlighted the importance of epistemically internalist warrant (justification) in the scientific practice. But in the case of public scientific testimony, it is important to recall epistemically externalist warrant (entitlement) which does not require that the recipient bases her belief on epistemic reasons, such as reasons for believing that the speaker is reliable. However, according to a social externalist picture, entitled belief through public scientific testimony requires that such testimony is normally reliable in the recipient's social environment (Gerken 2013b; Chapter 2.3.a, 2.4.b).

So, although warranted testimonial belief requires that the individual receivers of public scientific testimony exercise cognitive competence, testimonial entitlement does not require strong internalist conditions or individual vigilance. This is not to deny that laypersons' deliberations would be improved if they acquired testimonial justification, as opposed to mere entitlement. Recall that I have explicitly argued that discursive justification may improve communal deliberation (Chapter 3.5.c following Gerken 2013b, 2015a, 2020a, 2020c). However, this point must be juxtaposed with the point that testimonial entitlement is, for many deliberative purposes, an adequate epistemic position and, at any rate, the only viable one.

From the perspective of bounded rationality, a layperson is better off trusting public scientific testimony than trying to become a hobby scientist in the area in question. This is one reason why the folk conception of rational uptake as uptake that is based on critical autonomous understanding is a problematic descendent of the *Nullius in verba* tradition. In contrast to this picture, a key point of a *societal* division of cognitive labor is that scientists conduct the scientific work so that the public does not have to. From this point of view, it is paramount to consider the social conditions for rational uptake of public scientific testimony. Drawing on others, Slater et al. put the point as follows: "The practical reality is that the public is not—and likely will never be—in a position to vet scientific claims themselves (Anderson 2011: 144; Jasanoff 2014: 24; Keren 2018). They must instead rely on the division of epistemic labor and *trust* the scientific community as a source of intellectual authority, relying on the community itself to vet its own deliverances" (Slater et al. 2019: 253).

Consequently, it is reasonable to refocus from what individual recipients can do to what society as a whole can do. As *Democracy* suggests, the importance of an

[5] Anderson 2012; Michaelian 2013; Guerrero 2016; Keren 2018; Slater et al. 2019; Contessa forthcoming.

informed public in societies that pursue the ideals of deliberative democracy renders the focus on societal enabling of effective public scientific testimony especially pertinent.

7.4.c Entitlement through appreciative deference: So far, I have argued that even when individual conditions on testimonial entitlement must be met, it is often a social obligation to structure the social environment in a way that gives individuals a fair shot at fulfilling them. Recall the structural point that the laypersons who are most in need of public scientific testimony, due to lacking education and informational resources, are, for the same reasons, also typically the ones who are least capable of assessing it. In consequence, many societies that pursue ideals of deliberative democracy recognize the social obligation to provide education, access to information, etc. Likewise, such societies should also recognize social obligations to secure a social environment that allows its underprivileged members to acquire entitled testimonial belief through their *appreciative deference* to public scientific testimony.

I will approach a characterization of the idea of appreciative deference by illustrating its importance with a case from my spare time: I moonlight in a small NGO that assists small-scale artisanal goldminers in places such as Uganda and the Philippines with replacing the use of mercury in extracting gold from ore with a mercury-free method (Køster-Rasmussen et al. 2016). For that purpose, the miners simply need to be entitled in believing *that* mercury is highly toxic. They need not be justified in believing detailed biochemical details involved in the scientific justification for the assumption that mercury is toxic. Our program involves very basic toxicology training for some individuals—miner trainers, schoolteachers, etc. But teaching the scientific intricacies of mercury toxicology to miners who often lack any education would, if feasible at all, place an unreasonable burden on the shoulders of the recipients, who cannot afford investing their time in education.

The case represents ubiquitous circumstances of recipients of public scientific testimony who lack education, access to information, and even the privilege to pursue such niceties. Consequently, it illustrates the importance of providing public scientific testimony in a way that enables the epistemically impoverished laypersons to acquire the sort of entitled belief that furthers their basic participation in society. Such basic participation consists partly in the ability to pursue their ends without serious epistemic disadvantages. But it also consists in the ability to be a part of social deliberation. So, reflections on epistemic externalism, bounded rationality, and the societal division of cognitive labor suggest that one key function of public scientific testimony is to enlighten in the minimal sense of allowing laypersons to form entitled testimonial belief. The broad social externalist picture drawn here suggests that this is partly but largely a matter of securing a social environment that allows laypersons to acquire entitled testimonial belief by an appreciative deference to public scientific testimony (Anderson 2012; Goldberg 2017).

Appreciative deference to public scientific testimony is a type of uptake in virtue of a basic sensitivity to the fact that the testimony is warranted in virtue of being based on scientific justification. Thus, appreciative deference differs from Wagenknecht's notion of *complete trust*, which was characterized, roughly, as trust in the absence of epistemic reasons beyond the testimony (cf. Chapter 4.3.a; Wagenknecht 2015). In contrast, appreciative deference involves a more or less articulate reason akin to a variety of the idea that the testimony is credible in virtue of being scientific testimony. As mentioned, the appreciation need not involve a fully conceptualized reason in these terms. Rather, there is room for wide variation in how the reason is conceptualized. It only requires that the broad idea figures as a generic reason in the uptake of the testimony. Similarly, appreciative deference involves sensitivity to the norms of layperson uptake of public scientific testimony. Such sensitivity does not require the ability to characterize or conceptualize norms of scientific testimony. It merely requires that one's practice properly reflects those norms.

Importantly, appreciation comes in degrees. It may involve as little as deference on the basis of the broad idea that scientific testimony generally credible. However, a sophisticated degree of appreciative deference may involve the ability to articulate for oneself and others the central reasons why it is epistemically rational to trust scientific testimony. That said, it is important to emphasize that appreciative deference does not require epistemic vigilance, understanding of the relevant science, epistemic assessment of the testifier, etc. (Fricker 2002, Sperber et al. 2010). Consequently, appreciative deference remains a form of *deference*. The fact that one has a reason to defer to an epistemic superior does not mean that one does not defer to her when one forms a testimonial belief that *p* without any comprehension of the first-order warrant for *p*. Thus, appreciative deference is a mode of testimonial belief formation that is situated between complete trust and epistemic vigilance.

Given that appreciative deference does not require epistemic vigilance, it may yield little or no epistemic justification, although it may yield epistemic entitlement (cf. Chapter 2.3.a). In some cases, a minimal appreciation of scientific testimony as such may be a necessary condition on entitled testimonial belief. Assume, for illustration, that S testifies to me that Jupiter and Saturn contribute to the Earth's habitability by protecting it from asteroid impact. (Amazingly, they do: Giant planets have a "giant impact on giant impacts" (Lewin 2016).) Assume that I trust S despite having no idea whether her testimony was properly based on scientific justification rather than wild speculation on the part of S. In this case, I would probably not be entitled in my resulting testimonial belief. This exemplifies the general point that not all contents are equally apt for entitled testimonial belief in minimal background cases. On the other hand, if I am warranted in believing or presupposing that S's testimony ultimately rests on scientific justification, I would typically be entitled. At least, my epistemic position on the issue would be improved.

However, it remains important that the appreciation involved in appreciative deference is a matter of degree. A well-educated recipient may appreciate that a public scientific testimony that exposure to mercury may lead to birth defects is very trustworthy because it is based on an unequivocal meta-analysis. Such an educated recipient might acquire a justification, and perhaps even a *discursive* justification, for the relevant testimonial belief. In contrast, consider a small-scale gold miner who lacks even basic appreciation of the relevant science or the notion of a meta-analysis, etc. Given an appropriately presented testimony that takes his educational level into consideration, such an individual may nevertheless base his deference on some minimal appreciation that the testimony is a scientific testimony. Although this basic appreciation of the scientific testimony as such does not involve a grasp of the scientific justification, it may nevertheless contribute to the miner's entitled belief that exposure to mercury may lead to birth defects. Thus, appreciative deference may contribute to, and in some cases be required for, testimonial entitlement.

Appreciative deference goes hand in hand with an ability to *not* defer when there are indications that there are epistemic flaws in the testimony in question, that there are superior sources, etc. For example, recipients may regard it as a red flag if public scientific testimony lacks epistemic qualifications, although it concerns a domain in which even the best scientific investigations are associated with considerable uncertainty. A practice of being cautious when epistemic norms of public scientific testimony are clearly violated is the flip side of the appreciation for such norms that is involved in appreciative deference when the norms are met.

Appreciative deference reflects scientific practice insofar as both collaborating scientists and layperson recipients of public scientific testimony are wise to defer to epistemic experts about domains where they lack epistemic expertise. Thus, the idea of appreciative deference contributes to a positive *testimony-within-science* alternative to the misguided *science-before-testimony* picture. As such, it contributes to the overall project of replacing the damaging anti-testimonial folk conception of science.

One might think the idea of appreciative deference conflicts with the principles advocating for articulating the strength and nature of scientific justification—viz., JEN and its species JET and *Justification Reporting* (Chapters 5.4.c and 6.3.a, respectively). However, I do not assume that recipients acquire much more than testimonial entitlement when public scientific testimony is accompanied by indications of the strength and degree of scientific justification. Yet, the recipients may acquire an appreciation of the fact that the testimony is based on scientific justification. Testimony is not wholly incapable of providing understanding (Boyd 2017). Thus, appreciative deference does not require that laypersons may acquire intricate scientific justification but rather that they can recognize public scientific testimony as such, trust it for that reason, and be trusted when they appeal to it. In such a social environment, public scientific testimony may transmit epistemic authority, although it rarely transmits scientific justification. This idea reflects the principle *Non-Inheritance of Scientific Justification* according to which S's testimony that *p* does not typically transmit the kind or degree of scientific justification possessed by S to H (Chapter 2.2.b). But this principle is compatible with the idea that entitled testimonial belief is empowering in a social environment that is characterized by appreciative deference. This idea, in turn, is not only compatible with, but also reinforced by, principles such as JET *Norm* and *Justification Reporting*.

Hence, it is well worth exploring what society can do and ought to do about securing the type of social conditions that permit rational testimonial trust in public scientific testimony.

7.4.d Structuring the social environment: What are the aspects of the social environment that will allow for entitled testimonial belief on the basis of appreciative deference to public scientific testimony?

One important aspect is the absence of defeaters such as merely apparent scientific disagreement. If pseudo-scientists are easily able to present themselves as epistemic equals, this may defeat entitled testimonial beliefs (Guerrero 2016; Gerken 2020d). Addressing this concern requires societal efforts to harness scientifically illegitimate lobbyists and sponsors of pseudo-science. Likewise, social institutions and public media need principled no-platform policies for views that are so epistemically egregious that they will only do epistemic damage to the public debate (Simpson and Srinivasan 2018; Levy 2019; Peters and Nottelmann forthcoming). For example, I cannot think of a science communication context in which proponents of flat Earth and holocaust-denying hypotheses deserve a hearing. However, the issue is complex because it is also an important task to secure basic freedom of speech for scientists and appropriate platforms for public scientific testimony. An appropriate mix of such platforms and no-platforming policies may contribute to a social environment characterized by appreciative deference. This involves a general tendency to accept public scientific testimony and not too many non-scientific or pseudo-scientific challenges to it.

Of course, difficult questions in political philosophy concern how such a social environment is best achieved.[6] For example, the epistemological desideratum of limiting or eliminating certain platforms for outlandish and harmful views must be balanced with the moral and epistemic reasons for a high degree of freedom of expression. But while such tensions are tenacious, they underwrite that a social

[6] Grasswick 2010; de Melo-Martín and Intemann 2014, 2018; Lynch 2018; Oreskes 2019; de Cruz 2020; Boult 2021.

environment that is conducive to laypersons' ability to attain testimonial entitlement must be sustained at a societal level.

Another aspect of the social environment is the general reliability of providers of public scientific testimony. Although I have articulated norms for individual scientific testifiers, the broader society must help establish a social environment that allows individual scientists to comply with these norms. For example, personal and institutional incentives to minimize questionable research practices or outright fraud must be put in place on a social level. Since science is interwoven with the larger society, it is often unclear whether the relevant systems of incentives and sanctions are attributable to the scientific community or the broader society (Jasanoff 2004a, 2004b; Rolin 2020). Some factors are clearly imposed upon science from the outside, for example, legislation pertaining to funding or fraud. Other factors are fairly clearly imposed from within the scientific community, for example, efforts to address the replicability crisis in social psychology (Zwaan et al. 2018). Yet other cases are in a gray zone. An example might be the International Panel of Climate Change's decision to indicate "confidence levels" to their scientific testimony (for discussion, see Betz 2013; Parker 2014; John 2015a).

Another example is the underappreciated task of ensuring that scientists can communicate research in a manner that allows for a reliable uptake by the public. Some scientists should possess not merely epistemic and contributory expertise but also interactional and T-shaped expertise (Chapter 1.2.c). This aim may be pursued by providing basic science communication training to some scientists (Figdor 2017). Moreover, scientific experts and science reporters must adapt to a changing media landscape, and this includes developing (online) strategies to counter systematic misinformation (Iyengar and Massey 2019). One component in doing so is to ensure that scientific experts are visible as such (Goldman 2001; Leefmann and Lesle 2020; Gerken 2020e).

Finally, sustaining the environmental conditions that enable individual recipients to form entitled testimonial belief or acceptance on the basis of uptake of public scientific testimony is a societal matter (Rolin 2020). Despite local backfire effects and polarization effects, an educational system that emphasizes science literacy is a crucial social task (Chapter 6.4.c). In particular, laypersons may benefit from a broad understanding of the nature of science (Lombrozo et al. 2008; Douglas 2015). They may also benefit from some appreciation of science's role in the societal division of labor (Keren 2018). Indeed, an increased understanding that public scientific testimony is trustworthy because it is based on scientific justification that lies beyond the grasp of laypersons may contribute to a more general appreciative deference to public scientific testimony. Such an understanding may also diminish *discursive deception*—the phenomenon that one's epistemic position appears to be inadequate if one is unable to defend it against challenges (Chapter 5.3.d). Similarly, improved science education may diminish folk epistemological misconceptions about science and resulting biases in judgments about scientific hypotheses

(Chapter 5.2–3). For example, appreciating that a scientific hypothesis is fallible but epistemically superior to non-scientific alternatives may curb the negative impact of epistemic qualification effects—i.e., that indications of uncertainty may be detrimental to laypersons' uptake of it (Chapter 5.3.d).

So, generally, it is an important societal task to ensure that the public has an appreciative deference to public scientific testimony.[7] Such an appreciative deference involves that laypersons are aware of their own epistemic limitations and the need to accept public scientific testimony. Indeed, it may be epistemically irrational to personally assess a complex scientific matter. Hence, there may be a default epistemic norm for layperson recipients of public scientific testimony. So, in the spirit of exploration, consider the following suggestion, *Laypersons' Uptake Norm* or LUN for short:

> **LUN**
>
> In a context of public science communication, PSC, in which S's public scientific testimony that *p* conveys that *p*, the default attitude of a layperson recipient, H, should be to believe or accept that *p* if H has strong and undefeated warrant for believing that S's testimony that *p* is properly based on adequate scientific justification.

Simplified, LUN has it that *if H has strong and undefeated warrant for believing that S's public scientific testimony that p is epistemically adequate, then H should believe or accept that p.* The default norm, LUP, mirrors *Normative Intra-Scientific Uptake*, NISU (from Chapter 4.4.b), in that it only sets forth a sufficient condition but not a necessary one. After all, H may be warranted in believing or accepting that *p* on other bases than S's public scientific testimony. Moreover, LUP only sets forth a default sufficient condition since S might have excellent warrant for believing that *p* is false, that there is no scientific consensus about *p*, etc.

Although LUN concerns H's warrant for believing that S's testimony is properly based on adequate scientific justification, it is not a reductionist principle according to which H's testimonial warrant reduces to such antecedent warrant. Moreover, in many contexts LUN only requires *propositional* warrant and this may be an entitlement that is largely explained by epistemic dimensions of the social environment that are inaccessible to H. Such epistemic dimensions include facts about whether there is a lot of pseudo-scientific testimony, whether scientists generally meet or violate epistemic norms such as NEST, and whether they often engage in scientific expert trespassing testimony, etc. In other contexts, the relevant warrant may be a more demanding type of justification, of course. For

[7] Fricker 2006b; Turner 2014; Anderson 2011, 2012; Slater et al. 2019; Boult 2021; Contessa forthcoming.

example, there are cases where H may not be required to accept that *p* unless she has reason to think that S meets the relevant norms of public scientific testimony.

I will not develop or defend LUN here beyond the motivation for it that I have sketched above. Doing so requires specifying the epistemically relevant social factors, and this a major project. But it bears mention that stronger principles have been debated (Bicchieri et al. 2011; Zagzebski 2012; Jäger 2016). A proper defense of LUN involves addressing several big issues. One is whether it is compatible with the desideratum of an anti-authoritarian stance that I have always had sympathy for (as my parents can testify). So, like most of the issues and suggestions in this section, I put the putative norm, LUN, on the table to indicate the broader consequences of the main conclusions of the book. Developing these considerations is the task for an investigation that overlaps philosophy of science and political epistemology.

7.4.e The societal division of labor among testifiers, recipients, and society: *Enterprise* and *Democracy* help motivate an epistemically externalist conception of the societal division of cognitive labor. This conception emphasizes the general social environment over individual vigilance (Chapter 2.3.a, 2.4.b). Specifically, it emphasizes the general structures, environmental conditions, and social incentives that help constitute the social environment in which public scientific testimony occurs.[8] So, I will conclude with a few points about how this emphasis on the societal tasks relates to the individual norms for scientific experts and science reporters promoted in Chapters 5 and 6.

One might think that these societal obligations replace normative obligations on individual providers of public scientific testimony such as those expressed by JET and *Justification Reporting*. According to these norms, the strength and nature of the relevant scientific justification should be conveyed whenever feasible. It may appear that these norms do not align with the social obligation to establish a social environment in which laypersons may trust public scientific testimonies without trying to evaluate the underlying science. In response to this concern, note that JET and *Justification Reporting* do not require that the public should assess the details of scientific justification. This is one reason why the principles call for reporting the strength of the scientific justification. However, reporting the nature of the justification offers the recipient a feel for the relevant scientific basis or, more minimally, the fact that there is a scientific basis. Often this is enough to make it vivid how impoverished a non-scientific assessment of the hypothesis in question is. So, rather than being in conflict with a reasonable societal division of cognitive labor, JET and *Justification Reporting* reinforce it and contribute to a social environment characterized by an appreciative deference. Hence, the overall

[8] Anderson 2012; Turner 2014; Grasswick 2010; Rolin 2020; Contessa forthcoming.

societal division of labor that I have advocated for consists of the following aspects:

(i) Fairly minimal requirements for the *individual layperson consumers* of public scientific testimony.
(ii) Fairly demanding requirements for the *individual providers* of public scientific testimony.
(iii) Very comprehensive requirements for the *society* that enables the individual testifiers and recipients to meet the requirements that they are subject to.

This idea may be illustrated as a pyramid structure of the societal obligations involved in public scientific testimony (see Figure 7.1 below). In this *Testimonial Obligations Pyramid* (or TOP for short), the *base level* consists of the societal obligations to secure a social environment conducive to a reasonable societal division of cognitive labor. As discussed in the previous sections, this involves securing appropriate platforms for public scientific testimony, improving science education, promoting truth-conducive incentives and sanctions for scientists and science communicators, diminishing epistemic defeaters consisting of fake news, pseudoscientific testimony, epistemic trespassing testimony, false balance, etc. On top of the base level, there is a *middle level* of normative obligations for the producers of public scientific testimony. These consist in testifying in accordance with the relevant norms and guidelines such as NEST, JET, *Justification Reporting, Expert Trespassing Guideline, Epistemically Balanced Reporting*, etc. Finally, there is a yet smaller *top level* of obligations for the lay consumers of public scientific testimony. These consist in appreciative deference to public scientific testimony in accordance with LUN as well as general good citizenship that supports the base level.

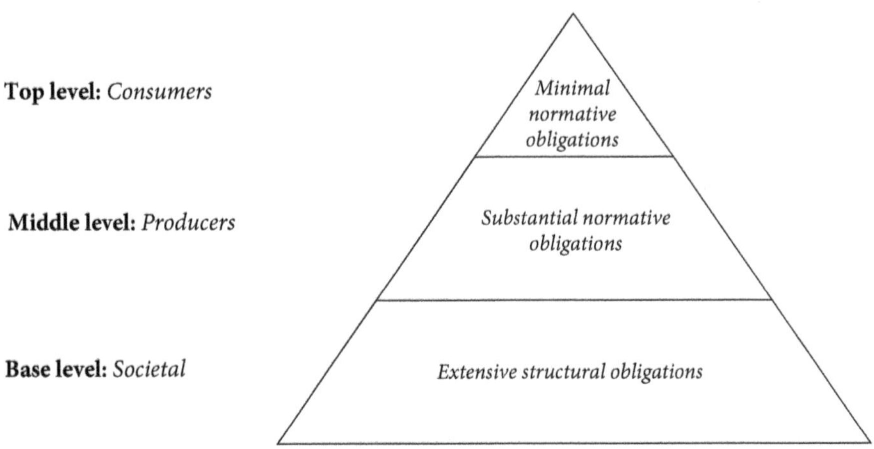

Figure 7.1 *Testimonial Obligations Pyramid* (TOP)

The *Testimonial Obligations Pyramid* illustrates that the bulk of the obligations are at the societal base, whereas the obligations that are imposed on individual recipients of public scientific testimony are far less demanding. Likewise, the TOP structure illustrates that the social environment is the base upon which the production and consumption of public scientific testimony take place. Without such a base, the individual producers and consumers of public scientific testimony may be eminently reasonable to no avail. For example, if impeccable public scientific testimony occurs in a social environment in which it is swamped by pseudo-scientific testimony that is indiscernible to laypersons, even vigilant individual recipients will typically be at a loss.

Finally, the societal division of labor that is illustrated by the *Testimonial Obligations Pyramid* provides a stark contrast to the *Nullius in verba* inspired picture. After all, this picture *inverts* the pyramid by placing weighty demands on the individual recipients of scientific testimony. According to this picture, recipients are supposed to be critical thinkers who autonomously check for themselves. This picture continues to be influential and is echoed in the contemporary slogan "Do your own research" and less dramatic calls for individual vigilance. However, it is misguided to pursue a society of autonomous science skeptics. It is far more reasonable to pursue a social structure, such as TOP, in which people can reap the epistemic benefits of collaborative science through appreciative deference to public scientific testimony. Laypersons may benefit from the intra-scientific division of cognitive labor in virtue of a societal division of cognitive labor. On this note, it is time to conclude.

7.5 The Significance of Scientific Testimony

So, what is the significance of scientific testimony? I would have loved to conclude my investigation with a simple answer. But what the investigation has shown is that there is no simple answer. More specifically, it has shown that varieties of scientific testimony are woven into the fabric of science and society in a myriad of ways. So, the best I can do is to draw the investigation to a close by highlighting some connections between my various conclusions. I will start by following the distinction between intra-scientific and public scientific testimony before concluding with some more general points and methodological remarks.

7.5.a The significance of intra-scientific testimony: One of my overarching ambitions has been to replace the *science-before-testimony* picture—crystalized in the Royal Society's slogan, *Nullius in verba*—with a positive *testimony-within-science* alternative that gives intra-scientific testimony its due place as a vital part of scientific practice.

Due to the increased focus on scientific collaboration, the *science-before-testimony* picture has been gradually relinquished by philosophers of science. But despite this relinquishment, testimony has not become a central topic in the philosophy of science. In particular, the relinquishment of the *Nullius in verba* tradition has not resulted in an alternative picture of the relationship between scientific practice and scientific testimony. One consequence is that aspects of the outdated picture may remain influential in folk conceptions of science. Thus, the project has been twofold. One part is the negative project of articulating explicit reasons for renouncing the *science-before-testimony* picture. Another part is the positive project of outlining an alternative *testimony-within-science* picture that may replace it.

Some general tenets of the positive picture are outlined in this chapter, but most of the details have been provided along the way. Examples include the characterizations of varieties of scientific testimony and the development of norms governing intra-scientific testimony. These are substantive contributions to an account of *how* scientific testimony is a part of science. If I know my colleagues well, the specific norms that I have proposed will prove to be controversial. But if such controversy concerns the specific norms rather than the idea that such norms are proper methodological norms that partly characterize scientific practice, the discussion will already have advanced a great deal.

7.5.b The significance of public scientific testimony: Another central ambition has been to articulate norms and guidelines for varieties of public scientific testimony. I have primarily been concerned with the norms and guidelines that apply to *producers* of public scientific testimony—i.e., scientists and science reporters. However, this concern demands that the psychology of layperson recipients' uptake of public scientific testimony is taken seriously. So, I have integrated my endeavor with the empirical research on laypersons' reception of public scientific testimony. Consequently, I have developed principles of science communication in a manner that takes into account empirical research on the biases, social pressures, and folk misconceptions about science.

Moreover, the proposals have been integrated with a social externalist framework, which is illustrated by the *Testimonial Obligations Pyramid* (TOP). According to TOP, the obligations concerning public scientific testimony fall less on the layperson recipients than on the providers and the society at large. As I have emphasized, this picture contrasts starkly with *Nullius in verba*-inspired pictures that emphasize individual autonomy and skepticism of epistemic authorities. This is not to belittle critical reasoning skills, science literacy, and healthy skepticism. Rather, the point is that the pursuit of the ideal of a scientifically informed public must involve pursuing structural social features beyond individual cognitive abilities. For example, it must involve the pursuit of scientific institutions and platforms for public scientific testimony as well as a social environment that is conducive to warranted public trust in public scientific testimony.

While this is easier said than done, and many societies are currently doing a rotten job of it, the ideal is worth pursuing. A society characterized by a general appreciation of science and an appreciative deference toward first-, second-, and third-hand scientific testimony enables science to help level the playing field in public debates. So, although heaps of practical and principled problems remain, public scientific testimony is able to play a privileged and authoritative role in a diverse society that is structured by a reasonable societal division of cognitive labor.

7.5.c The relationship between intra-scientific and public scientific testimony: The idea of the cognitive division of labor underlies my take on scientific testimony. For example, the theses *Parthood* and *Methodology* and the specific epistemic norms governing intra-scientific testimony highlight that it is a vital part of collaborative science based on the intra-scientific division of cognitive labor. Likewise, the theses *Enterprise* and *Democracy* highlight the interdependence between science and societies in the pursuit of a societal division of cognitive labor that promotes an informed public.

Although I have addressed intra-scientific testimony and public scientific testimony in a piecemeal manner, the resulting conclusions are deeply interrelated. From overlapping perspectives, they indicate the centrality of scientific testimony in science and society. Thus, they jointly contribute to a more general framework for theorizing about scientific testimony.

The interrelation is vivid when conclusions pertaining to intra-scientific testimony also bear on the ideal of a scientifically informed public. To see this, recall that the *Nullius in verba* tradition continues to influence folk conceptions of science such as the outdated idea that scientists must autonomously observe and understand everything themselves. This idea of autonomous scientists may encourage the unfortunate conception of the scientist as a solitary genius. This conception may, in turn, encourage a *great white man fetish*, which is not only racist, sexist, and elitist but moreover hampering public scientific testimony (Chapter 5.3.c). Thus, the effort to replace the *science-before-testimony* picture with a more explicit *testimony-within-science* alternative is not only important to understand the nature of science. It is also important to understand and ameliorate the role of science in society.

On the other side of the coin, conclusions regarding public scientific testimony may inform our approach to understanding the scientific practice. For example, the idea that public scientific testimony is governed by norms and guidelines informs a principled account of what it takes to be a member of the scientific enterprise. Permit me an anecdote for illustration: While writing this book, I did an interview in which I gave some reasons why reputable newspapers should be weary of platforming climate science deniers. What I found in my inbox the following morning was—you guessed it!—emails from upset climate science deniers. However, as a layperson in the relevant domain, I was unable to rebut their specific arguments, and it was not straightforward to assess their epistemic credentials. However, their public testimony blatantly violated reasonable norms for scientific expert testifiers. For example, they regarded uncertainties in existing research as devastating. But they made extremely strong claims of their own without any epistemic qualifications. So, their public testimony was an important red flag. They simply failed to give public scientific testimony in accordance with its appropriate norms. I am not suggesting that whether a testifier meets or violates norms of public scientific testimony determines whether she is a scientist. The point of the anecdote is that the norms and guidelines governing public scientific testimony can help to illuminate what it is to be a scientist and thereby the nature of science. Indeed, it can do so not only in the abstract but also in the context of concrete problems.

Thus, the discussion of intra-scientific testimony in the scientific process also sheds light on the role of science in society, and the discussion of public scientific testimony also sheds light on the nature of science. So, although the two investigations may be separated for some purposes, they are best seen as two overlapping perspectives on how scientific testimony situates science in society.

7.5.d Scientific testimony in science and society: The characterization of varieties of scientific testimony through their governing norms contributes to a traditional project in the philosophy of science: that of providing a principled characterization of the scientific enterprise that may help distinguish it from non-scientific and pseudo-scientific enterprises. As my anecdote illustrates, this may be helpful in pernicious situations such as the ones that laypersons find themselves in when faced with disagreements about scientific matters in the public debate. Although it is important to emphasize that a demarcation criterion cannot be expected, it is also important to emphasize that less may do (Lakatos 1978; Hansson 2017; Lombrozo 2017). Even contingent and inconclusive characterizations of science through an explication of the norms governing scientific testimony may increase our understanding of hitherto underappreciated aspects of scientific methods and practices. This in turn increases our ability to pinpoint systematic violations of scientific norms. Such an ability may help us determine whether someone is a scientist or a pseudo-scientist or something in between, such as a scientist who is engaged in questionable research practices. Indeed, many of the specific proposals of the present book—such as the epistemic norms of intra-scientific testimony—contribute to partial but principled characterizations of central aspects of scientific practice. Other contributions—such as the principles of science communication—build on this characterization for ameliorative purposes.

From a methodological point of view, the investigation exemplifies how scientific testimony makes for an extremely important but underappreciated and hence underexplored connection between largely separate debates and disciplines. One such connection is between the debates concerning collaboration in the philosophy of science and the debates about testimony in social epistemology. Another connection is between the rich body of empirical work on science communication and conceptual work in the philosophy of science. Likewise, scientific testimony intersects philosophy of science and political philosophy. Thus, a larger methodological conclusion is that scientific testimony provides an important perspective on the nature of science and its role in society.

This book is characterized by taking this perspective of scientific testimony. I have sought to make progress by developing a coherent general framework and by articulating some specific norms and principles that are central to it. But the main contribution of the present exploration may be that of revealing how scientific testimony is interwoven with a wide range of dimensions of science and society.

Coda

Scientific Testimony, Cognitive Diversity, and Epistemic Injustice

C.1 Scientific Testimony's Relationship to Cognitive Diversity and Epistemic Injustice

Given the approach to scientific testimony developed throughout the book, scientific testimony has important ramifications for the debates concerning (cognitive) diversity and epistemic injustice. So, in this brief Coda, I highlight some of these ramifications in a brief and selective manner. I do so in order to indicate relations to some broader debates, which are not thematized in the book, but which are too important to ignore. Thus, my aim is not to make substantive progress on the grand debates regarding cognitive diversity and epistemic injustice, but to highlight some ways in which scientific testimony is related to these debates.

Here is what I will do: Section C2 is devoted to characterizing cognitive diversity and epistemic injustice, respectively. In Section C3, I consider how these notions relate to intra-scientific testimony. In Section C4, I consider how they relate to public scientific testimony. In Section C5, I wrap up.

C.2 The Nature of Cognitive Diversity and Epistemic Injustice

To consider how varieties of scientific testimony bear on cognitive diversity and epistemic injustice, these notions should be characterized. So, I will briefly do so, starting with cognitive diversity.

Cognitive diversity may be characterized as consisting of variances in epistemic perspectives, worldviews, values, standards, or epistemically significant aspects of meaning (List 2006; Muldoon 2013). Cognitive diversity need not be attitudinal—i.e., it may consist in differences in values that are not represented in the diverse individuals' or groups' propositional attitudes. Hence, cognitive diversity need not be understood or appreciated by the agents manifesting it. One may be embedded in a *practice* that reflects certain cognitive values without having conceptualized these values. For example, two groups that differ about whether one should testify in a certain context may follow complex principles of socially appropriate

Scientific Testimony: Its roles in science and society. Mikkel Gerken, Oxford University Press. © Mikkel Gerken 2022.
DOI: 10.1093/oso/9780198857273.003.0009

testimony that they have not reflectively considered. Thus, cognitive diversity may be merely *presupposed* by a group's cognitive practices (see Gerken 2012c, 2013a on presupposition). Cognitive diversity may, but need not, arise from *demographic* diversity, which is diversity in terms of traits—such as gender, race, age, education, occupation, etc.—that are characteristic of a subset of a population (Ely and Thomas 2001; Phillips 2017). There is much more to be said about cognitive diversity (Steel et al. 2018; Gerken Ms b). However, the present broad characterization may subsume different species of cognitive diversity that may be distinguished between.

Meanwhile, the idea of epistemic injustice should also be characterized. One species of it—*distributive* epistemic injustice—concerns "the unfair distribution of epistemic goods such as education or information" (Fricker 2013: 1318). However, much attention has been devoted to another species: *discriminatory* epistemic injustice, which is paradigmatically explained by *identity prejudices* that pertain to gender, class, race, or social power. Following (Fricker 2007, 2013), I adopt the following sufficient condition of generic discriminatory epistemic injustice (*Generic DEI* via Gerken 2019: 1):

> ***Generic DEI***
>
> S suffers a discriminatory epistemic injustice if S is wronged specifically in her capacity as an epistemic subject.

Initially, Fricker characterized epistemic injustice in terms of knowledge (Fricker 2007: 1). But she has subsequently broadened the characterization (Fricker 2013: 1320, 2017). I explicitly argue that such a broadening is required (Gerken 2019). There are several subspecies of discriminatory epistemic injustice. The natural one to focus on here is *testimonial injustice*, which may be characterized as an epistemic injustice that occurs when S is wronged specifically in her capacity as a testifier (Fricker 2007: 35). Testimonial injustice may arise when a subject's testimony is not given the credit that it is due because she belongs to a group that suffers from a *credibility deficit* that is explained by a negative identity prejudice, social stereotype, or cognitive bias. Dotson has identified a related but subtly different phenomenon of "testimonial smothering." It consists in a variety of testimonial injustice that occurs in some cases in which "the speaker perceives one's immediate audience as unwilling or unable to gain the appropriate uptake of proffered testimony" (Dotson 2011: 244).

There is plenty more to be said about both cognitive diversity and epistemic injustice. But I think we have what we need for initiating an exploration of how they relate to scientific testimony. I will follow the book's overall structure of considering the relationships to intra-scientific testimony and public scientific testimony in turn. I will start with the former.

C.3 How Cognitive Diversity and Epistemic Injustice Relate to Intra-Scientific Testimony

In this section, I will argue that although cognitive diversity may, given inter-scientific testimony, have considerable epistemic benefits, it also introduces some communicative complications. Relatedly, intra-scientific testimony of cognitively diverse groups may be liable to epistemic injustice.

As mentioned in Chapter 1.4.b, different lines of research suggest that cognitive diversity in the scientific community can be epistemically beneficial. Strands of evidence come from empirical studies and formal models of scientific collaboration (see references in Chapter 1.4.b). Likewise, feminist philosophers have argued that cognitive diversity is epistemically beneficial in scientific collaboration. For example, Longino has argued that a diverse community of scientists is a desideratum in part because a scientifically objective process requires critical scrutiny from different perspectives (Longino 1990, 2002). Specifically, Longino argues that a cognitively diverse scientific community may minimize scientists' ubiquitous biases. But she also emphasizes that this requires that its members hold each other accountable and that norms of intra-scientific communication are met (Longino 1990, 2002). Relatedly, standpoint theorists argue that marginalized groups should be included because they are often in a privileged epistemic position—especially with regard to aspects of their social reality (Harding 1991, 2004). Generally, feminist philosophies of science often include some version of the thesis that cognitive diversity is not only morally required but also epistemically beneficial (Solomon 2006a).

I share much of the optimism that increasing cognitive diversity may be epistemically beneficial to scientific collaborations. But it is important to emphasize that these benefits require effective intra-scientific collaboration. Scientific collaboration could hardly reap the epistemic benefits of cognitive diversity without intra-scientific testimony. Moreover, the aim of reaping the epistemic benefits of cognitively diverse scientists may inform the norms of intra-scientific communication (cf. Longino 2002). So, initially, cognitive diversity and intra-scientific testimony appear to combine in epistemically beneficial ways.

Despite this good epistemic news, cognitive diversity may also complicate effective intra-scientific testimony. In particular, it may increase the risk of miscommunication due to linguistic and conceptual equivocation. This risk is especially high in multidisciplinary collaboration. To explain this risk, I will introduce a brand of epistemic diversity that I label *semantically significant cognitive diversity* (Gerken Ms b). It occurs when epistemic diversity leads to semantic differences among orthographically/phonetically identical terms that are not context-sensitive. In philosophy of science, the issue is often discussed in terms of *semantic incommensurability*, following Kuhn who famously held that theoretical terms such as 'species,' 'mass,' and 'element' differ in meaning across different scientific paradigms (Kuhn 2012/1962; Andersen et al. 2006; Sankey 1998, 2006). Semantically significant cognitive diversity may be *non-transparent*, as one can competently possess and use a linguistic term without fully grasping the conditions of its application.[1] Consequently, conceptual equivocation does not require a Kuhnian paradigm change but may occur in interdisciplinary and multidisciplinary collaboration. In particular, *multi*disciplinary collaboration, which characteristically lacks integration of terminology, may be prone to miscommunication due to equivocation (Johnson 2013; Nisbet and Fahy 2015). However, semantically significant cognitive diversity may also arise from differences in cognitive interests or values when they are terminologically tied to disciplinary conventions. Thus, semantically significant cognitive diversity may be a fairly widespread problem for intra-scientific testimony.

That said, the problem is not a devastating one but only a source of fallibility. Andersen argues that cross-disciplinary communicative difficulties are often better explained by lack of communication due to different cognitive interests than by semantic incommensurability (Andersen 2012: 273–4, 2016). However, many cases of inter- and multidisciplinary collaboration occur when different fields overlap in terms of area. In such cases, the risk of multidisciplinary miscommunication due to equivocation arises (Donovan et al. 2015; Politi 2017). More generally, communicative differences often reflect underlying cognitive diversity. For example, studies of cross-disciplinary collaboration suggest that the communicative obstacles are intertwined with diversity in methodology, cognitive interests, and values (MacLeod and Nersessian 2014; MacLeod 2018).

Since endeavors to overcome the challenges of miscommunication are required for the epistemic force of multidisciplinary collaboration, they are vital parts of the scientific process and method. For example, Klein notes that "Any interdisciplinary effort, then, requires analyzing terminology to improve understanding of phenomena and to construct an integrated framework with a common vocabulary" (Klein 2005: 43–4; see also Hall and O'Rourke 2014). Similarly, Galison has introduced the metaphor of "trading zones" for collaborations across scientific disciplines and has argued that they lead to the development of a common vocabulary (Galison 1997). A congenial approach is Campbell's "fish scale" model according to which skills and terms are partly coordinated with adjacent disciplines (Campbell 1969. While somewhat metaphorical, these models indicate transitions from multidisciplinary to interdisciplinary collaboration. Moreover, engaging in interdisciplinary collaboration—"entering trading zones"—may help scientists acquire interactional expertise and T-shaped expertise (Collins et al. 2007; Oskam 2009; Chapter 1.2.c). An upshot is that the challenges for intra-scientific testimony that may be seen as local obstacles may also be seen as

[1] Kripke 1980; Burge 1979; Sankey 2006; Gerken 2013a.

instrumental to scientific change from a more global perspective. Aligning technical terminology in order to enable effective collaboration is little different from calibrating an instrument in order to enable it to measure accurately. This point reinforces the book's overarching *testimony-within-science* approach and, in particular, the idea that intra-scientific testimony is a vital part of science.

In sum: Even if cognitive diversity remains a net plus in scientific collaboration—as I think it does—risks of miscommunication that arise from cognitive diversity raise challenges for intra-scientific testimony and scientific collaboration. I focus on such communicative challenges to draw the connection to inter-scientific testimony. But it has also been argued that cognitive diversity may yield challenges such as social tensions, conflict, avoidance of collaborators, decreased social cohesion, etc. (Ely and Thomas 2001; Galinsky et al. 2015; Eagly 2016). These challenges should figure in a full picture of the significance of cognitive diversity in scientific practice (Intemann 2011; Eagly 2016; Peters 2019).

So far, I have focused on how cognitive diversity has epistemic pros and cons vis-à-vis intra-scientific testimony. However, it must also be noted that testimonial injustice may rear its ugly head when the intra-scientific testimony of some cognitively diverse scientists is not taken as seriously as it should be. Some such cases may be diagnosed in terms of violation of the *Norm of Intra-Scientific Uptake*, NISU (Chapter 4.4.b). In those cases, the recipients have strong and undefeated warrant for believing that S's testimony is properly based on adequate scientific justification. But because of stereotypes pertaining to S's cognitively diverse group, there is no uptake of S's intra-scientific testimony (see Solomon 2006a; Intemann 2011 for interesting cases). Thus, NISU may find an application in helping to diagnose an important type of intra-scientific testimonial injustice. However, not all types of intra-scientific testimonial injustice have this structure. Other examples include epistemic injustice toward cognitively diverse groups that are characterized by differences in cognitive values and interests that are not shared by the majority of scientists.

The key concern is the testimonial injustice suffered by the individual testifiers and groups in such cases. But, in addition, testimonial injustice is problematic because it may undermine the epistemic benefits of critical assessment from diverse perspectives (Longino 1990, 2002). Given that epistemic injustice often arises from social stereotypes about out-groups, the problem is unlikely to be resolved *simply* by ensuring representation of cognitive diverse minorities into scientific institutions. Securing genuine cognitive diversity in science, and the associated epistemic benefits, includes sustaining a collaborative culture in which the cognitive diverse members may function *as such* in the intra-scientific division of cognitive labor (Eagly 2016). But, arguably, this requires that intra-scientific testimonies of cognitively diverse members of the community are appropriately weighted, epistemically speaking.

This task is highly complex because some exclusions of the intra-scientific testimony of some cognitively diverse groups are perfectly legitimate (Intemann 2011). For example, if cartographers exclude pseudo-scientific testimony from members of the flat Earth society, they are not committing any epistemic injustice. In less clear-cut cases, it may be reasonable to regard the intra-scientific testimony by certain group members as less interesting or less epistemically forceful. In general, there is a tricky balance to be struck when it comes to including or excluding purported intra-scientific testimony of cognitively diverse members of the scientific community. The epistemic benefits and moral obligation to include cognitively diverse voices within the scientific practice must be balanced with the need for retaining minimal epistemic standards (Simpson and Srinivasan 2018; Levy 2019). Alas, while retaining epistemic standards is legitimate *in principle*, it may *in practice* become a tool of oppression. Specifically, this may happen if the dominant groups identify the epistemically minimal standards with their own epistemic and value-driven standards in their uptake of cognitively diverse intra-scientific testimony.

However, understanding norms of intra-scientific testimony, such as the epistemic norm for providing it, NIST, and the norm for uptake of it, NISU, may help us develop a more principled account of when epistemic injustice toward cognitively diverse scientists occurs. As noted, testimonial injustice may occur when recipients violate NISU and fail to trust a clearly credible scientist's testimony due to social prejudices. More subtly, indirect testimonial injustice may occur if recipients trust a testifier who clearly violates NIST because they regard the testifier as an in-group member or the content of the testimony as supportive of their particular values. In this manner, abstract principles such as NIST and NISU may be of use in diagnosing cases of epistemic injustice more precisely. This is one way in which the themes of the book may intersect with the project of addressing epistemic injustice.

C.4 How Cognitive Diversity and Epistemic Injustice Relate to Public Scientific Testimony

At its best, public scientific testimony may help minimize distributive epistemic injustice by decreasing unjust distribution of informational and educational resources (Kurtulmus and Irzik 2017). Public scientific testimony in accordance with the overarching presentational norm, *Justification Explication Norm* (JEN), may have further beneficial educational side-effects. Consequently, it has some potential to help minimize *distributive* epistemic injustice. But public scientific testimony also has more subtle ramifications for discriminatory epistemic injustice and cognitive diversity. True to the social externalist approach illustrated by the *Testimonial Obligations Pyramid*, I will suggest that social and institutional initiatives should be central to combating epistemic injustice for cognitively diverse groups and in general (Chapter 7.4.e; Anderson 2012). Specifically, I will argue that it is an important societal task to enable that public scientific testimony is cast in a way that is sensitive to the cognitive situation of laypersons in epistemically compromised circumstances.

In Chapter 7.4, I argued that it is problematic to expect underprivileged recipients to engage in epistemic vigilance and to expect them to pursue justified (as opposed to entitled) testimonial belief from public scientific testimony. This point extends to cognitively diverse minorities in cases where their cognitive perspective impedes such vigilance. Recall the point that *those who are least epistemically equipped to make warranted epistemic judgments about the contents of public scientific testimony also tend to be those who are least epistemically equipped to assess the source of it* (Chapter 7.4.b). If the cognitive diversity of a group leads to epistemic unreliability about some domain, group members will also often be ill-equipped to critically assess sources of information about that domain—i.e., to be appropriately vigilant. Even if a group's particular cognitive interests or idiosyncratic ways of categorizing phenomena mark an epistemically legitimate perspective, they may interfere with uptake. For example, although people with cognitive disorders have an epistemically legitimate perspective on those disorders, it may interfere with their ability to assess public scientific testimony that reflects clinical hypotheses about the disorders.

In consequence, it may be a social responsibility to ensure a social environment in which members of cognitively diverse minorities have a decent opportunity for an uptake of public scientific testimony that results in entitled testimonial belief. This responsibility may involve presentational requirements, or at least desiderata, that go beyond the basic pedagogical requirement of making the public scientific testimony generally comprehensible. Such presentational requirements and desiderata may even go beyond the norm JEN's requirement of articulating aspects of the nature of the relevant scientific justification (Chapter 5.4.c). For example, it may be that a cognitively diverse group's perspective or epistemic values must be addressed in some public scientific testimony. It may also be the case that public scientific testimony should rebut public misconceptions that would otherwise defeat the entitled belief of members of cognitively diverse groups. In many cases, however, good public scientific testimony must simply be sensitive to the distinctive, and in some cases compromised, epistemic position of some recipients.

Recall, for example, the case of communicating the perils of mercury to uneducated *artisanal small-scale gold miners* discussed in Chapter 7.4.c. In this case, many recipients may only stand a chance of acquiring entitled testimonial beliefs if the perils of mercury toxicity are also related to their own cognitive perspective. Public scientific testimony that does so need not pander to the recipients' non-cognitive values. It may rather be a matter of selecting examples, using terminology, and articulating explanations in a manner that aligns with the recipients' cognitive perspective, concepts, and way of thinking. More generally, pursuing a social environment that is characterized by an appreciative deference to public scientific testimony may enable cognitively diverse individuals to generate entitled belief from it. In an environment in which public scientific testimony is standardly accepted, and rarely challenged by opining laypersons or pseudo-scientists, members of cognitively diverse groups will encounter less misleading defeaters to such testimony. This is especially important even when the public scientific testimony involves cognitive values, concepts, or perspectives that are foreign to them.

Securing such a social environment may also play a part in addressing testimonial injustice by aligning a skewed credibility economy. In a social environment characterized by an *appreciative deference* to public scientific testimony, the epistemic authority of science may help elevate the credibility of the relevant laypersons' testimony and, thereby, diminish unjust credibility deficits. That is, someone with a cognitive perspective that is not shared by the majority may have an improved avenue of communication insofar as scientific justification aligns with the perspective. Consider, for example, a poor person who attempts to convey something she has an epistemically privileged perspective on—namely, how living in poverty has ramifications for mental health. Assume that she can point to the vast scientific research on this issue. Of course, this is far from always the case. But if it is, she may be less likely to suffer testimonial injustices in a society characterized by an appreciative deference to public scientific testimony than in one characterized by selective uptake of it.

Similarly, a social environment characterized by an appreciative deference to science may also help minimize testimonial smothering (Dotson 2011). After all, socially marginalized individuals, who do not generally expect uptake of their testimony, may be more likely to engage in public deliberation if they can reasonably expect to be recognized as conveying scientific authority when they do so. For example, consider a member of a racial minority who considers citing recent public scientific testimony that racial bias affects job hiring. If he can reasonably expect an appreciative deference to the public scientific testimony that he cites, he may be more likely to raise the point than if he reasonably expects that it will be dismissed as complaining or airing of grievances. So, if a society is characterized by an appreciative deference to public scientific testimony, people who themselves suffer from a credibility deficit may have an opportunity to gain authority by way of pointing to public scientific testimony. This may help diminish testimonial smothering even though it will surely not eliminate it. For example, it should not be overlooked that, given *distributive* epistemic injustice, not everyone has the same opportunity to point to public scientific testimony.

Cognitively diverse groups who suffer credibility deficiencies precisely because their cognitive perspective or cognitive interests are not shared by the majority may also benefit from a society characterized by appreciative deference to public scientific testimony. Recall that an important aspect of social cognition is that individuals are sometimes assessed on the basis of the social stereotypes of the groups that they belong to.[2] For example, negative epistemic assessment of members of cognitive diverse groups may occur when they are taken to be an out-group. More specifically, recall the thesis *Epistemic Underestimation* according to which both accurate and inaccurate social stereotypes may lead evaluators to underestimate a subject's epistemic position (Chapter 5.3.e). Thus, stereotyped testifiers may suffer a credibility deficit which, in turn, results in testimonial injustice—at least when the stereotypes are inaccurate (Gerken 2019, 2022). As with the case of *socially* marginalized groups, the prevalence of such testimonial injustice may also lead members of *cognitively diverse* groups to be more likely to cede from public debates. If one does not think that one's cognitive interest or perspective will receive a reasonable uptake in a public debate, one is less likely to speak up. Thus, the result may be testimonial smothering (Dotson 2011).

Again, the problem may be partly alleviated by a social environment characterized by an appreciative deference to public scientific testimony and by a minimal obligation for the recipients in accordance with the *Testimonial Obligations Pyramid* (Chapter 7.4.e). In such a society, cognitively diverse members may, at least in some instances, form entitled beliefs from public scientific testimony, and this may give them some credibility that they would otherwise lack. In such cases, the authority of public scientific testimony may allow cognitively diverse groups to take part in aspects of public debate that they might otherwise have trouble contributing to or be reluctant to enter.

In general, a social environment which permits testimonial entitlement through appreciative deference to public scientific testimony may help to level the discursive playing field for members of groups which are marginalized due to being epistemically impoverished or cognitively diverse. I believe that this point amounts to a first step toward meeting the challenge of "integrating expertise with democratic values" (Kitcher 2011: 11). However, it is a first step on a long and winding road.

C.5 Concluding Remarks on Cognitive Diversity and Epistemic Injustice

If nothing else, this brief Coda makes it clear that both intra-scientific testimony and public scientific testimony stand in complex but important relationships to

[2] Ames et al. 2012; Balliet et al. 2014; Carter and Phillips 2017; Spaulding 2018; O'Connor and Weatherall 2019.

cognitive diversity and epistemic injustice. My aim has primarily been to highlight some of these relationships in order to indicate avenues for further research.

With regard to intra-scientific testimony, I have argued that cognitive diversity has both epistemically beneficial and epistemically problematic consequences, some of which lead to epistemic injustice. With regard to public scientific testimony, I have argued that it should be a societal task to alleviate structural challenges concerning epistemic injustice for marginalized groups—including cognitively diverse ones (Chapter 7.4). In both cases, I suggested that the problematic aspects may be diminished by pursuing a social environment that permits for testimonial entitlement by a general appreciative deference to scientific testimony. Contra the picture echoing the *Nullius in verba* slogan, this is a task for society at large rather than for individual providers and consumers of public scientific testimony.

These issues concerning the relationship between cognitive diversity and intra-scientific testimony are important in their own right. But they are also central aspects of scientific practice. Thus, this brief reflection on cognitive diversity and epistemic injustice reinforces the overarching theme of the book that scientific testimony, and the conditions for its uptake, are important aspects of science.

APPENDIX

List of Principles

The main principles discussed in the book are listed in the order in which they occur. The charitable reader will note that since some of the enlisted principles are contradictory, I do not endorse all of them.

1.2.a ***Epistemic Expertise*** p. 20
S possesses epistemic expertise in a domain, D, that consists of a set of propositions iff S has acquired a specialized competence in virtue of which she is likely to possess or be able to form, in suitable conditions, extraordinarily reliable judgments about members of D.

1.3.b ***Collaboration's Contribution*** p. 32
Scientific collaboration contributes immensely to the epistemic force of science.

1.3.c ***Testimony's Contribution*** p. 33
Intra-scientific testimony is an epistemically vital part of scientific collaboration.

1.5.b ***Distinctive Norms*** p. 41
The epistemic contribution of scientific collaboration depends on distinctive norms of intra-scientific testimony.

2.1.a ***Testimony*** p. 46
Testimony that *p* is an assertion that *p*, which is offered as a ground for belief or acceptance that *p* on its basis.

2.2.a ***Transmission of Epistemic Properties-N*** p. 51
For every speaker, S, and hearer, H, H's belief that *p* is warranted (justified, known) on the basis of S's testimony that *p* only if S's belief that *p* is warranted (justified, known).

2.2.a ***Testimonial Knowledge Requires Testifier Knowledge*** p. 51
H acquires testimonial knowledge that *p* through S's testimony that *p* only if S knows that *p*.

2.2.b ***Inheritance of Warrant*** p. 53
Whenever H acquires testimonial warrant through S's testimony that *p*, S's testimony that *p* transmits the kind or degree of warrant possessed by S to H.

2.2.b ***Non-Inheritance of Warrant*** p. 54
Unless the nature of the warrant that S possesses for believing that *p* is articulated, or otherwise clear, to H, S's testimony that *p* does not transmit the kind or degree of warrant possessed by S to H.

2.2.b ***Non-Inheritance of Scientific Justification*** p. 55
Unless the scientific justification that S possesses for believing that *p* is articulated or independently clear to H, S's testimony that *p* does not transmit the kind or degree of scientific justification possessed by S to H.

2.3.a *Reason Criterion (Justification)* p. 57
S's warrant, W, for her belief that *p* is a justification if and only if W constitutively depends, for its warranting force, on the competent exercise of S's faculty of reason.

2.3.a *Reason Criterion (Entitlement)* p. 57
S's warrant, W, for her belief that *p* is an entitlement if and only if W does not constitutively depend, for its warranting force, on the competent exercise of S's faculty of reason.

2.3.a *Discursive Justification* p. 58
S's warrant for believing that *p* is a discursive justification if and only if S is able to articulate some epistemic reasons for believing that *p*.

2.3.b *Reductionism* p. 61
Testimonial warrant requires, and reduces to, other types of warrant that are ultimately non-testimonial.

2.3.b *Anti-Reductionism* p. 62
Testimony is a basic, and hence irreducible, source of warranted belief and knowledge.

2.3.b *Acceptance Principle* p. 62
A person is entitled to accept as true something that is presented as true and that is intelligible to him, unless there are stronger reasons not to do so.

3.1.b *Testifier Characterization* p. 78
A testimony is a scientific testimony iff the testifier is a scientist testifying *qua* scientist.

3.1.c *Content Characterization* p. 80
A testimony is a scientific testimony iff its content is scientific.

3.1.d *Justification Characterization* p. 81
A testimony is a scientific testimony iff it is properly based on scientific justification.

3.3 *Hallmark I* p. 85
In many domains, scientific justification is generally epistemically superior to non-scientific types of warrant.

3.4 *Hallmark II* p. 93
Scientific justification generally comes in degrees of epistemic strength.

3.5 *Hallmark III* p. 95
Scientific justification generally involves a high degree of *discursive* justification.

4.2.b WASA p. 107
In a conversational context, CC, in which S's assertion that *p* conveys that *p*, S meets the epistemic conditions on appropriate assertion that *p* (if and) only if S's assertion is properly based on a degree of warrant for believing that *p* that is adequate relative to CC.

4.2.b NIST p. 109
In a context of intra-scientific communication, CISC, in which S's intra-scientific testimony that *p* conveys that *p*, S meets the epistemic conditions on appropriate intra-scientific testimony that *p* only if S's intra-scientific testimony is properly based on a degree of scientific justification for believing or accepting that *p* that is adequate relative to CISC.

4.3.a *Hardwig's Dictum* p. 116
"a scientific community has no alternative to trust"

4.4.b NISU p. 124
In a context of intra-scientific communication, CISC, in which S's intra-scientific testimony that *p* conveys that *p*, the default attitude of a collaborating scientist, H, should be to believe or accept that *p* if H has strong and undefeated warrant for believing that S's testimony that *p* is properly based on adequate scientific justification.

4.4.c WISU p. 126
In a context of intra-scientific communication, CISC, in which S's intra-scientific testimony that *p* conveys that *p*, a collaborating scientist, H, is, as a default, warranted in believing or accepting that *p* if H has strong and undefeated warrant for believing that S's testimony that *p* is properly based on adequate scientific justification.

5.2.b *The Challenge of Selective Uptake* p. 142
Laypersons who generally accept public scientific testimony fail to accept public scientific testimony concerning select, equally well warranted, scientific hypotheses.

5.3.e *Epistemic Overestimation* p. 153
Both accurate and inaccurate social stereotypes may lead evaluators to overestimate a subject's epistemic position.

5.3.e *Epistemic Underestimation* p. 153
Both accurate and inaccurate social stereotypes may lead evaluators to underestimate a subject's epistemic position.

5.4.b NEST p. 156
In a science communication context, SCC, in which S's scientific expert testimony that *p* conveys that *p*, S meets the epistemic conditions on appropriate scientific expert testimony that *p* only if S's scientific expert testimony is properly based on a degree of scientific justification for believing or accepting that *p* that is adequate relative to SCC.

5.4.c *Justification Explication Norm* (JEN) p. 158
Public scientific testifiers should, whenever feasible, include appropriate aspects of the nature and strength of scientific justification, or lack thereof, for the scientific hypothesis in question.

5.4.c *Justification Expert Testimony* (JET) p. 158
Scientific expert testifiers should, whenever feasible, include appropriate aspects of the nature and strength of scientific justification, or lack thereof, for the scientific hypothesis in question.

5.5.a *Expert Trespassing Testimony* p. 164
S's testimony that *p* is expert trespassing testimony iff
(i) S is an expert in D1 where D1 is a domain of expertise.
(ii) S is not an expert in D2 where D2 is a purported domain of expertise.
(iii) $p \notin$ D1.
(iv) $p \in$ D2.

5.5.a *Expert Trespassing Context* p. 164
S's conversational context is an expert trespassing context iff
a significant subset of the audience is likely or reasonable to regard S's expert trespassing testimony as expert testimony.

5.5.c *Expert Trespassing Guideline* p. 167
When S provides expert trespassing testimony in a context where it may likely and/or reasonably be taken to be expert testimony, S should qualify her testimony to indicate that it does not amount to expert testimony.

6.2.a *Deficit Reporting* p. 175
Science reporters should, whenever feasible, merely report the scientific hypotheses that meet a reasonable epistemic threshold.

6.2.b *Consensus Reporting* p. 177
Science reporters should, whenever feasible, report the scientific consensus, or lack thereof, for a reported scientific hypothesis.

6.2.e *Value-Based Reporting* p. 182
Science reporters should, whenever feasible, report a scientific hypothesis in a manner that appeals to the social values of the intended recipients.

6.3.a *Justification Reporting* p. 185
Science reporters should, whenever feasible, report appropriate aspects of the nature and strength of scientific justification, or lack thereof, for a reported scientific hypothesis.

6.5.a *Balanced Reporting* p. 203
Science reporters should, whenever feasible, report opposing hypotheses in a manner that does not favor any one of them.

6.5.a *Reliable Reporting* p. 203
Science reporters should, whenever feasible, report the most reliably based hypotheses and avoid reporting hypotheses that are not reliably based.

6.5.a *The Question of Balance* p. 204
How should *Balanced Reporting* and *Reliable Reporting* be balanced in science reporting?

6.5.b *Inclusive Reliable Reporting* p. 205
Science reporters should, whenever feasible, report hypotheses in a manner that favors the most reliably based ones by indicating the nature and strength of their respective scientific justifications.

6.5.b *Epistemically Balanced Reporting* p. 205
Science reporters should, whenever feasible, report opposing hypotheses in a manner that reflects the nature and strength of their respective scientific justifications or lack thereof.

7.1.b *Methodology* p. 212
The distinctive norms governing intra-scientific testimony are vital to the scientific methods of collaborative science.

7.1.b *Parthood* p. 212
Intra-scientific testimony is a vital part of collaborative science.

7.3.a *Enterprise* p. 223
Public scientific testimony is critical for the scientific enterprise in societies pursuing the ideals of deliberative democracy.

7.3.a *Democracy* p. 223
Public scientific testimony is a critical part of societies pursuing the ideals of deliberative democracy.

7.4.d LUN p. 240
In a context of public science communication, PSC, in which S's public scientific testimony that *p* conveys that *p*, the default attitude of a layperson recipient, H, should be to believe or accept that *p* if H has strong and undefeated warrant for believing that S's testimony that *p* is properly based on adequate scientific justification.

C2 *Generic DEI* p. 249
S suffers a discriminatory epistemic injustice if S is wronged specifically in her capacity as an epistemic subject.

Literature

Aad, G., Abbott, B., Abdallah, J., Abdinov, O., Aben, R., Abolins, M., ... and Abulaiti, Y. (2015). Combined measurement of the Higgs Boson mass in p p collisions at s = 7 and 8 TeV with the ATLAS and CMS experiments. *Physical review letters* 114 (19): 191803.

ABC News (2020). Fauci throws cold water on Trump's declaration that malaria drug chloroquine is a "game changer." https://abcnews.go.com/Politics/fauci-throws-cold-water-trumps-declaration-malaria-drug/story?id=69716324.

Achen, C. H., and Bartels, L. M. (2017). *Democracy for Realists: Why Elections Do Not Produce Responsive Government*, Vol. 4. Princeton University Press.

Adam, D. (2019). A solution to psychology's reproducibility problem just failed its first test. *Science*, https://www.sciencemag.org/news/2019/05/solution-psychology-s-reproducibility-problem-just-failed-its-first-test.

Adler, J. (2002). *Belief's Own Ethics*. MIT Press.

Aksnes, D. W. (2006). Citation rates and perceptions of scientific contribution. *Journal of the American Society for Information Science and Technology* 57 (2): 169–85.

Alexander, J., Gonnerman, C., and Waterman, J. (2014). Salience and epistemic egocentrism: An empirical study. In (ed. Beebe, J.) *Advances in Experimental Epistemology*. Bloomsbury: 97–118.

Alexander, J., Himmelreich, J., and Thompson, C. (2015). Epistemic landscapes, optimal search, and the division of cognitive labor. *Philosophy of Science* 82 (3): 424–53.

Alfano, M., and Klein, C. (2019). Trust in a social and digital world. *Social Epistemology Review and Reply Collective* 8 (10): 1–8.

Alfano, M., and Sullivan, E. (2021). Online trust and distrust. In (eds. Hannon, M. and de Ridder, J.), *The Routledge Handbook of Political Epistemology*. Routledge: 480–91.

Allchin, D. (2003). Scientific myth-conceptions. *Science Education* 87 (3): 329–51.

Allen, B. L. (2018). Strongly participatory science and knowledge justice in an environmentally contested region. *Science, Technology, and Human Values* 43 (6): 947–71.

American Psychological Association (2009). *Publication Manual of the American Psychological Association*, 6th edition.

Ames, D. (2004). Inside the mind-reader's toolkit: Projection and stereotyping in mental state inference. *Journal of Personality and Social Psychology* 87: 340–53.

Ames, D., Weber, E. U., and Zou, X. (2012). Mind-reading in strategic interaction: The impact of perceived similarity on projection and stereotyping. *Organizational Behavior and Human Decision Processes* 117 (1): 96–110.

Andersen, H. (2012). Conceptual development in interdisciplinary research. In (eds. Feest, U. and Steinle, F.) *Scientific Concepts and Investigative Practice*. Walter de Gruyter: 271–92.

Andersen, H. (2016). Collaboration, interdisciplinarity, and the epistemology of contemporary science. *Studies in History and Philosophy of Science Part A* 56: 1–10.

Andersen, H., Barker, P., and Chen, X. (2006). *The Cognitive Structure of Scientific Revolutions*. Cambridge University Press.

Andersen, H., and Wagenknecht, S. (2013). Epistemic dependence in interdisciplinary groups. *Synthese* 190 (11): 1881–98.

Anderson, E. (2006). The epistemology of democracy. *Episteme* 3 (1–2): 8–22.
Anderson, E. (2011). Democracy, public policy, and lay assessments of scientific testimony. *Episteme* 8 (2): 144–64.
Anderson, E. (2012). Epistemic justice as a virtue of social institutions. *Social Epistemology* 26 (2): 163–73.
Angler, M. (2017). *Science Journalism: An Introduction*. Routledge.
Aschengrau, A., and Seage, G. R. (2020). *Essentials of Epidemiology in Public Health*, 4th edition. Jones and Bartlett Publishers.
Audi, R. (1997). The place of testimony in the fabric of knowledge and justification. *American Philosophical Quarterly* 34 (4): 405–22.
Audi, R. (2006). Testimony, credulity, and veracity. In (eds. Lackey, J. and Sosa, E.) *The Epistemology of Testimony*. Oxford University Press: 25–49.
Bach, K. (1994). Conversational impliciture. *Mind and Language* 9: 124–62.
Bach, K., and Harnish, R. (1979). *Linguistic Communication and Speech Acts*. MIT Press.
Bacon, F. (1620) [2008]. *Novum Organum*. In (ed. Vickers, B.) *Francis Bacon: The Major Works*. Oxford University Press.
Baghramian, M., and Croce, M. (2021). Experts, public policy, and the question of trust. In (eds. Hannon, M. and de Ridder, J.), *The Routledge Handbook of Political Epistemology*. Routledge: 446–57.
Bailer-Jones, D. M. (2009). *Scientific Models in Philosophy of Science*. University of Pittsburgh Press.
Ballantyne, N. (2019). Epistemic trespassing. *Mind* 128 (510): 367–95.
Balliet, D., Wu, J., and De Dreu, C. K. (2014). Ingroup favoritism in cooperation: A meta-analysis. *Psychological Bulletin* 140 (6): 1556.
Bargh J. A. (2007). *Social Psychology and the Unconscious: The Automaticity of Higher Mental Processes*. Psychological Press.
Baron, J., and Hershey, J. C. (1988). Outcome bias in decision evaluation. *Journal of Personality and Social Psychology* 54 (4): 569.
Bauer, M. (1998). The medicalization of science news: From the "rocket-scalpel" to the "gene-meteorite complex." *Social Science Information* 37 (4): 731–51.
BBC (2018a). BBC editorial guidelines. BBC Online, https://www.bbc.co.uk/editorialguidelines/guidelines.
BBC (2018b). Trump: Climate change scientists have "political agenda." BBC Online, October 15, https://www.bbc.com/news/world-us-canada-45859325.
Beatty, J., and Moore, A. (2010). Should we aim for consensus? *Episteme* 7 (3): 198–214.
Beauchamp, T. (2011). Informed consent: Its history, meaning, and present challenges. *Cambridge Quarterly of Healthcare Ethics* 20 (4): 515–23.
Beaver, D. (2004). Does collaborative research have greater epistemic authority? *Scientometrics* 60: 399–408.
Bedford, D. (2015). Does climate literacy matter? A case study of U.S. students' level of concern about anthropogenic global warming. *Journal of Geography* 115 (5): 187–97.
Benjamin, D. J., Berger, J. O., Johannesson, M., Nosek, B. A., Wagenmakers, E. J., Berk, R., . . . and Cesarini, D. (2018). Redefine statistical significance. *Nature Human Behaviour* 2 (1): 6.
Bennett, J., and Higgitt, R. (2019). London 1600–1800: Communities of natural knowledge and artificial practice. *British Journal for the History of Science* 52 (2): 183–96.
Benton, M. (2016). Expert opinion and second-hand knowledge. *Philosophy and Phenomenological Research* 92 (2): 492–508.
Berlin, J. A., and Golub, R. M. (2014). Meta-analysis as evidence: Building a better pyramid. *Jama* 312 (6): 603–6.

Bernecker, S., Flowerree, A. K., and Grundmann, T. (eds.) (2021). *The Epistemology of Fake News*. Oxford University Press.
Betz, G. (2013). In defence of the value free ideal. *European Journal for Philosophy of Science* 3: 207–20.
Bicchieri, C. (2006). *The Grammar of Society: The Nature and Dynamics of Social Norms*. Cambridge University Press.
Bicchieri, C. (2017). *Norms in the Wild: How to Diagnose, Measure, and Change Social Norms*. Oxford University Press.
Bicchieri, C., Xiao, E., and Muldoon, R. (2011). Trustworthiness is a social norm, but trusting is not. *Politics, Philosophy and Economics* 10 (2): 170–87.
Biddle, J., and Leuschner, A. (2015). Climate scepticism and the manufacture of doubt. *European Journal for Philosophy of Science* 5 (3): 261–78.
Bird, A. (2010). Social knowing: The social sense of "scientific knowledge." *Philosophical Perspectives* 24: 23–56.
Bird, A. (2014). When is there a group that knows? Distributed cognition, scientific knowledge, and the social epistemic subject. In (ed. Lackey, J.) *Essays in Collective Epistemology*. Oxford University Press: 42–63.
Bird, A. (2019). Systematicity, knowledge, and bias: How systematicity made clinical medicine a science. *Synthese* 196 (3): 863–79.
Bird, A. (forthcoming). Understanding the replication crisis as a base rate fallacy. *The British Journal for the Philosophy of Science*.
Blair, A. (1990). Tycho Brahe's critique of Copernicus and the Copernican system. *Journal of the History of Ideas* 51 (3): 355–77.
Boguná, M., Pastor-Satorras, R., Díaz-Guilera, A., and Arenas, A. (2004). Models of social networks based on social distance attachment. *Physical review E* 70 (5): 056122.
Bolsen, T., and Druckman, J. (2018). Do partisanship and politicization undermine the impact of a scientific consensus message about climate change? *Group Processes and Intergroup Relations* 21 (3): 389–402.
Bonjour, L. (1985). *The Structure of Empirical Knowledge*. Harvard University Press.
Bonjour, L. (1992). Internalism/externalism. In (eds. Dancy, J. and Sosa, E.) *A Companion to Epistemology*. Blackwell: 132–6.
Boult, C. (2021). The epistemic responsibilities of citizens in a democracy. In (eds. De Ridder, J. and Hannon, M.) *The Routledge Handbook of Political Epistemology*. Routledge: 407–18.
Boyd, K. (2017). Testifying understanding. *Episteme* 14 (1): 103–27.
Boyd, K. (2018). Epistemically pernicious groups and the groupstrapping problem. *Social Epistemology* 33 (1): 61–73.
Boyd, K. (forthcoming). Group epistemology and structural factors in online group polarization. *Episteme*.
Boyer-Kassem, T., Mayo-Wilson, C., and Weisberg, M. (eds.) (2017). *Scientific Collaboration and Collective Knowledge*. Oxford University Press.
Boykoff, M. T. (2007). Flogging a dead norm? Newspaper coverage of anthropogenic climate change in the United States and United Kingdom from 2003 to 2006. *Area* 39 (4), 470–81.
Boykoff, M. T., and Boykoff, J. M. (2004). Balance as bias: Global warming and the US prestige press. *Global Environmental Change* 14 (2): 125–36.
Brady, M. S., and Fricker, M. (eds.) (2016). *The Epistemic Life of Groups: Essays in the Epistemology of Collectives*. Oxford University Press.
Bramson, A., Grim, P., Singer, D., Berger, W., Sack, G., Fisher, S., Flocken, C., and Holman, B. (2017). Understanding polarization: Meanings, measures, and model evaluation. *Philosophy of Science* 84: 115–59.

Brewer, M. B. (2001). Ingroup identification and intergroup conflict. *Social Identity, Intergroup Conflict, and Conflict Reduction* 3: 17–41.

Brier, G. W. (1950). Verification of forecasts expressed in terms of probability. *Monthey Weather Review* 78 (1): 1–3.

Bright, L. K. (2017). Decision theoretic model of the productivity gap. *Erkenntnis* 82 (2): 421–42.

Brown, J., and Cappelen, H. (eds.) (2011). *Assertion*. Oxford University Press.

Brown, M. (2013). Review of *Science in a Democratic Society* by Philip Kitcher. *Minerva* 51: 389–97.

Brown, M. J. (2020). *Science and Moral Imagination: A New Ideal for Values in Science*. University of Pittsburgh Press.

Bruner, Justin P., and Holman, Bennett (2019). Self-correction in science: Meta-analysis, bias and social structure. *Studies in History and Philosophy of Science Part A* 78: 93–7.

Brüning, O., Burkhardt, H., and Myers, S. (2012). The large hadron collider. *Progress in Particle and Nuclear Physics* 67 (3): 705–34.

Buckwalter, W. (2014). The mystery of stakes and error in ascriber intuitions. In (ed. Beebe, J.) *Advances in Experimental Epistemology*. Bloomsbury Academic: 145–73.

Buckwalter, W., and Schaffer, J. (2015). Knowledge, stakes, and mistakes. *Noûs* 49 (2): 201–34.

Budescu, D. V., Por, H. H., Broomell, S. B., and Smithson, M. (2014). The interpretation of IPCC probabilistic statements around the world. *Nature Climate Change* 4 (6): 508–12.

Bullock, J. G., and Lenz, G. (2019). Partisan bias in surveys. *Annual Review of Political Science* 22: 325–42.

Burge, T. (1979). Individualism and the mental. *Midwest Studies in Philosophy* 4 (1): 73–122.

Burge, T. (1993). Content preservation. *The Philosophical Review* 102: 457–88.

Burge, T. (1997). Interlocution, perception, and memory. *Philosophical Studies* 86: 21–47.

Burge, T. (2003). Perceptual entitlement. *Philosophy and Phenomenological Research* 67 (3): 503–48.

Burge, T. (2011). Lecture II: Self and constitutive norms. *Journal of Philosophy* 108 (6–7): 316–38.

Burge, T. (2013). Postscript: "Content preservation." In *Cognition through Understanding: Self-Knowledge, Interlocution, Reasoning, Reflection*. Oxford University Press: 254–84.

Burge, T. (2020). Entitlement: The basis of empirical warrant. In (eds. Graham, P. and Pedersen, N.) *Epistemic Entitlement*. Oxford University Press: 37–142.

Burns, T. W., O'Connor, D. J., and Stocklmayer, S. M. (2003). Science communication: A contemporary definition. *Public Understanding of Science* 12 (2): 183–202.

Buzzacott, P. (2016). *DAN Annual Diving Report 2016 Edition: A Report on 2014 Data on Diving Fatalities, Injuries, and Incidents*. Divers Alert Network.

Campbell, D. (1969). Ethnocentrism of disciplines and the fish-scale model of omniscience. In (eds. Sherif, M. and Sherif, C. W.) *Interdisciplinary Relationships in the Social Sciences*. Routledge: 328–48.

Campbell, T., and Kay, A. (2014). Solution aversion: On the relation between ideology and motivated disbelief. *Journal of Personality and Social Psychology* 107: 809–24.

Carruthers, P. (2011). *The Opacity of Mind: An Integrative Theory of Self-Knowledge*. Oxford University Press.

Carston, R. (2002). *Thoughts and Utterances: The Pragmatics of Explicit Communication*. Blackwell.

Carter, A. (2020). On behalf of a bi-level account of trust. *Philosophical Studies* 177 (8), 2299–322.

Carter, A., and Nickel, P. (2014). On testimony and transmission. *Episteme* 11 (2): 145–55.
Carter, A. B., and Phillips, K. W. (2017). The double-edged sword of diversity: Toward a dual pathway model. *Social and Personality Psychology Compass* 11 (5): e12313.
Cartwright, N. (2010). What are randomised controlled trials good for? *Philosophical Studies* 147 (1): 59–70.
Cartwright, N. (2020). Why trust science? Reliability, particularity and the tangle of science. *Proceedings of the Aristotelian Society* 120 (3): 237–52.
Castanho Silva, B., Vegetti, F., and Littvay, L. (2017). The elite is up to something: Exploring the relation between populism and belief in conspiracy theories. *Swiss Political Science Review* 23 (4): 423–43.
Chakravartty, A. (2011). Scientific realism. In (ed. Zalta, E. N.) *The Stanford Encyclopedia of Philosophy*.
Cheon, H. (2014). In what sense is scientific knowledge collective knowledge? *Philosophy of the Social Sciences* 44 (4): 407–23.
Cho, A. (2011). Particle physicists' new extreme teams. *Science* 333 (6049): 1564–7.
Christensen, D. (2007). Epistemology of disagreement: The good news. *Philosophical Review* 116 (2): 187–217.
Christensen, D., and Kornblith, H. (1997). Testimony, memory and the limits of the a priori. *Philosophical Studies* 86 (1): 1–20.
Clarke, C. E., Weberling McKeever, B., Holton, A., and Dixon, G. N. (2015). The influence of weight-of-evidence messages on (vaccine) attitudes: A sequential mediation model. *Journal of Health Communication* 20 (11): 1302–9.
Clement, F. (2010). To trust or not to trust? Children's social epistemology. *Review of Philosophical Psychology* 1: 531–49.
CNet (2020). Coronavirus treatments: Everything you need to know about chloroquine and vaccines. CNet Online, March 23, https://www.cnet.com/how-to/coronavirus-treatments-everything-you-need-to-know-about-chloroquine-and-vaccines/.
CNN (2019). MMR vaccine does not cause autism, another study confirms. https://edition.cnn.com/2019/03/04/health/mmr-vaccine-autism-study/index.html. October 26.
CNN (2020). The healthiest way to brew your coffee—and possibly lengthen your life. https://edition.cnn.com/2020/04/22/health/healthiest-coffee-brew-wellness/index.html. April 23.
Coady, C. A. J. (1992), *Testimony: A Philosophical Study*. Clarendon Press.
Coady, D. (2012). *What to Believe Now: Applying Epistemology to Contemporary Issues*. Wiley-Blackwell.
Cohen, J. (1989). Deliberation and democratic legitimacy. In (eds. Matravers, D. and Pike, J.) *Debates in Contemporary Political Philosophy*, 342–60.
Cohen, L. J. (1989). Belief and acceptance. *Mind 98* (391): 367–89.
Cohen, L. J. (1992). *An Essay on Belief and Acceptance*. Clarendon Press.
Cole, C., Harris, P., and Koenig, M. (2012). Entitled to trust? Philosophical frameworks and the evidence from children. *Analyse and Kritik* 2: 195–216.
Collins, H. (2004). Interactional expertise as a third kind of knowledge. *Phenomenology and the Cognitive Sciences* 3 (2): 125–43.
Collins, H., and Evans, R. (2007). *Rethinking Expertise*. University of Chicago Press.
Collins, H., and Evans, R. (2015). Expertise revisited, part I: Interactional expertise. *Studies in History and Philosophy of Science Part A* 54: 113–23.
Collins, H., Evans, R., and Gorman, M. (2007). Trading zones and interactional expertise. *Studies in History and Philosophy of Science Part A* 38 (4): 657–66.
Collins, H., Evans, R., and Weinel, M. (2016). Expertise revisited II: Contributory expertise. *Studies in History and Philosophy of Science* 56: 103–10.

Conley, S. N., Foley, R. W., Gorman, M. E., Denham, J., and Coleman, K. (2017). Acquisition of T-shaped expertise: An exploratory study. *Social Epistemology* 31 (2): 165–83.

Contessa, G. (forthcoming). It takes a village to trust science: Towards a (thoroughly) social approach to social trust in science. *Erkenntnis.*

Cook, J., and Lewandowsky, S. (2016). Rational irrationality: Modeling climate change belief polarization using Bayesian networks. *Topics in Cognitive Science* 8 (1): 160–79.

Corner, A., Whitmarsh, L., and Xenias, D. (2012). Uncertainty, scepticism and attitudes towards climate change: Biased assimilation and attitude polarisation. *Climatic Change* 114 (3–4): 463–78.

Cowburn, J. (2013). *Scientism: A Word We Need.* Wipf and Stock.

Creswell, J. W., and Clark, V. L. P. (2017). *Designing and Conducting Mixed Methods Research.* Sage Publications.

Crichton, M. (2003). Available on: http://stephenschneider.stanford.edu/Publications/PDF_Papers/Crichton2003.pdf.

Croce, M. (2019a). For a service conception of epistemic authority: A collective approach. *Social Epistemology* 33 (2): 172–82.

Croce, M. (2019b). On what it takes to be an expert. *Philosophical Quarterly* 69 (274): 1–21.

Cullison, A. (2010). On the nature of testimony. *Episteme* 7 (2): 114–27.

Dang, H. (2019). Do collaborators in science need to agree? *Philosophy of Science* 86 (5): 1029–40.

Dang, H., and Bright, L. K. (forthcoming). Scientific conclusions need not be accurate, justified, or believed by their authors. *Synthese.*

De Cruz, Helen (2020). Believing to belong: Addressing the novice-expert problem in polarized scientific communication. *Social Epistemology* 34 (5): 440–52.

de Melo-Martín, I., and Intemann, K. (2014). Who's afraid of dissent? Addressing concerns about undermining scientific consensus in public policy developments. *Perspectives on Science* 22 (4): 593–615.

de Melo-Martín, I., and Intemann, K. (2018). *The Fight against Doubt: How to Bridge the Gap between Scientists and the Public.* Oxford University Press.

de Ridder, J. (2014a). Epistemic dependence and collective scientific knowledge. *Synthese* 191 (1): 1–17.

de Ridder, J. (2014b). Science and scientism in popular science writing. *Social Epistemology Review and Reply Collective* 3 (12): 23–39.

de Ridder, J., Peels, R., and van Woudenberg, R. (eds.) (2018). *Scientism: Prospects and Problems.* Oxford University Press.

De Waal, F. (2017). The surprising science of alpha males. Retrieved from https://www.ted.com/talks/frans_de_waal_the_surprising_science_of_alpha_males?referrer=playlist-unexpected_lessons_from_the_animal_world.

DeJesus, J. M., Callanan, M. A., Solis, G., and Gelman, S. A. (2019). Generic language in scientific communication. *Proceedings of the National Academy of Sciences* 116 (37): 18370–7.

Del Vicario, M., Vivaldo, G., Bessi, A., Zollo, F., Scala, A., Caldarelli, G., and Quattrociocchi, W. (2016). Echo chambers: Emotional contagion and group polarization on facebook. *Scientific Reports* 6: 37825.

Dellsén, F. (2018a). Scientific progress: Four accounts. *Philosophy Compass* 13 (11): e12525.

Dellsén, F. (2018b). When expert disagreement supports the consensus. *Australasian Journal of Philosophy* 96 (1): 142–56.

Dellsén, F. (2020). The epistemic value of expert autonomy. *Philosophy and Phenomenological Research* 100 (2): 344–61.

Deryugina, T., and Shurchkov, O. (2016). The effect of information provision on public consensus about climate change. *PloS One* 11 (4): e0151469

Descartes, R. (1628) [1985]. Rules for the direction of the mind. In (eds. Cottingham, J., Stoothoff, R., and Murdoch, D.) *The Philosophical Writings of Descartes*, Vol. 1. Cambridge University Press.

Descartes, R. (1641) [1985]. *Meditations on First Philosophy: With Selections from the Objections and Replies.* (Transl. J. Cottingham, Ed. B. Williams). Cambridge University Press.

Dickinson, J. L., Zuckerberg, B., and Bonter, D. N. (2010). Citizen science as an ecological research tool: Challenges and benefits. *Annual Review of Ecology, Evolution, and Systematics* 41: 149–72.

DiPaolo, J. (forthcoming). What's wrong with epistemic trespassing? *Philosophical Studies.*

Dixon, G. N., and Clarke, C. E. (2013). Heightening uncertainty around certain science: Media coverage, false balance, and the autism-vaccine controversy. *Science Communication* 35 (3): 358–82.

Dixon, G., Hmielowski, J., and Ma, Y. (2017). Improving climate change acceptance among US conservatives through value-based message targeting. *Science Communication* 39 (4): 520–34.

Donovan, S. M., O'Rourke, M., Looney, C. (2015). Your hypothesis or mine? Terminological and conceptual variation across disciplines. *SAGE Open* 5 (2): 1–13.

Dotson, K. (2011). Tracking epistemic violence, tracking practices of silencing. *Hypatia* 26 (2): 236–57.

Douglas, H. (2009). *Science, Policy, And The Value-Free Ideal.* University of Pittsburgh Press.

Douglas, H. (2013). Review of *Science in a Democratic Society* by Philip Kitcher. *British Journal of Philosophy of Science* 64: 901–5.

Douglas, H. (2015). Politics and science: Untangling values, ideologies, and reasons. *The Annals of the American Academy of Political and Social Science* 658 (1): 296–306.

Douven, I., and Cuypers, S. (2009). Fricker on testimonial justification. *Studies in History and Philosophy of Science Part A* 40 (1): 36–44.

Drummond, C., and Fischhoff, B. (2017a). Individuals with greater science literacy and education have more polarized beliefs on controversial science topics. *Proceedings of the National Academy of Sciences* 114 (36): 9587–92.

Drummond, C., and Fischhoff, B. (2017b). Development and validation of the scientific reasoning scale. *Journal of Behavioral Decision Making* 30 (1): 26–38.

Dunlap, R. E., Norgaard, R. B., and McCright, A. M. (2011). Organized climate change denial. In (eds. Dryzek, J. S., Norgaard, R. B., and Schlosberg, D.) *The Oxford Handbook of Climate Change and Society.* Oxford University Press: 144–60.

Dunwoody S. (2005). Weight-of-evidence reporting: What is it? Why use it? *Nieman Reports* 54 (4): 89–91.

Dunwoody, S. (2014). Science journalism. In (eds. Bucchi, M., and Trench, B.) *Routledge Handbook of Public Communication of Science and Technology.* Routledge: 27–39.

Dunwoody, S., and Kohl, P. A. (2017). Using weight-of-experts messaging to communicate accurately about contested science. *Science Communication* 1075547017707765.

Dutilh Novaes, C. (2015). A dialogical, multi-agent account of the normativity of logic. *dialectica* 69 (4): 587–609.

Eagly, H. (2016). When passionate advocates meet research on diversity, does the honest broker stand a chance? *Journal of Social Issues* 72 (1): 199–222.

Elgin, C. (2001). Word giving, word taking. In (eds. Byrne, A., Stalnaker, R., and Wedgwood, R.) *Fact and Value: Essays for Judith Jarvis Thomson.* MIT Press: 97–116.

Elgin, C. (2017). *True Enough*. MIT Press.
Elster, J. (1989). *Nuts and Bolts for the Social Sciences*. Cambridge University Press.
Ely, R., and Thomas, D. (2001). Cultural diversity at work: The effects of diversity perspectives on work group processes and outcomes. *Admin. Sci. Q.* 46: 229–73.
Enders, J., and de Weert, E. (2009). Towards a T-shaped profession: Academic work and career in the knowledge society. In (eds. Enders, J., De Weert, E., and de Weert, E.) *The Changing Face of Academic Life*. Palgrave Macmillan: 251–72.
Engel, P. (ed.) (2000). *Believing and Accepting*, Vol. 83. Springer Science and Business Media.
Entman, R. (1989). *Democracy without Citizens: Media and the Decay of American Politics*. Oxford University Press.
Estlund, D. (2008). *Democratic Authority: A Philosophical Framework*. Princeton University Press.
Estlund, D., and Landemore, H. (2018). The epistemic value of democratic deliberation. In (eds. Bächtiger, A., Dryzek, J. S., Mansbridge, J., and Warren, M. E.) *The Oxford Handbook of Deliberative Democracy*. Oxford University Press: 113–32.
European Commission (2019). *Group of Chief Scientific Advisors*. Retrieved in April 2019 from https://ec.europa.eu/research/sam/index.cfm?pg=hlg).
Evans, J. (2010). *Thinking Twice: Two Minds in One Brain*. Oxford University Press.
Eveland, W. P., Nathanson, A. I., Detenber, B. H., and McLeod, D. M. (1999). Rethinking the social distance corollary: Perceived likelihood of exposure and the third-person perception. *Communication Research* 26: 275–302.
Fagan, M. (2012). Collective scientific knowledge. *Philosophy Compass* 7 (12): 821–31.
Fahrbach, L. (2011). Theory change and degrees of success. *Philosophy of Science* 78 (5): 1283–92.
Fallis, D. (2006). The epistemic costs and benefits of collaboration. *The Southern Journal of Philosophy* 44 (S1): 197–208.
Faulkner, P. (2000). The social character of testimonial knowledge. *Journal of Philosophy* 97 (11): 581–601.
Faulkner, P. (2011). *Knowledge on Trust*. Oxford University Press.
Faulkner, P. (2018). Collective testimony and collective knowledge. *Ergo, an Open Access Journal of Philosophy* 5: 104–26.
Fidler, F., and Wilcox, J. (2018). Reproducibility of scientific results. In (ed. Zalta, E. N.) *The Stanford Encyclopedia of Philosophy* (Winter 2018 edition). <https://plato.stanford.edu/archives/win2018/entries/scientific-reproducibility/>.
Figdor, C. (2010). Is objective news possible? *Journalism Ethics: A Philosophical Approach*: 153–64.
Figdor, C. (2013). New skepticism about science. *Philosophers' Magazine* 60 (1): 51–6.
Figdor, C. (2017). (When) is scientific reporting ethical? The case for recognizing shared epistemic responsibility in science journalism. *Frontiers in Communication* 2: 1–7.
Figdor, C. (2018). Trust me: News, credibility deficits, and balance. In (eds. Fox, C. and Saunders, J.) *Media Ethics, Free Speech, and the Requirements of Democracy*. Routledge: 69–86.
Fischhoff, B. (2012). Communicating uncertainty: Fulfilling the duty to inform. *Issues in Science and Technology* 28: 63–70.
Fischhoff, B. (2013). The sciences of science communication. *Proceedings of the National Academy of Sciences* 110 (Supplement 3): 14033–9.
Fischhoff, B. (2019). Evaluating science communication. *Proceedings of the National Academy of Sciences* 116 (16): 7670–5.

Fishkin, J. S., and Luskin, R. C. (2005). Experimenting with a democratic ideal: Deliberative polling and public opinion. *Acta Politica* 40 (3): 284–98.
Fiske, S., Gilbert, D., and Lindzey, G. (eds.) (2010). *Handbook of Social Psychology*, Vol. 1. John Wiley and Sons.
Fleisher, W. (2021). Endorsement and assertion. *Noûs 55* (2): 363–84.
Franceschet, M., and Costantini, A. (2010). The effect of scholar collaboration on impact and quality of academic papers. *Journal of Informetrics* 4 (4): 540–53.
Franco, Paul, L. (2017). Assertion, non-epistemic values, and scientific practice. *Philosophy of Science* 84 (1): 160–80.
Fricker, E. (1994). Against gullibility. In (eds. Chakrabarti, A. and Matilal, B. K..) *Knowing from Words*. Kluwer Academic Publishers: 125–61.
Fricker, E. (1995). Telling and trusting: Reductionism and anti-reductionism in the epistemology of testimony. *Mind* 104: 393–411.
Fricker, E. (2002). Trusting others in the sciences: A priori or empirical warrant? *Studies in History and Philosophy of Science Part A* 33 (2): 373–83.
Fricker, E. (2006a). Second-hand knowledge. *Philosophy and Phenomenological Research* 73: 592–618.
Fricker, E. (2006b). Testimony and epistemic autonomy. In (eds. Lackey, J. and Sosa, E.) *The Epistemology of Testimony*. Oxford University Press: 225–53.
Fricker, E. (2006c). Varieties of anti-reductionism about testimony: A reply to Goldberg and Henderson. *Philosophy and Phenomenological Research* 72: 618–28.
Fricker, E. (2012). Stating and insinuating. *Aristotelian Society Supplementary Volume* 86 (1): 61–94.
Fricker, E. (2017). Norms, constitutive and social, and assertion. *American Philosophical Quarterly* 54 (4): 397–418.
Fricker, M. (2007). *Epistemic Injustice*. Oxford University Press.
Fricker, M. (2012). Group testimony? The making of a collective good informant. *Philosophy and Phenomenological Research* 84 (2): 249–76.
Fricker, M. (2013). Epistemic justice as a condition of political freedom? *Synthese* 190 (7): 1317–32.
Frigg, R., and Nguyen, J. (2021). Mirrors without warnings. *Synthese* 198 (3), 2427–47.
Frimer, J. A., Skitka, L. J., and Motyl, M. (2017). Liberals and conservatives are similarly motivated to avoid exposure to one another's opinions. *Journal of Experimental Social Psychology* 72: 1–12.
Frost-Arnold, K. (2013). Moral trust and scientific collaboration. *Studies in History and Philosophy of Science Part A* 44 (3): 301–10.
Frost-Arnold, K. (2014). Trustworthiness and truth: The epistemic pitfalls of internet accountability. *Episteme* 11 (1): 63–81.
Fumerton, R. A. (1995). *Metaepistemology and Skepticism*. Rowman and Littlefield Publishers.
Funk, C., and Kennedy, B. (2016). *The Politics of Climate*. Pew Research Center, 4.
Galinsky, A., Todd, A., Homan, A., Phillips, K., Apfelbaum, E., Sasaki, S., . . . Maddux, W. (2015). Maximizing the gains and minimizing the pains of diversity: A policy perspective. *Perspectives on Psychological Science* 10 (6): 742–8.
Galison, P. (1997). *Image and Logic: A Material Culture of Microphysics*. The University of Chicago Press.
Gauchat, G. (2012). Politicization of science in the public sphere: A study of public trust in the United States, 1974 to 2010. *American Sociological Review* 77 (2): 167–87.
Gaynes, R. P. (2011). *Germ Theory: Medical Pioneers in Infectious Diseases*. John Wiley and Sons.

Gelbspan, R. (1998). *The Heat Is On: The Climate Crisis, the Cover-up, the Prescription.* Perseus Books.

Gelfert, A. (2009). Indefensible middle ground for local reductionism about testimony. *Ratio* 22 (2): 170–90.

Gelfert, A. (2011). Who is an epistemic peer? *Logos and Episteme* 2 (4): 507–14.

Gelfert, A. (2013). Climate scepticism, epistemic dissonance, and the ethics of uncertainty. *Philosophy and Public Issues* 13 (1): 167–208.

Gelfert, A. (2014). *A Critical Introduction to Testimony.* A and C Black.

Gerken, M. (2011). Warrant and action. *Synthese* 178 (3): 529–47.

Gerken, M. (2012a). Discursive justification and skepticism. *Synthese* 189 (2): 373–94.

Gerken, M. (2012b). Critical study of Goldberg's *Relying on Others. Episteme* 9 (1): 81–8.

Gerken, M. (2012c). Univocal reasoning and inferential presuppositions. *Erkenntnis* 76 (3): 373–94.

Gerken, M. (2013a). *Epistemic Reasoning and the Mental.* Palgrave Macmillan.

Gerken, M. (2013b). Internalism and externalism in the epistemology of testimony. *Philosophy and Phenomenological Research* 87 (3): 532–57.

Gerken, M. (2013c). Epistemic focal bias. *Australasian Journal of Philosophy* 91 (1): 41–61.

Gerken, M. (2014a). Same, same but different: The epistemic norms of assertion, action and practical reasoning. *Philosophical Studies* 168 (3): 725–44.

Gerken, M. (2014b). Outsourced cognition. *Philosophical Issues* 24 (1): 127–58.

Gerken, M. (2015a). The epistemic norms of intra-scientific testimony. *Philosophy of the Social Sciences* 45 (6): 568–95.

Gerken, M. (2015b). The roles of knowledge ascriptions in epistemic assessment. *European Journal of Philosophy* 23 (1): 141–61.

Gerken, M. (2015c). Philosophical insights and modal cognition. In (eds. Collins, J. and Fischer, E.) *Experimental Philosophy, Rationalism, and Naturalism.* Routledge: 110–31.

Gerken, M. (2017a). *On Folk Epistemology: How We Think and Talk about Knowledge.* Oxford University Press.

Gerken, M. (2017b). Against knowledge-first epistemology. In (eds. Gordon, E. and Carter, J.) *Knowledge-First Approaches in Epistemology and Mind.* Oxford University Press: 46–71.

Gerken, M. (2018a). Expert trespassing testimony and the ethics of science communication. *Journal for General Philosophy of Science* 49 (3): 299–318.

Gerken, M. (2018b). Metaepistemology. *Routledge Encyclopedia of Philosophy.* doi:10.4324/0123456789-P076-1; https://www.rep.routledge.com/articles/thematic/metaepistemology/v-1.

Gerken, M. (2018c). Pragmatic encroachment on scientific knowledge? In (eds. McGrath, M. and Kim, B.) *Pragmatic Encroachment.* Routledge: 116–40.

Gerken, M. (2018d). The new evil demon and the devil in the details. In (ed. Mitova, V.) *The Factive Turn in Epistemology.* Cambridge University Press: 102–22.

Gerken, M. (2019). Pragmatic encroachment and the challenge from epistemic injustice. *Philosophers' Imprint* 19 (15): 1–19.

Gerken, M. (2020a). Epistemic entitlement: Its scope and limits. In (eds. Graham, P. and Pedersen, N.) *Epistemic Entitlement.* Oxford University Press: 150–78.

Gerken, M. (2020b). Truth-sensitivity and folk epistemology. *Philosophy and Phenomenological Research* 100 (1): 3–25.

Gerken, M. (2020c). Public scientific testimony in the scientific image. *Studies in History and Philosophy of Science A* 80: 90–101.

Gerken, M. (2020d). How to balance balanced reporting and reliable reporting. *Philosophical Studies* 177 (10): 3117–42.

Gerken, M. (2020e). Disagreement and epistemic injustice from a communal perspective. In (eds. Broncano-Berrocal, F. and Carter. A.) *The Epistemology of Group Disagreement*. Routledge: 139–62.

Gerken, M. (2021). Representation and misrepresentation of knowledge. *Behavioural and Brain Science, 44, E153.*

Gerken, M. (2022). Salient alternatives and epistemic injustice in folk epistemology. In (ed. Archer, S.) *Salience: A Philosophical Inquiry*. Routledge: 213-233.

Gerken, M. (forthcoming a). Dilemmas in science communication. In (ed. Hughes, N.) *Epistemic Dilemmas*. Oxford University Press.

Gerken, M. (Ms a). Trespassing testimony in scientific collaboration.

Gerken, M. (Ms b). Cognitive diversity and epistemic injustice.

Gerken, M., Alexander, J., Gonnerman, C., and Waterman, J. (2020). Salient alternatives in perspective. *Australasian Journal of Philosophy* 98 (4): 792–810.

Gerken, M., and Beebe, J. R. (2016). Knowledge in and out of contrast. *Noûs* 50 (1): 133–64.

Gerken, M., and Petersen, E. N. (2020). Epistemic norms of assertion and action. In (ed. Goldberg, S.) *Oxford Handbook of Assertion*. Oxford University Press: 683–706.

Gertler, B. (2015). Self-knowledge. In (ed. Zalta, E. N.) *The Stanford Encyclopedia of Philosophy*.

Giere, R. N. (2002). Discussion note: Distributed cognition in epistemic cultures. *Philosophy of Science* 69 (4): 637–44.

Giere, R. N. (2007). Distributed cognition without distributed knowing. *Social Epistemology: A Journal of Knowledge, Culture and Policy* 21: 313–20.

Gigerenzer, G. (2008). *Rationality for Mortals: How People Cope with Uncertainty*. Oxford University Press.

Gilbert, D. T., King, G., Pettigrew, S., and Wilson, T. D. (2016). Comment on "Estimating the Reproducibility of Psychological Science." *Science* 351: 1037. doi: 0.1126/science. aad7243.

Gilbert, M. (1989). *On Social Facts*. Princeton University Press.

Gilbert, M. (2000). *Sociality and Responsibility: New Essays in Plural Subject Theory*. Rowman and Littlefield.

Gilbert, M. (2002). Belief and acceptance as features of groups. *Protosociology: An International Journal of Interdisciplinary Research* 16: 35–69.

Goddiksen, M. (2014). Clarifying interactional and contributory expertise. *Studies in History and Philosophy of Science Part A* 47: 111–17.

Godfrey-Smith, P. (2003). *Theory and Reality: An Introduction to the Philosophy of Science*. University of Chicago Press.

Goethe, J. (1808/2014). *Faust: A Tragedy, Parts One and Two*. (Transl. M. Greenberg). Yale University Press.

Goldberg, S. (2008). Testimonial knowledge in early childhood, revisited. *Philosophy and Phenomenological Research* 76 (1): 1–36.

Goldberg, S. (2010a). *Relying on Others: An Essay in Epistemology*. Oxford University Press.

Goldberg, S. (2010b). Assertion, testimony, and the epistemic significance of speech. *Logos and Episteme* 1 (1): 59–65.

Goldberg, S. (2011). The division of epistemic labor. *Episteme* 8 (1): 112–25.

Goldberg, S. (2017). Should have known. *Synthese* 194 (8): 2863–94.

Goldberg, S., and Henderson, D. (2006). Monitoring and anti-reductionism in the epistemology of testimony. *Philosophy and Phenomenological Research* 72: 600–17.

Goldman, A. I. (1999). *Knowledge in a Social World*. Oxford University Press.

Goldman, A. I. (2001). Experts: Which ones should you trust? *Philosophy and Phenomenological Research* 63 (1): 85–110.

Goldman, A. I. (2018). Expertise. *Topoi* 37 (1): 3–10.
Gordon, E. C. (2016). Social epistemology and the acquisition of understanding. In (eds. Ammon, S., Baumberger, C., and Grimm, S.) *Explaining Understanding: New Perspectives from Epistemology and Philosophy of Science.* Routledge: 293–317.
Graham, P. (1997). What is testimony? *Philosophical Quarterly* 47 (187): 227–32.
Graham, P. (2000). Transferring knowledge. *Nous* 34: 131–52.
Graham, P. (2006). Can testimony generate knowledge? *Philosophica* 78: 105–27.
Graham, P. (2010). Testimonial entitlement and the function of comprehension. In (eds. Haddock, A., Millar, A., and Pritchards, D.) *Social Epistemology.* Oxford University Press: 148–74.
Graham, P. (2012a). Epistemic entitlement. *Noûs* 46 (3): 449–82.
Graham, P. (2012b). Testimony, trust, and social norms. *Abstracta* 6 (S6): 92–116.
Graham, P. (2015a). Testimony as speech act, testimony as source. In (eds. Mi, C. Sosa, E., and Slote, M.) *Moral and Intellectual Virtues in Western and Chinese Philosophy: The Turn toward Virtue.* Routledge: 121–44.
Graham, P. (2015b). Epistemic normativity and social norms. In (eds. Henderson, D. and Greco, J.) *Epistemic Evaluation: Purposeful Epistemology.* Oxford University Press: 247–73.
Graham, P. (2016). Testimonial knowledge: A unified account. *Philosophical Issues* 26 (1): 172–86.
Graham, P. (2018a). Formulating reductionism about testimonial warrant and the challenge from childhood testimony. *Synthese* 195 (7): 3013–33.
Graham, P. (2018b). Sincerity and the reliability of testimony: Burge on the a priori basis of testimonial warrant. In (eds. Michaelson, E. and Stokke, A.) *Lying: Knowledge, Language, Ethics, Politics.* Oxford University Press: 85–112.
Graham, P. (2020). What is epistemic entitlement? Reliable competence, reasons, inference, access. In (eds. Greco, J. and Kelp, C.) *Virtue-Theoretic Epistemology: New Methods and Approaches.* Cambridge University Press: 93–123.
Grasswick, H. (2010). Scientific and lay communities: Earning epistemic trust through knowledge sharing. *Synthese* 177 (3): 387–409.
Greco, J. (2012). Recent work on testimonial knowledge. *American Philosophical Quarterly* 49 (1): 15–28.
Grice, P. (1989). *Studies in the Way of Words.* Harvard University Press.
Griggs, R. A., and Cox, J. R. (1982). The elusive thematic-materials effect in Wason's selection task. *British Journal of Psychology* 73 (3): 407–20.
Grim, P., Singer, D., Bramson, A., Holman, B., McGeehan, S., and Berger, W. (2019). Diversity, ability, and expertise in epistemic communities. *Philosophy of Science* 86 (1): 98–123.
Grimm, Stephen R. (2006). Is understanding a species of knowledge? *British Journal for the Philosophy of Science* 57 (3): 515–35.
Grindrod, J., Andow, J., and Hansen, N. (2019). Third-person knowledge ascriptions: A crucial experiment for contextualism. *Mind and Language* 34 (2): 158–82.
Grundmann, Thomas (forthcoming). Experts: What are they and how can laypeople identify them? In (eds. Lackey, J. and McGlynn, A.) *Oxford Handbook of Social Epistemology.* Oxford University Press.
Guerrero, A. (2016). Living with ignorance in a world of experts. In (ed. Peels, R.) *Perspectives on Ignorance from Moral and Social Philosophy.* Routledge: 168–97.
Gundersen, T. (2018). Scientists as experts: A distinct role? *Studies in History and Philosophy of Science Part A* 69: 52–9.

Gustafson, A., and Rice, R. E. (2020). A review of the effects of uncertainty in public science communication. *Public Understanding of Science* 29 (6): 614–33.

Guy, S., Kashima, Y., Walker, I., and O'Neill, S. (2014). Investigating the effects of knowledge and ideology on climate change beliefs. *European Journal of Social Psychology* 44 (5): 421–9.

Hackett, E. (2005). Essential tensions: Identity, control, and risk in research. *Social Studies of Science* 35 (5): 787–826.

Hahn, U., and Harris, A. J. (2014). What does it mean to be biased: Motivated reasoning and rationality. Psychology of Learning and Motivation 61: 41–102. (Academic Press).

Hájek, A., and Hall, N. (2002). Induction and probability. *The Blackwell Guide to the Philosophy of Science*: 149–72.

Hall, B. H., Jaffe, A., and Trajtenberg, M. (2005). Market value and patent citations. *RAND Journal of Economics*: 16–38.

Hall, T. E., and O'Rourke, M. (2014). Responding to communication challenges in transdisciplinary sustainability science. In (eds. Huutoniemi, K. and Tapio, P.) *Heuristics for Transdisciplinary Sustainability Studies: Solution-Oriented Approaches to Complex Problems*. Routledge: 119–39.

Hallen, B. L., Bingham, C. B., Hill, C., Carolina, N., and Cohen, S. L. (2017). At least bias is bipartisan: A meta-analytic comparison of partisan bias in liberals and conservatives. *SSRN:* https://ssrn. com/abstract, 2952510.

Hallsson, B. G. (2019). The epistemic significance of political disagreement. *Philosophical Studies* 176 (8): 2187–202.

Hamilton, L. C. (2016). Public awareness of the scientific consensus on climate. *Sage Open* 6 (4): 2158244016676296.

Hansson, S. O. (2017). Science and pseudo-science. In (ed. Zalta, E.) *The Stanford Encyclopedia of Philosophy*, <https://plato.stanford.edu/archives/sum2017/entries/pseudo-science/>.

Harding, S. (1991). *Whose Science? Whose Knowledge? Thinking from Women's Lives.* Cornell University Press.

Harding, S. (2004). A socially relevant philosophy of science? Resources from standpoint theory's controversiality. *Hypatia* 19 (1): 25–47.

Hardwig, J. (1985). Epistemic dependence. *Journal of Philosophy* 82 (7): 335–49.

Hardwig, J. (1991). The role of trust in knowledge. *The Journal of Philosophy* 88 (12): 693–708.

Hardwig, J. (1994). Toward an ethics of expertise. In (ed. Wueste, D. E.) *Professional Ethics and Social Responsibility*. Rowman and Littlefield: 83–101.

Hart, P. S., and Nisbet, E. C. (2012). Boomerang effects in science communication: How motivated reasoning and identity cues amplify opinion polarization about climate mitigation policies. *Communication Research* 39 (6): 701–23.

Hart, W., Albarracín, D., Eagly, A. H., Brechan, I., Lindberg, M. J., and Merrill, L. (2009). Feeling validated versus being correct: A meta-analysis of selective exposure to information. *Psychological Bulletin* 135 (4): 555–88.

Hassin, R. R., Uleman, J. S., and Bargh, J. A. (eds.) (2005). *The New Unconscious*. Oxford University Press.

Hawley, K. (2012). *Trust: A Very Short Introduction*. Oxford University Press.

Hawley, K. (2019). *How to Be Trustworthy*. Oxford University Press.

Heesen, R., Bright, L., and Zucker, A. (2019). Vindicating methodological triangulation. *Synthese* 196 (8): 3067–81.

Henderson, D., and Graham, P. (2017a). Epistemic norms and the "epistemic game" they regulate: The basic structured epistemic costs and benefits. *American Philosophical Quarterly* 54 (4): 367–82.

Henderson, D., and Graham, P. (2017b). A refined account of the "epistemic game": epistemic norms, temptations, and epistemic cooperation. *American Philosophical Quarterly* 54 (4): 383–96.

Hendriks, F., Kienhues, D., and Bromme, R. (2015). Measuring laypeople's trust in experts in a digital age: The Muenster Epistemic Trustworthiness Inventory (METI). *PloS One* 10 (10): e0139309.

Hendriks, F., Kienhues, D., and Bromme, R. (2016). Disclose your flaws! Admission positively affects the perceived trustworthiness of an expert science blogger. *Studies in Communication Sciences* 16 (2): 124–31.

Hergovich, A., Schott, R., and Burger, C. (2010). Biased evaluation of abstracts depending on topic and conclusion: Further evidence of a confirmation bias within scientific psychology. *Current Psychology* 29 (3): 188–209.

Hinchman, E. (2005). Telling as inviting to trust. *Philosophy and Phenomenological Research* 70 (3): 562–87.

Holbrook, J. B. (2013). What is interdisciplinary communication? Reflections on the very idea of disciplinary integration. *Synthese* 190 (11): 1865–79.

Holst, C., and Molander, A. (2019). Epistemic democracy and the role of experts. *Contemporary Political Theory* 18 (4): 541–61.

Hong, L., and Page, S. (2004). Groups of diverse problem solvers can outperform groups of high-ability problem solvers. *Proceedings of the National Academy of Sciences* 101: 16385–9.

Hopkins, E. J., Weisberg, D. S., and Taylor, J. C. V. (2016). The seductive allure is a reductive allure: People prefer scientific explanations that contain logically irrelevant reductive information. *Cognition* 155: 67–76.

Hopkins, E. J., Weisberg, D. S., and Taylor, J. C. V. (2019). Expertise in science and philosophy moderates the seductive allure of reductive explanations. *Acta Psychologica* 198: 102890.

Hoyningen-Huene, P. (2016). *Systematicity: The Nature of Science*. Oxford University Press.

Huebner, B., Kukla, R., and Winsberg, E. (2018). Making an author in radically collaborative research. In (eds. Boyer-Kassem, T., Mayo-Wilson, C., and Wiesberg, M.) *Scientific Collaboration and Collective Knowledge: New Essays*. Oxford University Press: 95–116.

Hull, D. L. (1988). *Science as a Process: An Evolutionary Account of the Social and Conceptual Development of Science*. University of Chicago Press.

Huxster, J. K., Slater, M. H., . . . and Hopkins, M. (2018). Understanding "understanding" in public understanding of science. *Public Understanding of Science* 27 (7): 756–71.

Hvidtfeldt, R. (2018). *The Structure of Interdisciplinary Science*. Palgrave Macmillan.

Hviid, A., Hansen, J. V., Frisch, M., and Melbye, M. (2019). Measles, mumps, rubella vaccination and autism: A nationwide cohort study. *Annals of Internal Medicine* 170 (8): 513–20.

Hyman, J. (2015). *Action, Knowledge, and Will*. Oxford University Press.

Intemann, K. (2011). Diversity and dissent in science: Does democracy always serve feminist aims? In (ed. Grasswick, H.) *Feminist Epistemology and Philosophy of Science*. Springer: 111–32.

Intemann, K. (2017). Who needs a consensus anyway? Addressing manufactured doubt and increasing public trust in climate science. *Public Affairs Quarterly* 31 (3): 189–208.

Ioannidis, J. (2005). Why most published research findings are false. *PLoS Med* 2 (8): e124.

Ioannidis J. (2018). All science should inform policy and regulation. *PLoS Med* 15 (5): e1002576.

Irwin, A. (2015). Citizen science and scientific citizenship: Same words, different meanings? *Science Communication Today*: 29–38.

Irzik, G., and Kurtulmus, F. (2019). What is epistemic public trust in science? *The British Journal for the Philosophy of Science* 70 (4): 1145–66.

Irzik, G., and Kurtulmus, F. (2021). Well-ordered science and public trust in science. *Synthese* 198: 4731–48.

Isenberg, D. (1986). Group polarization: A critical review and meta-analysis. *Journal of Personality and Social Psychology* 50 (6): 1141–51.

Iyengar, S., and Massey, D. S. (2019). Scientific communication in a post-truth society. *Proceedings of the National Academy of Sciences* 116 (16): 7656–61.

Jäger, C. (2016). Epistemic authority, preemptive reasons, and understanding. *Episteme* 13 (2): 167–85.

Jamieson, K. H., Kahan, D., and Scheufele, D. A. (eds.) (2017). *The Oxford Handbook of the Science of Science Communication*. Oxford University Press.

Jasanoff, S. (1990). *The Fifth Branch: Science Advisers as Policymakers*. Harvard University Press.

Jasanoff, S. (ed.) (2004a). *States of Knowledge: The Co-Production of Science and the Social Order*. Routledge.

Jasanoff, S. (2004b). Ordering knowledge, ordering society. In (ed. Jasanoff, S.) *States of Knowledge: The Co-Production of Science and the Social Order*. Routledge: 13–45.

Jasanoff, S. (2014). A mirror for science. *Public Understanding of Science* 23 (1): 21–6.

Jellison, J., and Riskind, J. (1970). A social comparison of abilities interpretation of risk taking behavior. *Journal of Personality and Social Psychology* 15 (4): 375–90.

Jensen, J. D. (2008). Scientific uncertainty in news coverage of cancer research: Effects of hedging on scientists' and journalists' credibility. *Human Communication Research* 34 (3): 347–69.

Jensen, J. D., and Hurley, R. J. (2012). Conflicting stories about public scientific controversies: Effects of news convergence and divergence on scientists' credibility. *Public Understanding of Science* 21: 689–704.

Jensen, J. D., Pokharel, M., Scherr, C. L., King, A. J., Brown, N., and Jones, C. (2017). Communicating uncertain science to the public: How amount and source of uncertainty impact fatalism, backlash, and overload. *Risk Analysis* 37 (1): 40–51.

John, L. K., Loewenstein, G., and Prelec, D. (2012). Measuring the prevalence of questionable research practices with incentives for truth telling. *Psychological Science* 23 (5): 524–32.

John, S. (2015a). The example of the IPCC does not vindicate the Value Free Ideal: A reply to Gregor Betz. *European Journal for Philosophy of Science* 5 (1): 1–13.

John, S. (2015b). Inductive risk and the contexts of communication. *Synthese* 192 (1): 79–96.

John, S. (2018). Epistemic trust and the ethics of science communication: Against transparency, openness, sincerity and honesty. *Social Epistemology* 32 (2): 75–87.

Johnson, C. (2015). Testimony and the constitutive norm of assertion. *International Journal of Philosophical Studies* 23 (3): 356–75.

Johnson, D. R. (2017). Bridging the political divide: Highlighting explanatory power mitigates biased evaluation of climate arguments. *Journal of Environmental Psychology* 51: 248–55.

Johnson, N. (2013). The GM safety dance: What's rule and what's real. 10 July, *Grist.org*.

Jordan, T. H., Chen, Y. T., Gasparini, P., Madariaga, R., Main, I., Marzocchi, W., . . . and Zschau, J. (2011). Operational earthquake forecasting: State of knowledge and guidelines for utilization. *Annals of Geophysics* 54 (4): 315–91.

Jung, A. (2012). Medialization and credibility: Paradoxical effect or (re)-stabilization of boundaries? Epidemiology and stem cell research in the press. In (eds. Rödder, S., Franzen, M., and Weingart, P.) *The Sciences' Media Connection: Public Communication and Its Repercussions*. Springer: 107–30.

Kahan, D. M. (2013). Ideology, motivated reasoning, and cognitive reflection. *Judgment and Decision Making* 8 (4): 407–24.

Kahan, D. (2015a). What is the "science of science communication." *Journal of Science Communication* 14 (3): 1–10.

Kahan, D. (2015b). Climate-science communication and the measurement problem. *Political Psychology 36*: 1–43.

Kahan, D. (2016). The politically motivated reasoning paradigm, part 1: What politically motivated reasoning is and how to measure it. *Emerging Trends in the Social and Behavioral Sciences*: 1–16.

Kahan, D. (2017). The "gateway belief" illusion: Reanalyzing the results of a scientific-consensus messaging study. *Journal of Science Communication* 16 (05): A03.

Kahan, D., Braman, D., Cohen, G., Gastil, J., and Slovic, P. (2010). Who fears the HPV vaccine, who doesn't, and why? An experimental study of the mechanisms of cultural cognition. *Law and Human Behavior* 34 (6): 501–16.

Kahan, D., Braman, D., Gastil, J., Slovic, P., and Mertz, C. K. (2007). Culture and identity-protective cognition: Explaining the white-male effect in risk perception. *Journal of Empirical Legal Studies* 4 (3): 465–505.

Kahan, D. M., and Corbin, J. C. (2016). A note on the perverse effects of actively open-minded thinking on climate-change polarization. *Research and Politics* 3 (4): 2053168016676705.

Kahan, D., Jenkins-Smith, H., and Braman D. (2011). Cultural cognition of scientific consensus. *Journal of Risk Research* 14: 147–74. doi:10.1080/136 69877.2010.511246.

Kahan, D. M., Peters, E., Dawson, E. C., and Slovic, P. (2017). Motivated numeracy and enlightened self-government. *Behavioural Public Policy* 1 (1): 54–86.

Kahan, D. M., Peters, E., Wittlin, M., Slovic, P., Ouellette, L. L., Braman, D., and Mandel, G. (2012). The polarizing impact of science literacy and numeracy on perceived climate change risks. *Nature Climate Change* 2 (10): 732–5.

Kahneman, D. (2011). *Thinking Fast and Slow*. Farrar, Straus and Giroux.

Kahneman, D. (2012). A proposal to deal with questions about priming effects. *Nature*, September 26, http://www.nature.com/polopoly_fs/7.6716.1349271308!/suppinfoFile/Kahneman%20Letter.pdf.

Kallestrup, J. (2019). Groups, trust and testimony. In (ed. Dormandy, K.) *Trust in Epistemology*. Routledge: 136–58.

Kallestrup, J. (2020). The epistemology of testimonial trust. *Philosophy and Phenomenological Research* 101 (1): 150–74.

Kampourakis, K., and McCain, K. (2019). *Uncertainty: How It Makes Science Advance*. Oxford University Press.

Kant, I. (1784) [1991]. An answer to the question: What is enlightenment? In (ed. Reiss, H.) *Political Writings*, 2nd edition. Cambridge University Press: 54–60.

Kappel, K., and Holmen, S. (2019). Why science communication, and does it work? A taxonomy of science communication aims and a survey of the empirical evidence. *Frontiers in Communication* 4: 55.

Kappel, K., and Zahle, J. (2019). The epistemic role of science and experts in democracy. In (eds. Graham, P., Fricker, M., Henderson, D., Pedersen, N.) *The Routledge Handbook of Social Epistemology*. Routledge: 397–405.

Karmarkar, U. R., and Tormala, Z. L. (2010). Believe me, I have no idea what I'm talking about: The effects of source certainty on consumer involvement and persuasion. *Journal of Consumer Research 36* (6): 1033–49.

Keohane, R. O., Lane, M., and Oppenheimer, M. (2014). The ethics of scientific communication under uncertainty. *Politics, Philosophy and Economics* 13: 343–68.

Keren, A. (2007). Epistemic authority, testimony and the transmission of knowledge. *Episteme* 4 (3): 368–81.

Keren, A. (2013). Kitcher on well-ordered science: Should science be measured against the outcomes of ideal democratic deliberation? *Theoria: An International Journal for Theory, History and Foundations of Science* 28 (2): 233–44.

Keren, A. (2018). The public understanding of what? Laypersons' epistemic needs, the division of cognitive labor, and the demarcation of science. *Philosophy of Science* 85 (5): 781–92.

Khalifa, K. (2013). The role of explanation in understanding. *British Journal for the Philosophy of Science* 64 (1): 161–87.

Khalifa, K. (2017). *Understanding, Explanation, and Scientific Knowledge.* Cambridge University Press.

Khalifa, K. (2020). Understanding, truth, and epistemic goals. *Philosophy of Science* 87 (5): 944–56.

Khalifa, K., and Millson, J. (2020). Explanatory obligations. *Episteme* 17 (3): 384–401.

Kitcher, P. (1990). The division of cognitive labor. *Philosophy of Science* 87 (1): 5–22.

Kitcher, P. (1993). *The Advancement of Science.* Oxford University Press.

Kitcher, P. (2003). *Science, Truth, and Democracy.* Oxford University Press.

Kitcher, P. (2011). *Science in a Democratic Society.* Prometheus Books.

Klausen, S. H. (2017). No cause for epistemic alarm: Radically collaborative science, knowledge and authorship. *Social Epistemology* 6 (3): 38–61.

Klein, J. T. (2005). Interdisciplinary teamwork: The dynamics of collaboration and integration. In (eds. Derry, S. J., Schunn, C. D., and Gernsbacher, M. A.) *Interdisciplinary Collaboration: An Emerging Cognitive Science.* Lawrence Erlbaum: 23–50.

Klein, J. T. (2010). A taxonomy of interdisciplinarity. In (eds. Frodeman, R., Klein, J. T., and Mitcham, M.) *The Oxford Handbook of Interdisciplinarity.* Oxford University Press: 15–30.

König, M., and Harris, P. (2004). Trust in children's use of true and false statements. *Psychological Science* 15: 694–8.

König, M., and Harris, P. (2007). The basis of epistemic trust. *Episteme* 4: 264–84.

König, M., and Stephens, E. (2014). Characterizing children's responsiveness to cues of speaker trustworthiness: two proposals. In (eds. Robinson, E. and Einav, S.) *Trust and Skepticism: Children's Selective Learning from Testimony.* Psychology Press: 21–35.

Korevaar, J., and Moed, H. (1996). Validation of bibliometric indicators in the field of mathematics. *Scientometrics* 37 (1): 117–30.

Kornblith, H. (2002). *Knowledge and Its Place in Nature.* Oxford University Press.

Køster-Rasmussen, R., Westergaard, M. L., Brasholt, M., Gutierrez, R., Jørs, E., and Thomsen, J. F. (2016). Mercury pollution from small-scale gold mining can be stopped by implementing the gravity-borax method: A two-year follow-up study from two mining communities in the Philippines. *NEW SOLUTIONS: A Journal of Environmental and Occupational Health Policy* 25 (4): 567–87.

Kovaka, K. (2019). Climate change denial and beliefs about science. *Synthese* 198 (3): 2355–74.

Kraft, P. W., Lodge, M., and Taber, C. S. (2015). Why people "don't trust the evidence": Motivated reasoning and scientific beliefs. *Annals of the American Academy of Political and Social Science* 658 (1): 121–33.

Kripke, S. (1979) [2011]. A puzzle about belief. In *Philosophical Troubles*. Oxford University Press: 125–61.
Kripke, S. (1980). *Naming and Necessity*. Harvard University Press.
Kuhn, T. (1977). *The Essential Tension*. University of Chicago Press.
Kuhn, T. (1962) [2012]. *The Structure of Scientific Revolutions*. University of Chicago Press.
Kukla, R. (2014). Performative force, convention, and discursive injustice. *Hypatia* 29 (2): 440–57.
Kullenberg, C., and Kasperowski, D. (2016). What is citizen science? A scientometric meta-analysis. *PloS One* 11 (1): e0147152.
Kunda, Z. (1987). Motivated inference: Self-serving generation and evaluation of causal theories. *Journal of Personality and Social Psychology* 53 (4): 636–47.
Kunda, Z. (1990). The case for motivated reasoning. *Psychological Bulletin* 108 (3): 480.
Kuorikoski, J., and Marchionni, C. (2016). Evidential diversity and the triangulation of phenomena. *Philosophy of Science* 83 (2): 227–47.
Kurtulmus, F., and Irzik, G. (2017). Justice in the distribution of knowledge. *Episteme* 14 (2): 129–46.
Kusch, M. (2002). *Knowledge by Agreement: The Programme of Communitarian Epistemology*. Oxford University Press.
Lackey, J. (1999). Testimonial knowledge and transmission. *Philosophical Quarterly* 49 (197): 471–90.
Lackey, J. (2005). Testimony and the infant/child objection. *Philosophical Studies 126*: 163–90.
Lackey, J. (2006a). Learning from words. *Philosophy and Phenomenological Research* 73 (1): 77–101.
Lackey, J. (2006b). The nature of testimony. *Pacific Philosophical Quarterly* 87: 177–97.
Lackey, J. (2008). *Learning from Words: Testimony as a Source of Knowledge*. Oxford University Press.
Lackey, J. (2011). Assertion and isolated second-hand knowledge. In (eds. Brown, J. and Cappelen, H.) *Assertion: New Philosophical Essays*. Oxford University Press: 251–76.
Lackey, J. (2014). *Essays in Collective Epistemology*. Oxford University Press.
Lackey, J. (2015). A deflationary account of group testimony. In (ed. Lackey, J.) *Essays in Collective Epistemology*. Oxford University Press: 64–94.
Lackey, J. (2016). What is justified group belief? *Philosophical Review* 125 (3): 341–96.
Lackey, J. (2021). *The Epistemology of Groups*. Oxford University Press.
Ladyman, J., Ross, D., Collier, J., Spurrett, D., and Collier, J. G. (2007). *Every Thing Must Go: Metaphysics Naturalized*. Oxford University Press.
Lakatos, I. (1974). The role of crucial experiments in science. *Studies in the History and Philosophy of Science* 4: 309–25.
Lakatos, I. (1978). Science and pseudoscience. In *The Methodology of Scientific Research Programmes: Philosophical Papers*, Vol. 1. Cambridge University Press: 1–7.
Lamb, E. (2012). 5 sigma what's that? *Scientific American*. https://blogs.scientificamerican.com/observations/five-sigmawhats-that/.
Landemore, H. (2012). *Democratic Reason: Politics, Collective Intelligence, and the Rule of the Many*. Princeton University Press.
Landemore, H. (2013). Deliberation, cognitive diversity, and democratic inclusiveness. *Synthese* 190 (7): 1209–31.
Latour, B., and Woolgar, S. (1979) [2013]. *Laboratory Life: The Construction of Scientific Facts*. Princeton University Press.
Laudan, L. (1981). A confutation of convergent realism. *Philosophy of Science* 48: 19–49.

Laudan, L. (1983). The demise of the demarcation problem. In (eds. Cohan, R. S. and Laudan, L.) *Physics, Philosophy, and Psychoanalysis*. Reidel: 111–27.
Lazer, D., Baum, M., Benkler, Y., et al. (2018). The science of fake news. *Science* 359 (6380): 1094–6.
Leefmann, J., and Lesle, S. (2020). Knowledge from scientific expert testimony without epistemic trust. *Synthese* 197 (8): 3611–41.
Lerman, A. E., Sadin, M. L., and Trachtman, S. (2017). Policy uptake as political behavior: Evidence from the Affordable Care Act. *American Political Science Review* 111: 755–70.
Leung, W., et al. (2015). Drosophila Muller F elements maintain a distinct set of genomic properties over 40 million years of evolution. *G3: Genes| Genomes| Genetics* 5 (5): 719–40.
Levy, N. (2018). Taking responsibility for health in an epistemically polluted environment. *Theoretical Medicine and Bioethics* 39 (2): 123–41.
Levy, N. (2019). Due deference to denialism: Explaining ordinary people's rejection of established scientific findings. *Synthese* 196 (1): 313–27.
Levy, N., and Alfano, M. (2019). Knowledge from vice: Deeply social epistemology. *Mind* 129 (515): 887–915.
Levy, N., and Ross, R. (2021). The cognitive science of fake news. In (eds. Hannon, M. and de Ridder, J.) *Routledge Handbook of Political Epistemology*. Routledge: 181–91.
Lewandowsky, S., Cook, J., and Lloyd, E. (2018). The "Alice in Wonderland" mechanics of the rejection of (climate) science: Simulating coherence by conspiracism. *Synthese* 195 (1): 175–96.
Lewin, S. (2016). Life on Earth can thank its lucky stars for Jupiter and Saturn. https://www.space.com/31577-earth-life-jupiter-saturn-giant-impacts.html.
Lewis, P. J. (2001). Why the pessimistic induction is a fallacy. *Synthese* 129 (3): 371–80.
Lipton, P. (1998). The epistemology of testimony. *Studies in History and Philosophy of Science Part A* 29 (1): 1–31.
Lipton, P. (2003). Inference to the Best Explanation. Routledge.
List, C. (2006). Special issue on Epistemic Diversity, Episteme 3 (3).
List, C., and Pettit, P. (2011). *Group Agency: The Possibility, Design, and Status of Corporate Agents*. Oxford University Press.
Littlejohn, C., and Turri, J. (eds.) (2014). *Epistemic Norms: New Essays on Action, Belief and Assertion*. Oxford University Press.
Locke, J. (1690) [1975]. *An Essay concerning Human Understanding*. (Ed. Nidditch, P.). Clarendon Press.
Lombrozo, T. (2017). What is pseudo-science? *NPR Cosmos and Culture* 13 (7), https://www.npr.org/sections/13.7/2017/05/08/527354190/what-is-pseudoscience.
Lombrozo, T., Thanukos, A., and Weisberg, M. (2008). The importance of understanding the nature of science for accepting evolution. *Evolution: Education and Outreach* 1: 290–8.
Longino, H. (1990). *Science as Social Knowledge*. Princeton University Press.
Longino, H. E. (2002). *The Fate of Knowledge*. Princeton University Press.
Lord, E., and Sylvan, K. (2019). Prime time (for the basing relation). In (eds. Carter, J. A., and Bondy, P.) *Well-Founded Belief* Routledge: 141–73.
Lynch, M. P. (2018). Epistemic arrogance and the value of political dissent. In (ed. Johnson, C.), *Voicing Dissent*. Routledge: 129–39.
Machamer, P. (2005). Galileo Galilei. In (ed. Zalta, E. N.) *The Stanford Encyclopedia of* Philosophy, https://plato.stanford.edu/archives/sum2017/entries/galileo/.
Maibach, E., Roser-Renouf, C., and Leiserowitz, A. (2008). Communication and marketing as climate change intervention assets: A public health perspective. *American Journal of Preventive Medicine* 35: 488–500.

Malmgren, A. S. (2006). Is there a priori knowledge by testimony? *The Philosophical Review* 115 (2): 199–241.
Mann, T. (1947/1999). *Doctor Faustus*. Vintage International. (Transl. J. E. Woods).
Manson, N. (2007). *Rethinking Informed Consent in Bioethics*. Cambridge University Press.
Martini, C. (2014). Experts in science: A view from the trenches. *Synthese* 191 (1): 3–15.
Martinson, B. C., Anderson, M. S., and De Vries, R. (2005). Scientists behaving badly. *Nature* 435 (7043): 737.
Mayo-Wilson, C. (2014). Reliability of testimonial norms in scientific communities. *Synthese* 191 (1): 55–78.
MacLeod, M. (2018). What makes interdisciplinarity difficult? Some consequences of domain specificity in interdisciplinary practice. *Synthese* 195 (2): 697–720.
MacLeod, M., and Nersessian, N. J. (2014). Strategies for coordinating experimentation and modeling in integrative systems biology. *Journal of Experimental Zoology (Molecular and Developmental Evolution)* 9999: 1–10.
McAllister, L., Daly, M., Chandler, P., McNatt, M., Benham, A., and Boykoff, M. (2021). Balance as bias, resolute on the retreat? Updates and analyses of newspaper coverage in the United States, United Kingdom, New Zealand, Australia and Canada over the past 15 years. *Environmental Research Letters* 16 (9): 094008.
McCain, K. (2015). Explanation and the nature of scientific knowledge. *Science and Education* 24 (7–8): 827–54.
McCain, K., and Poston, T. (2014). Why explanatoriness is evidentially relevant. *Thought: A Journal of Philosophy* 3 (2): 145–53.
McCright, A. M., Dunlap, R. E., and Xiao, C. (2013). Perceived scientific agreement and support for government action on climate change in the USA. *Climatic Change* 119 (2): 511–18.
McHugh, C., Way, J., and Whiting, D. (eds.) (2018). *Normativity: Epistemic and Practical*. Oxford University Press.
McKenna, R. (2019). Irrelevant cultural influences on belief. *Journal of Applied Philosophy* 36 (5): 755–68.
Merkley, E. (2020). Are experts (news)worthy? Balance, conflict, and mass media coverage of expert consensus. *Political Communication* 37 (4): 530–49.
Merton, R. K. (1973). *The Sociology of Science: Theoretical and Empirical Investigations*. University of Chicago Press.
Michaelian, K. (2010). In defence of gullibility: The epistemology of testimony and the psychology of deception detection. *Synthese* 176 (3): 399–427.
Michaelian, K. (2013). The evolution of testimony: Receiver vigilance, speaker honesty, and the reliability of communication. *Episteme* 10 (1): 37–59.
Milgram, E. (2015). *The Great Endarkenment*. Oxford University Press.
Miller, B. (2009). What does it mean that PRIMES is in P? Popularization and distortion revisited. *Social Studies of Science* 39 (2): 257–88.
Miller, B. (2013). When is consensus knowledge based? Distinguishing shared knowledge from mere agreement. *Synthese* 190 (7): 1293–316.
Miller, B. (2015). Why (some) knowledge is the property of a community and possibly none of its members. *The Philosophical Quarterly* 65 (260): 417–41.
Miller, B. (2016). Scientific consensus and expert testimony in courts: Lessons from the bendectin litigation. *Foundations of Science* 21 (1): 15–33.
Miller, B., and Freiman, O. (2020). Trust and distributed epistemic labor. In (ed. Simon, J.) *The Routledge Handbook on Trust and Philosophy*. Routledge: 341–53.

Miller, S. (2001). Public understanding of science at the crossroads. *Public Understanding of Science* 10: 115–20.
Mizrahi, M. (2013). The pessimistic induction: A bad argument gone too far. *Synthese* 190 (15): 3209–26.
Mizrahi, M. (2017). What's so bad about scientism? *Social Epistemology* 31 (4): 351–67.
Mooney, C., and Nisbet, M. C. (2005). Undoing Darwin. *Columbia Journalism Review* 44 (3): 30–9.
Moore, R. (2017). Gricean communication and cognitive development. *Philosophical Quarterly* 67 (267): 303–26.
Moore, R. (2018). Gricean communication, language development, and animal minds. *Philosophy Compass* 13 (12): e12550.
Moxham, N. (2019). Natural Knowledge, Inc.: The Royal Society as a metropolitan corporation. *British Journal for the History of Science* 52 (2): 249–71.
Muldoon, R. (2013). Diversity and the division of cognitive labor. *Philosophy Compass* 8 (2): 117–25.
Müller, F. (2015). The pessimistic meta-induction: Obsolete through scientific progress? *International Studies in the Philosophy of Science* 29 (4): 393–412.
Murad, M. H., Asi, N., Alsawas, M., and Alahdab, F. (2016). New evidence pyramid. *BMJ Evidence-Based Medicine* 21 (4): 125–7.
Mynatt, C. R., Doherty, M. E., and Tweney, R. D. (1977). Confirmation bias in a simulated research environment: An experimental study of scientific inference. *The Quarterly Journal of Experimental Psychology* 29 (1): 85–95.
Nagel, J., San Juan, V., and Mar, R. A. (2013). Lay denial of knowledge for justified true beliefs. *Cognition* 129 (3): 652–61.
Nelkin, D. (1987). *Selling Science: How the Press Covers Science and Technology*. W. H. Freeman and Co.
Neta, R. (2019). The basing relation. *Philosophical Review* 128 (2): 179–217.
Neurath O. (1921) [1973]. Anti-Spengler. In (eds. Neurath M. and Cohen, R. S.) *Empiricism and Sociology*. Vienna Circle Collection, Vol 1. Springer: 158–213.
Nguyen, C. T. (2020). Echo chambers and epistemic bubbles. *Episteme* 17 (2): 141–61.
Nickel, P. (2013). Norms of assertion, testimony and privacy. *Episteme* 10 (02): 207–17.
Nickerson, R. S. (1998). Confirmation bias: A ubiquitous phenomenon in many guises. *Review of General Psychology* 2: 175–220
Nisbet, E. C., Cooper, K. E., and Garrett, R. K. (2015). The partisan brain: How dissonant science messages lead conservatives and liberals to (dis) trust science. *The Annals of the American Academy of Political and Social Science* 658 (1): 36–66.
Nisbet, M. C., and Fahy, D. (2015). The need for knowledge-based journalism in politicized science debates. *The Annals of the American Academy of Political and Social Science* 658: 223–34.
Nosek, B., Alter, G., Banks, G., Borsboom, D. Bowman, S. Breckler, S., Buck, S., et al. (2015). Promoting an open research culture. *Science* 348 (6242): 1422–5.
Nosek, B. A., Ebersole, C. R., DeHaven, A. C., and Mellor, D. T. (2018). The preregistration revolution. *Proceedings of the National Academy of Sciences*: 201708274.
NPR (2019). When it comes to vaccines and autism, why is it hard to refute misinformation. *NPR Podcast*. Retrieved Sept. 19, 2020 from https://www.npr.org/2019/07/22/744023623/when-it-comes-to-vaccines-and-autism-why-is-it-hard-to-refute-misinformation?t=1583509766257andt=1600520332782.
Nyhan, B., and Reifler, J. (2010). When corrections fail: The persistence of political misperceptions. *Political Behavior* 32 (2): 303–30.

O'Connor, C., and Weatherall, J. O. (2019). *The Misinformation Age: How False Beliefs Spread.* Yale University Press.
Olsson, E., and Vallinder, A. (2013). Norms of assertion and communication in social networks. *Synthese* 190 (13): 2557–71.
Open Science Collaboration (2015). Estimating the reproducibility of psychological science. *Science* 349 (6251): aac4716.
Oreskes, N. (1999). *The Rejection of Continental Drift: Theory and Method in American Earth Science.* Oxford University Press.
Oreskes, N. (2019). *Why Trust Science?* Princeton University Press.
Oreskes, N., and Conway, E. M. (2010). *Merchants of Doubt: How a Handful of Scientists Obscured the Truth on Issues from Tobacco Smoke to Global Warming.* Bloomsbury Publishing.
Origgi, G. (2015). What is an expert that a person may trust her? Towards a political epistemology of expertise. *Humana Mente* 8 (28): 159–68.
O'Rourke, M., Crowley, S., and Gonnerman, C. (2016). On the nature of cross-disciplinary integration: A philosophical framework. *Studies in History and Philosophy of Biological and Biomedical Sciences* 56: 62–70.
Oskam, I. F. (2009). T-shaped engineers for interdisciplinary innovation: An attractive perspective for young people as well as a must for innovative organisations. In *37th Annual Conference—Attracting Students in Engineering* (14), Rotterdam, The Netherlands: 1–10.
Osman, M., Heath, A. J., and Löfstedt, R. (2018). The problems of increasing transparency on uncertainty. *Public Understanding of Science* 27: 131–8.
Owens, D. (2000). *Reason without Freedom: The Problem of Epistemic Normativity.* Routledge.
Papachrisanthou, M. M., and Davis, R. L. (2019). The resurgence of measles, mumps, and pertussis. *The Journal for Nurse Practitioners* 15 (6): 391–5.
Pariser, E. (2011). *The Filter Bubble: What the Internet Is Hiding from You.* Penguin.
Park, S. (2011). A confutation of the pessimistic induction. *Journal for General Philosophy of Science* 42 (1): 75–84.
Parker, W. (2014). Values and uncertainties in climate prediction, revisited. *Studies in History and Philosophy of Science Part A* 46: 24–30.
Pedersen, N. J. L. L., and Kallestrup, J. (2013). The epistemology of absence-based inference. *Synthese* 190 (13): 2573–93.
Peels, R. (2016). The empirical case against introspection. *Philosophical Studies* 173 (9): 2461–85.
Peels, R. (2018). A conceptual map of scientism. In (eds. de Ridder, P. and van Woudenberg, R.) *Scientism: Prospects and Problems.* Oxford University Press: 28–56.
Peet, A. (2019). Knowledge-yielding communication. *Philosophical Studies* 176 (12): 3303–27.
Peet, A., and Pitcovski, E. (2017). Lost in transmission: Testimonial justification and practical reason. *Analysis* 77 (2): 336–44.
Pellechia, M. G. (1997). Trends in science coverage: A content analysis of three US newspapers. *Public Understanding of Science* 6 (1): 49–68.
Pennycook, G., and Rand, D. G. (2019). Lazy, not biased: Susceptibility to partisan fake news is better explained by lack of reasoning than by motivated reasoning. *Cognition* 188: 39–50.
Persson, J., Sahlin, N. E., and Wallin, A. (2015). Climate change, values, and the cultural cognition thesis. *Environmental Science and Policy 52*: 1–5.

Peter, F. (2008). *Democratic Legitimacy*. Routledge.
Peters, U. (2019). Implicit bias, ideological bias, and epistemic risks in philosophy. *Mind and Language* 34 (3): 393–419.
Peters, U. (forthcoming a). Illegitimate values, confirmation bias, and Mandevillian cognition in science. *The British Journal for the Philosophy of Science*.
Peters, U. (forthcoming b). Values in science: Assessing the case for mixed claims. *Inquiry*.
Peters, U., and Nottelmann, N. (forthcoming). Weighing the costs: The epistemic dilemma of no-platforming. *Synthese*.
Peterson, E., and Iyengar, S. (2019). Partisan gaps in political information and information-seeking behavior: Motivated reasoning or cheerleading?
Phillips, J., Buckwalter, W., Cushman, F., Friedman, O., Martin, A., Turri, J., Santos, L., and Knobe, J. (forthcoming). Knowledge before belief. *Behavioral and Brain Sciences*: 1–37.
Phillips, K. (2017). What is the real value of diversity in organizations? Questioning our assumptions. In (ed. Page, S.) *The Diversity Bonus: How Great Teams Pay off in the Knowledge Economy*. Princeton University Press: 223–45.
Pinillos, N. Á. (2018). Knowledge, ignorance and climate change (New York Times, Nov. 26).
Politi, V. (2017). Specialisation, interdisciplinarity, and incommensurability. *International Studies in the Philosophy of Science* 31 (3): 301–17.
Polyani, M. (1962) [2000]. The Republic of Science: Its political and economic theory. *Minerva* 38: 1–21.
Popper, K. R. (1934) [2002]. *Logik der Forschung*. Akademie Verlag. English translation as *The Logic of Scientific Discovery*, London: Routledge.
Popper, K. R. (1963). *Conjectures and Refutations: The Growth of Scientific Knowledge*. Harper.
Porritt, J., et al. (2018). Climate change is real: We must not offer credibility to those who deny it. *The Guardian*, Aug. 26, https://www.theguardian.com/environment/2018/aug/26/climate-change-is-real-we-must-not-offer-credibility-to-those-who-deny-it.
Porter S., England, L., Juodis, M., ten Brinke, L., and Wilson, K. (2008). Is the face a window to the soul? Investigation of the accuracy of intuitive judgments of the trustworthiness of human faces. *Canadian Journal of Behavioral Science* 40: 171–7.
Pritchard, D. (2004). The epistemology of testimony. *Philosophical Issues* 14: 326–48.
Psillos, S. (1999). *Scientific Realism: How Science Tracks Truth*. Routledge.
Putnam, H. (1978). *Meaning and the Moral Sciences*. Routledge.
Quast, C., and Seidel, M. (2018). The philosophy of expertise: What is expertise? *Topoi* 37 (1): 1–2.
Ranney, M. A., and Clark, D. (2016). Climate change conceptual change: Scientific information can transform attitudes. *Topics in Cognitive Science* 8 (1): 49–75.
Read, R. (2018). I won't go on the BBC if it supplies climate change deniers as "balance." *The Guardian*, Aug. 2, https://www.theguardian.com/commentisfree/2018/aug/02/bbc-climate-change-deniers-balance.
Reichenbach, H. (1938). *Experience and Prediction: An Analysis of the Foundations and the Structure of Knowledge*. The University of Chicago Press.
Rescorla, M. (2009). Assertion and its constitutive norms. *Philosophy and Phenomenological Research* 79 (1): 98–130.
Resnik, D. B., and Shamoo, A. E. (2011). The Singapore statement on research integrity. *Accountability in Research* 18 (2): 71–5.
Reyes-Galindo, L., and Duarte, T. (2015). Bringing tacit knowledge back to contributory and interactional expertise: A reply to Goddiksen. *Studies in History and Philosophy of Science Part A* 49: 99–102.

Rini, R. (2017). Fake news and partisan epistemology. *Kennedy Institute of Ethics Journal* 27 (S2): E43–64.

Robbins J. M., and Krueger, J. I. (2005). Social projection to ingroups and outgroups: A review and meta-analysis. *Personal. Soc. Psychol. Rev.* 9 (3): 2–47.

Rolin, K. (2008). Science as collective knowledge. *Cognitive Systems Research* 9: 115–24.

Rolin, K. (2010). Group justification in science. *Episteme* 7 (3): 215–31.

Rolin, K. (2020). Trust in science. In (ed. Simon, J.) *The Routledge Handbook of Trust and Philosophy*. Routledge: 354–66.

Rossini, F. A., and Porter, A. L. (1979). Frameworks for integrating interdisciplinary research. *Research Policy* 8 (1): 70–9.

Roush, S. (2009). Randomized controlled trials and the flow of information: Comment on Cartwright. *Philosophical Studies* 143 (1): 137–45.

Rowbottom, D. (2011). Kuhn vs. Popper on criticism and dogmatism in science: A resolution at the group level. *Studies in History and Philosophy of Science Part A* 42 (1): 117–24.

Royal Society (2020). Retrieved Sept. 8, 2020 from https://royalsociety.org/about-us/history/.

Rule, N. O., Krendl, A. C., Ivcevic, Z., and Ambady, N. (2013). Accuracy and consensus in judgments of trustworthiness from faces: Behavioral and neural correlates. *Journal of Personality and Social Psychology* 104: 409–26.

Salzberg, S. (2015). Ted Cruz uses the Galileo gambit to deny global warming. Forbes. Retrieved Sept. 15, 2020 from http://www.forbes.com/sites/stevensalzberg/2015/03/30/ted-cruz-uses-the-galileo-gambit-to-deny-global-warming/.

Sankey, H. (1998). Taxonomic incommensurability. *International Studies in the Philosophy of Science* 12 (1): 7–16.

Sankey, H. (2006). Incommensurability. In (eds. Sarkar, S. and Pfeifer, J.) *The Philosophy of Science: An Encyclopedia*. Routledge: 370–3.

Satta, M. (forthcoming). Epistemic trespassing and expert witness testimony. *Journal of Ethics and Social Philosophy*.

Saul, J. (2002). Speaker meaning, what is said, and what is implicated. *Noûs* 36 (2): 228–48.

Schaffer, J., and Knobe, J. (2012). Contrastive knowledge surveyed. *Noûs* 46 (4): 675–708.

Schickore, J. (2018). Scientific discovery. In (ed. Zalta, E. N.) *The Stanford Encyclopedia of Philosophy*, <https://plato.stanford.edu/archives/sum2018/entries/scientific-discovery/>.

Schickore, J., and Steinle, F. (2006). *Revisiting Discovery and Justification: Historical and Philosophical Perspectives on the Context Distinction*. Springer.

Schmitt, F. (2006). Testimonial justification and transindividual reasons. In (eds. Lackey, J. and Sosa, E.) *The Epistemology of Testimony*. Oxford University Press: 193–224.

Schmoch, U., and Schubert, T. (2008). Are international co-publications an indicator for quality of scientific research? *Scientometrics* 74 (3): 361–77.

Selinger, E., and Crease, R. (2006). *The Philosophy of Expertise*. Columbia University Press.

Shapin, S. (1994). *A Social History of Truth Civility and Science in Seventeenth-Century England*. University of Chicago Press.

Shi, J., Visschers, V. H., and Siegrist, M. (2015). Public perception of climate change: The importance of knowledge and cultural worldviews. *Risk Analysis* 35 (12): 2183–201.

Shi, J., Visschers, V. H., Siegrist, M., and Arvai, J. (2016). Knowledge as a driver of public perceptions about climate change reassessed. *Nature Climate Change* 6 (8): 759.

Shieber, J. (2015). *Testimony: A Philosophical Introduction*. Routledge.

Shogenji, T. (2006). A defense of reductionism about testimonial justification of beliefs. *Noûs* 40 (2): 331–46.

Sherman, D. K., and Cohen, G. L. (2006). The psychology of self-defense: Self-affirmation theory. *Advances in Experimental Social Psychology* 38: 183–242.
Simion, M. (2017). Epistemic norms and "he said/she said" reporting. *Episteme* 14 (4): 413–22.
Simion, M. (forthcoming). Testimonial contractarianism. *Noûs*.
Simion, M., and Kelp, C. (2020a). How to be an anti-reductionist. *Synthese* 197 (7): 2849–66.
Simion, M., and Kelp, C. (2020b). The constitutive norm view of assertion. In (ed. Goldberg, S.) *The Oxford Handbook of Assertion*. Oxford University Press: 59–74.
Simpson, R. M., and Srinivasan, A. (2018). No platforming. *Academic Freedom*: 186–210.
Sinatra, G. M., Kienhues, D., and Hofer, B. K. (2014). Addressing challenges to public understanding of science: Epistemic cognition, motivated reasoning, and conceptual change. *Educational Psychologist* 49 (2): 123–38.
Singer, D. J. (2019). Diversity, not randomness, trumps ability. *Philosophy of Science* 86 (1): 178–91.
Slater, M., Huxster, J., and Bresticker, J. (2019). Understanding and trusting science. *Journal for General Philosophy of Science* 50 (2): 247–61.
Smart, P. (2018). Mandevillian intelligence. *Synthese* 195 (9): 4169–200.
Smith, A. (1977) [1776]. *An Inquiry into the Nature and Causes of the Wealth of Nations*. University of Chicago Press.
Smith, T. W., and Son, J. (2013). Trends in public attitudes and confidence in institutions. *General Social Survey Final Report*.
Solomon, M. (1992). Scientific rationality and human reasoning. *Philosophy of Science* 59 (3): 439–55.
Solomon, M. (2006a). Norms of epistemic diversity. *Episteme* 3: 23–36.
Solomon, M. (2006b). Groupthink versus the wisdom of crowds: The social epistemology of deliberation and dissent. *The Southern Journal of Philosophy* 44 (S1): 28–42.
Sonnenwald, D. H. (2007). Scientific collaboration. *Annual Review of Information Science and Technology* 41 (1): 643–81.
Sosa, E. (1991). *Knowledge in Perspective: Selected Essays in Epistemology*. Cambridge University Press.
Spaulding, S. (2016). Mind misreading. *Philosophical Issues* 26 (1): 422–40.
Spaulding, S. (2018). *How We Understand Others: Philosophy and Social Cognition*. Routledge.
Sperber, D. (2013). Speakers are honest because hearers are vigilant: Reply to Kourken Michaelian. *Episteme 10* (1): 61–71.
Sperber, D., and Wilson, D. (1986) [1995]. *Relevance: Communication and Cognition*. Blackwell. (2nd revised edition Wiley-Blackwell, 1995.)
Sperber, D., Clément, F., Heintz, C., Mascaro, O., Mercier, H., Origgi, G., and Wilson, D. (2010). Epistemic vigilance. *Mind and Language* 25 (4): 359–93.
Stanovich, K. (2011). *Rationality and the Reflective Mind*. Oxford University Press.
Steel, D., Fazelpour, S., Crewe, B., Gillette, K., Crewe, B., and Burgess, M. (2018). Multiple concepts of diversity and their ethical-epistemic implications for science. *European Journal for Philosophy of Science* 8 (3): 761–80.
Steele, K. (2012). The scientist qua policy advisor makes value judgments. *Philosophy of Science* 79 (5): 893–904.
Stenega, J. (2016). Three criteria for consensus conferences. *Foundations of Science* 21 (1): 35–49.
Stephan, E., Liberman, N., and Trope, Y. (2011). The effects of time perspective and level of construal on social distance. *Journal of Experimental Social Psychology* 47: 397–402.

Storage, D., Horne, Z., Cimpian, A., and Leslie, S.-J. (2016). The frequency of "brilliant" and "genius" in teaching evaluations predicts the representation of women and African Americans across fields. *PloS One* 11 (3): e0150194.

Strandberg, K., Himmelroos, S., and Grönlund, K. (2019). Do discussions in like-minded groups necessarily lead to more extreme opinions? Deliberative democracy and group polarization. *International Political Science Review* 40 (1): 41–57.

Strevens, M. (2003). The role of the priority rule in science. *Journal of Philosophy* 100 (2): 55–79.

Strevens, M. (2011). Economic approaches to understanding scientific norms. *Episteme* 8 (2): 184–200.

Strevens, M. (2013). No understanding without explanation. *Studies in History and Philosophy of Science Part A* 44 (3): 510–15.

Strevens, M. (2017). Scientific sharing: Communism and the social contract. *Scientific Collaboration and Collective Knowledge*: 1–50.

Sturgis P., and Allum, N. (2004). Science in society: Re-evaluating the deficit model of public attitudes. *Public Understanding of Science* 13 (1): 55–74.

Sullivan, E., Sondag, M., Rutter, I., Meulemans, W., Cunningham, S., Speckmann, B., and Alfano, M. (2020). Vulnerability in social epistemic networks. *International Journal of Philosophical Studies* 28 (5): 731–53.

Sunstein, C. (2002). The law of group polarization. *Journal of Political Philosophy* 10 (2): 175–95.

Sunstein, C. (2009). *Going to Extremes: How Like Minds Unite and Divide*. Oxford University Press.

Taber, C., and Lodge, M. (2006). Motivated skepticism in the evaluation of political beliefs. *American Journal of Political Science* 50 (3): 755–69.

Thagard, P. (1997). Collaborative knowledge. *Noûs* 31 (2): 242–61.

Thagard, P. (1999). *How Scientists Explain Disease*. Princeton University Press.

Thagard, P. (2006). How to collaborate: Procedural knowledge in the cooperative development of science. *The Southern Journal of Philosophy* 44 (S1): 177–96.

Thoma, J. (2015). The epistemic division of labor revisited. *Philosophy of Science* 82 (3): 454–72.

Thomm, E., and Bromme, R. (2012). "It should at least seem scientific!" Textual features of "scientificness" and their impact on lay assessments of online information. *Science Education* 96 (2): 187–211.

Thomson, J. J. (2008). *Normativity*. Open Court.

Todorov, A., Olivola, C. Y., Dotsch, R., and Mende-Siedlecki, P. (2015). Social attributions from faces: Determinants, consequences, accuracy, and functional significance. *Annual Review of Psychology* 66: 519–45.

Tollefsen, D. (2007). Group testimony. *Social Epistemology* 21 (3): 299–311.

Tollefsen, D. (2015). *Groups as Agents*. John Wiley and Sons.

Tong, E., and Glantz, S. (2007). Tobacco industry efforts undermining evidence linking secondhand smoke with cardiovascular disease. *Circulation* 116: 1845–54.

Tuomela, R. (2000). Belief versus acceptance. *Philosophical Explorations* 3 (2): 122–37.

Tuomela, R. (2004). Group knowledge analyzed. *Episteme* 1 (2): 109–27.

Tuomela, R. (2013). *Cooperation: A Philosophical Study*, Vol. 82. Springer Science and Business Media.

Turner, J. (1982). Towards a cognitive redefinition of the social group. In (ed. Tajfel, H.) *Social Identity and Intergroup Relations*. Praeger: 59–84.

Turner, J., Wetherell, M., and Hogg, M. (1989). A referent informational influence explanation of group polarization. *British Journal of Social Psychology* 28: 135–48.
Turner, S. (2001). What is the problem with experts? *Social Studies of Science* 31 (1): 123–49.
Turner, S. (2014). *The Politics of Expertise.* Routledge.
Turri, J. (2015a). Evidence of factive norms of belief and decision. *Synthese* 192 (12): 4009–30.
Turri, J. (2015b). Skeptical appeal: The source-content bias. *Cognitive Science* 38 (5): 307–24.
Turri, J. (2016). The radicalism of truth-insensitive epistemology: Truth's profound effect on the evaluation of belief. *Philosophy and Phenomenological Research* 93 (2): 348–67.
Turri, J. (2017). Epistemic contextualism: An idle hypothesis. *Australasian Journal of Philosophy* 95 (1): 141–56.
Turri, J., and Buckwalter, W. (2017). Descartes's schism, Locke's reunion: Completing the pragmatic turn in epistemology. *American Philosophical Quarterly* 54 (1): 25–46.
UCLA Statistical Consulting Group (2021). Choosing the correct statistical test in SAS, STATE, SPSS and R. Retrieved June 28, 2021, from https://stats.idre.ucla.edu/other/mult-pkg/whatstat/.
Uleman, J. S., Adil Saribay, S., and Gonzalez, C. M. (2008). Spontaneous inferences, implicit impressions, and implicit theories. *Annual Review of Psychology* 59: 329–60.
Valenti, J. (2000). Improving the scientist/journalist conversation. *Science and Engineering Ethics* 6 (4): 543–8.
van Bavel, J. J., and Pereira, A. (2018). The partisan brain: An identity-based model of political belief. *Trends in Cognitive Sciences* 20: 1–12.
van der Bles, A. M., van der Linden, S., Freeman, A. L., Mitchell, J., Galvao, A. B., Zaval, L., and Spiegelhalter, D. J. (2019). Communicating uncertainty about facts, numbers and science. *Royal Society Open Science* 6 (5): 181870.
van der Bles, A. M., van der Linden, S., Freeman, A. L., and Spiegelhalter, D. J. (2020). The effects of communicating uncertainty on public trust in facts and numbers. *Proceedings of the National Academy of Sciences* 117 (14): 7672–83.
van der Linden, S. L., Leiserowitz, A. A., Feinberg, G. D., and Maibach, E. W. (2015). The scientific consensus on climate change as a gateway belief: Experimental evidence. *PloS One* 10 (2): e0118489.
van der Linden, S. L., Leiserowitz, A., and Maibach, E. W. (2016). Communicating the scientific consensus on human-caused climate change is an effective and depolarizing public engagement strategy: Experimental evidence from a large national replication study. *SSRN Electronic Journal.* DOI: 10.2139/ssrn.2733956.
van der Linden, S. L., Leiserowitz, A., and Maibach, E. (2017). Gateway illusion or cultural cognition confusion? *Journal of Science Communication* 16 (05): A04.
Van Fraassen, B. C. (1980). *The scientific Image.* Oxford University Press.
Van Prooijen, J. W., and Jostmann, N. B. (2013). Belief in conspiracy theories: The influence of uncertainty and perceived morality. *European Journal of Social Psychology* 43 (1): 109–15.
van't Veer, A. E., and Giner-Sorolla, R. (2016). Pre-registration in social psychology: A discussion and suggested template. *Journal of Experimental Social Psychology* 67: 2–12.
Wagenknecht, S. (2014). Opaque and translucent epistemic-dependence in collaborative scientific practice. *Episteme* 11 (4): 475–92.
Wagenknecht, S. (2015). Facing the incompleteness of epistemic trust: Managing dependence in scientific practice. *Social Epistemology* 29 (2): 160–84.

Wagenknecht, S. (2016). *A Social Epistemology of Research Groups: Collaboration in Scientific Practice*. Palgrave Macmillan.

Wagenmakers, E. J., and Dutilh, G. (2016). Seven selfish reasons for preregistration. *APS Observer* 29 (9).

Wason, P. (1966). Reasoning. In (ed. Foss, B.) *New Horizons in Psychology*. Penguin.

Wason, P. (1968). Reasoning about a rule. *Quarterly Journal of Experimental Psychology* 20: 273–81.

Wason, P., and Johnson-Laird, P. (1972). *Psychology of Reasoning: Structure and Content*. Harvard University Press.

Waterman, J., Gonnerman, C., Yan, K., and Alexander, J. (2018). Knowledge, subjective certainty, and philosophical skepticism: A cross-cultural study. In (eds. Mizumoto, M., Stich, S., and McCready, E.) *Epistemology for the Rest of the World: Linguistic and Cultural Diversity and Epistemology*. Oxford University Press: 187–214.

Weatherall, J., O'Connor, C., and Bruner, J. (2020). How to beat science and influence people: Policymakers and propaganda in epistemic networks. *The British Journal for the Philosophy of Science* 71 (4): 1157–86.

Weber, E. U., and Stern, P. C. (2011). Public understanding of climate change in the United States. *American Psychologist* 66 (4): 315.

Weisberg, D. S., Hopkins, E. J., and Taylor, J. C. V. (2018). People's explanatory preferences for scientific phenomena. *Cognitive Research: Principles and Implications* 3 (44): 1–14.

Weisberg, D. S., Landrum, A. R., Metz, S. E., and Weisberg, M. (2018). No missing link: Knowledge predicts acceptance of evolution in the United States. *BioScience* 68 (3): 212–22.

Weisberg, M., and Muldoon, R. (2009). Epistemic landscapes and the division of cognitive labor. *Philosophy of Science* 76 (2): 225–52.

Wetsman, N. (2020). We have no idea how dangerous football really is. Is there a scientific case for banning the sport? *Popular Science*, https://www.popsci.com/how-dangerous-is-football-cte/.

WHO (2019). Ten threats to global health in 2019. Retrieved March 2019 from https://www.who.int/emergencies/ten-threats-to-global-health-in-2019.

WHO (2000). *Tobacco Company Strategies to Undermine Tobacco Control Activities at the World Health Organization*. World Health Organization.

Whyte, K. P., and Crease, R. P. (2010). Trust, expertise, and the philosophy of science. *Synthese 177* (3): 411–25.

Wilholt, T. (2013). Epistemic trust in science. *British Journal for the Philosophy of Science* 64 (2): 233–53.

Wilholt, T. (2016). Collaborative research, scientific communities, and the social diffusion of trustworthiness. In (eds. Brady, M. and Fricker, M.) *The Epistemic Life of Groups: Essays in the Epistemology of Collectives*. Oxford University Press: 218–33.

Williamson, T. (2000). *Knowledge and Its Limits*. Oxford University Press.

Winsberg, E. (2001). Simulations, models, and theories: Complex physical systems and their representations. *Philosophy of Science* 68 (S3): S442–54.

Winsberg, E. (2012). Values and uncertainties in the predictions of global climate models. *Kennedy Institute of Ethics Journal* 22 (2): 111–37.

Winsberg, E. (2018). Communicating uncertainty to policymakers: The ineliminable role of values. In (eds. Loyd, E. and Winsberg, E.) *Climate Modelling*. Palgrave Macmillan: 381–412.

Winsberg, E., Huebner, B., and Kukla, R. (2014). Accountability and values in radically collaborative research. *Studies in History and Philosophy of Science Part A* 46: 16–23.

Wood, T., and Porter, E. (2019). The elusive backfire effect: Mass attitudes' steadfast factual adherence. *Political Behavior* 41 (1): 135–63.

Woolston, C. (2015). Fruit-fly paper has 1,000 authors. *Nature 521* (7552). Online version: https://www.nature.com/news/fruit-fly-paper-has-1-000-authors-1.17555.

Worrall, J. (2007). Why there's no cause to randomize. *British Journal for the Philosophy of Science* 58 (3): 451–88.

Worsnip, A. (2021). The skeptic and the climate change skeptic. In (eds. Hannon, M. and de Ridder, J.), *The Routledge Handbook of Political Epistemology*. Routledge: 469–79.

Wray, K. B. (2001). Collective belief and acceptance. *Synthese* 129 (3): 319–33.

Wray, K. B. (2002). The epistemic significance of collaborative research. *Philosophy of Science* 69 (1): 150–68.

Wray, K. B. (2015). History of epistemic communities and collaborative research. In (ed. Wright, J.) *International Encyclopedia of the Social and Behavioral Sciences*, 2nd edition, Vol. 7. Elsevier: 867–72.

Wright, S. (2016). The transmission of knowledge and justification. *Synthese* 193 (1): 293–311.

Wright, S. (2018). *Knowledge Transmission.* Routledge.

Wuchty, S., Jones, B. F., and Uzzi, B. (2007). The increasing dominance of teams in production of knowledge. *Science* 316 (5827): 1036–9.

Zagzebski, L. T. (2012). *Epistemic Authority: A Theory of Trust, Authority, and Autonomy in Belief.* Oxford University Press.

Zahle, J. (2019). Data, epistemic values, and multiple methods in case study research. *Studies in History and Philosophy of Science Part A* 78: 32–9.

Zamir, E., Ritov, I., and Teichman, D. (2014). Seeing is believing: The anti-inference bias. *Indiana Law Journal* 89: 195–229.

Zhou, J. (2016). Boomerangs versus javelins: How polarization constrains communication on climate change. *Environmental Politics* 25 (5): 788–811.

Zollman, K. (2010). The epistemic benefit of transient diversity. *Erkenntnis* 72: 17–35.

Zollman, K. (2015). Modeling the social consequences of testimonial norms. *Philosophical Studies* 172 (9): 2371–83.

Zuckerman, H., and Merton, R. K. (1973). Age, aging, and age structure in science. In (ed. Merton, R. K.) *The Sociology of Science: Theoretical and Empirical Investigations.* University of Chicago Press: 497–559.

Zwaan, R. A., Etz, A., Lucas, R. E., and Donnellan, M. B. (2018). Making replication mainstream. *Behavioral and Brain Sciences* 41: 1–50.

Author Index

For the benefit of digital users, table entries that span two pages (e.g., 52–53) may, on occasion, appear on only one of those pages.

Subject Index

Note: The italicized entries indicate basic explanations or substantive discussion. Thus, a reader may use the index to look up basic points by following the italicized entries. Note that the principles occurring in the book are collected in the appendix 'List of Principles' rather than in the present index.

Figures are indicated by an italic "*f*", respectively, following the page number.

For the benefit of digital users, table entries that span two pages (e.g., 52–53) may, on occasion, appear on only one of those pages.